Natural Resources and National Welfare

edited by
Ann Seidman

The Praeger Special Studies program—utilizing the most modern and efficient book production techniques and a selective worldwide distribution network—makes available to the academic, government, and business communities significant, timely research in U.S. and international economic, social, and political development.

Natural Resources and National Welfare

The Case of Copper

PRAEGER SPECIAL STUDIES IN INTERNATIONAL ECONOMICS AND DEVELOPMENT

Praeger Publishers   New York   Washington   London

Library of Congress Cataloging in Publication Data
Main entry under title:

Natural resources and national welfare.

  (Praeger special studies in international
economics and development)
  Based on papers presented at a conference held
in Lusaka, Zambia, July 3-9, 1974.
  Includes index.
  1. Copper industry and trade—Congresses.
2. Copper mines and mining—Congresses. I. Seidman,
Ann Willcox, 1926-
HD9539. C6N34   333. 8'5   75-60
ISBN 0-275-05450-0

PRAEGER PUBLISHERS
111 Fourth Avenue, New York, N.Y. 10003, U.S.A.

Published in the United States of America in 1975
by Praeger Publishers, Inc.

Printed in the United States of America

CONTENTS

PART IV: COPPER MINING AND DEVELOPMENT
IN ZAIRE

x

LIST OF TABLES AND FIGURES

Natural Resources and National Welfare

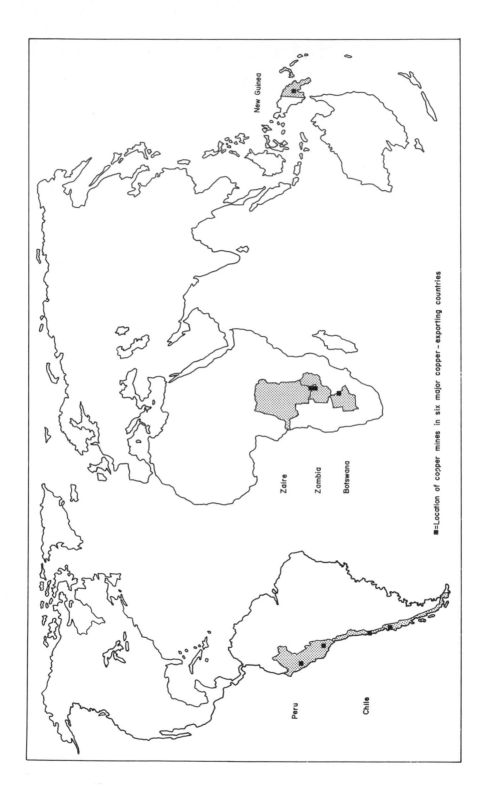

New Guinea

Zaire

Zambia

Botswana

Peru

Chile

■=Location of copper mines in six major copper-exporting countries

INTRODUCTION:
WHY THE COPPER COUNTRIES?
Ann Seidman

This volume consists of studies presented to a conference of
researchers held in Lusaka, Zambia, in July 1974, on Natural Resources and National Welfare: The Case of Copper. The idea of the
conference was born in 1972 at the Law and Development Institute,
which met at the University of Wisconsin during the summers of 1971
and 1972 and where university teachers and government personnel
from about 23 Third World countries were studying how laws and
working rules shape efforts to attain development. A commissioner
of mines raised the question as to what kind of mining legislation his
country should adopt to ensure that it benefited from newly discovered
mineral deposits. The institute participants concluded that a case
study evaluating the consequences of alternative development strategies might be useful to all governments seeking to maximize the national welfare to be obtained from rich mineral wealth.

The four major copper-exporting countries—Chile, Peru, Zambia, and Zaire—were chosen for the study, in part because, by chance,
the Law and Development Institute included participants from each.
More important, these four developing countries on two continents

---

The original essays presented at the Lusaka conference were
considerably longer than those here published. To keep publication
costs down and avoid duplication, it was necessary to eliminate several and edit others severely, reducing them in length by about a
third. This was rendered especially difficult by the unsatisfactory
state of communications across the five continents to which we all
scattered when the conference ended. I wish to express my sincere
appreciation to my colleagues for their helpful cooperation in this
painful process.

produce a major share of the world's copper exports. A few large mining companies were responsible for the expansion of their mines and the worldwide marketing network for their exports. These conditions, combined, rendered it feasible to analyze in depth the issues and potentially conflicting and perhaps overlapping interests of the countries and companies in a single comparative research program.

THE MAJOR COPPER-EXPORTING COUNTRIES

The economies of the main exporters of copper—Chile, Peru, Zaire, and Zambia—have been fundamentally influenced by the opening of vast mines to produce copper for sale to the factories of Europe and North America. These mines, which produced the bulk of their copper exports, have been developed over the last half-century, using the most modern technologies available. Although in the 20 years since the mid-1950s, these countries' relative shares of the copper market have shifted, they have remained the four largest copper exporters in the world.

Table 1.1 presents basic data indicating the relative size of the four countries and the significance of copper to their economies. The conversion table (Table 1.2) indicates the official values of the currencies in use in each country during the period under consideration.

The four major exporters of copper are at historically different stages of political and economic development. Chile and Peru attained independence from the Spanish Empire in the early 1800s, more than a century and a quarter before Zambia and Zaire. Both countries inherited a powerful domestic class of large landlords. Local businessmen emerged, especially in trade. Some of them, particularly in Chile, began to invest in small copper mines before the turn of the century and in manufacturing especially since the Great Depression of the 1930s. Yet in both countries, the development of large, technologically advanced copper mines was achieved in the 20th century by a handful of giant, primarily U.S.-owned copper companies: Anaconda and Kennecott in Chile; and Cerro de Pasco and, more recently, the Southern Peru Copper Company in Peru.

In Zambia and Zaire, on the other hand, the initial investments of the big copper companies were made under the umbrella of colonial rule, British in the former case and Belgian in the latter. Taxes on Africans and labor recruitment policies compelled Africans to migrate from remote rural areas to furnish the cheap labor needed to dig the mines. Taxes on company profits were low. Anglo-American Corporation and Roan Selection Trust (which became a subsidiary of American Metal Climax) together with the associated financial and trading institutions, shipped out an estimated half to two-thirds of the investable surpluses produced in then Northern Rhodesia.

TABLE 1.1

Some (Estimated) Background Data for Four Major Copper-Exporting Countries: Chile, Peru, Zambia, and Zaire, 1970

	Chile	Peru	Zambia	Zaire
Land area (km.^{2})	741,767	1,285,216	752,000	2,335,409
Population	10,000,000	13,171,800	4,500,000	21,637,876
Urban percent of total	75	52	50	30
Mining output as percent of exports	73	46*	95	67
Percent of GDP	10	5.5	42	26.1 (1969)
Per capita income (dollars)	450	291	299	70-74
World copper position: percent production, 1970 (metric tons)	13.2	4.1	13.4	7.4
Known reserves (millions of tons/copper content)	53.8	18.1	27.2	18.1
Percent reserves (1964): world reserves assessed at 50.5 million tons at same grade as currently produced	21.1	5.9	11.9	9.5
Average grade of ore (percent copper)	1.53	1.14	3.38	4.2
Cost per pound of mining copper: cents per pound	30.88	26.48	30.16	25.32
Range of mining cost: cents per pound	26.30-52.00	25.60-32.00	25.80-35.00	20.10-38.50
Proportion nationally owned	Major foreign-owned companies expropriated 1971; new investments by foreign corporations encouraged	Peruvian government recovered most of undeveloped copper properties; developed properties mainly owned by foreign firms; Cerro de Pasco mine purchased by government	51 percent equity acquired by government	100 percent equity acquired by government. Foreign consortiums opening new mines with government owning minority of shares
Linkages				
Forward				
Percent refined	40 (1966)	20 (1966)	85 (1970) 79 (1966)	65 (1970) 50 (1966)
Percent fabricated	15	—	—	—
Backward				
Proportion of machines, plant, and equipment imported	High	High	Almost all	Almost all
Percent of wage employment	3(mining) About one-half copper	2(mining)	14(mining)	4 (1969)

*Copper = 20 percent.

Note: The comparability of the data is necessarily affected by the fluctuations of domestic and international prices, exchange rates, and errors of estimation, so that the data can only be considered comparable in terms of rough orders of magnitude.

Source: Compiled by the author with the assistance of Nora Hamilton, University of Wisconsin, Madison, Wisconsin.

TABLE 1.2

Exchange Rates: Trade Conversion Factor (Average for Annual Period), 1967–73, for Major Copper– Exporting Nations

Country	1967	1968	1969	1970	1971	1972	1973
Botswana (using South African Rand): U.S. dollar per Rand	1.400	1.400	1.400	1.400	1.400	1.294	1.437
Chile: Escudos per U.S. dollar	5.068	6.877	9.099	11.665	12.225	18.865	93.750
Papua New Guinea (using Australian dollar): U.S. dollar per Australian dollar	1.120	1.120	1.120	1.120	1.134	1.190	1.417
Peru: Soles per U.S. dollar	29.79	38.70	38.71	38.71	38.69	38.70	38.70
Zaire: Congo francs/Zaires per U.S. dollar	0.329	0.500	0.500	0.500	0.500	0.500	0.500
Zambia: U.S. dollar per Zambian kwachas	1.40	1.40	1.40	1.40	1.40	1.40	1.53

Note: Conversion factor is always expressed in terms of national currency unit currently in circulation for all back periods without regard to changes in the unit of account over time.

Source: International Monetary Fund, International Financial Statistics, September 1974.

6

Only after independence, in the 1960s, could African governments attempt to capture the investable surpluses produced by the mines and reallocate them to a diversified range of productive activities. Africanization programs accelerated the entrance of a few Africans into top ministerial and civil service posts, enabling them to enjoy the high salaries and living standards previously reserved for whites. The bulk of the middle-level jobs, however, continue to be manned by expatriates. A range of government measures facilitated the entry of African entrepreneurs into trade. A handful of Africans began to accumulate wealth and invest in large farms and real estate. Few private individuals had the know-how or capital to enter into industrial production.

Despite the apparent differences in their historical experiences and the detailed characteristics of their economies, all four countries may be said to exhibit surprisingly similar features. A relatively small proportion of the total population in each of these countries is engaged in mining, but the mines still produce an important share of their exports, their gross domestic product (GDP), and their government revenues. Copper exports appear to be somewhat less significant in Zaire and Peru, where larger populations and land areas have facilitated production of a somewhat more diversified range of exports. Copper produces a smaller proportion of the GDP and revenues in Chile and Peru, where other sectors of the domestic economy have been somewhat more monetized through the combined efforts of domestic entrepreneurs, government measures, and foreign investment in the last century and a half of political independence.

The distribution of income in all four countries is sharply skewed, with perhaps 10 percent of the local inhabitants, associated with the narrow "modern" enclave, obtaining half to three-fourths of the national income. In Zambia and Zaire, this still includes a high proportion of non-Africans. Over the years, a major share of the income produced by the mines has been pumped out of all the countries in the form of profits, interest, and high salaries of foreign high-level personnel. The incomes of the masses of the population, whether peasants or urban wage-earners, have been barely more than sufficient to support life. Growing numbers of rural-dwellers have crowded into peri-urban slums clustered at the foot of modern city skyscrapers, swelling the ranks of unskilled, unemployed workers desperately seeking jobs at bare subsistence wages.

These are the features of underdevelopment that must be overcome by the formulation of appropriate development strategies. Expansion of appropriate kinds of industrial and agricultural output in all regions should spread productive employment opportunities for the vast majority of the peasants and wage-earners to raise their levels of living. In a sense, the success achieved in attaining these goals provides the criterion for evaluating the development strategies devised.

THE BIG COPPER COMPANIES

The international copper market in which the four copper-exporting countries have traditionally sold their copper has, historically, been characterized by a high degree of concentration and control.[1] The two largest companies, Anaconda and Kennecott, built their world-wide operations around powerful, vertically integrated, corporate structures controlling the entire copper production process, from the mines in the United States and Chile, through the final sales of the end products in which copper is used. To a somewhat lesser extent, the other big firms asserted their influence through their ownership of the mines, smelters, and refineries, as well as their worldwide marketing systems. They showed a distinct tendency to erect smelters, and more particularly refineries, in their home countries, rather than in those regions where the mines themselves were located, as Table 1.3 shows.

In 1966 about half of all the mine capacity was located in the developing areas, over 40 percent of it in Africa and Latin America. Only about a third of the smelter capacity and barely 20 percent of the refinery capacity was located there. Forecasts suggest that this trend is likely to continue.

In 1966 a slightly larger proportion of the smelting and refinery capacity was located in Africa than in the Latin American countries, perhaps reflecting a tendency for mining companies in Africa, particularly in Zambia, to consider their investments (prior to 1966) "safer." The proportion of copper refined in Chile had actually declined since the mid-1950s.

Mine output in Asia is forecast to increase more rapidly than elsewhere, primarily due to the opening of the new mine in Bougainville, Papua New Guinea. In the post-World War II period, a significant share of the world's mining, smelting, and refining of copper has expanded in the socialist countries, which, by the mid-1960s, produced almost a fifth of the world copper output. Under the impact of all these factors, the dominant position of the biggest multinational copper companies appears to have decreased somewhat, as indicated in Table 1.4.

As might be expected, all the copper companies have developed new strategies and perspectives in response to the changes in world copper demand, technological requirements of the industry, and the emergence of more nationally oriented policies of the host-country governments where their mines had been located. This has been in-

TABLE 1.3

World Copper Production Capacity,[a] 1966 and 1976[b]

	Mine				Smelter				Refined			
	1966		1976[b]		1966		1976[b]		1966		1976[b]	
	1,000 Tons	Per-cent	1,000 Tons	Per-cent	1,000 Tons	Per-cent	1,000 Tons	Per-cent	1,000 Tons	Per-cent	1,000 Tons	Per-cent
Developed:												
North America	1,750	39.4	2,927	35.7	1,819	39.1	2,417	33.2	2,391	45.8	3,164	38.6
Western Europe	144	3.2	340	4.1	415	8.9	881	11.7	1,161	22.2	1,733	21.1
Other developed[c]	348	7.8	573	7.0	576	12.3	1,418	18.9	534	10.2	1,541	18.8
Total developed	2,242	50.5	3,840	46.9	2,810	60.4	4,716	62.8	4,086	78.2	6,438	78.7
Developing:												
Latin America	930	20.9	1,656	20.2	816	17.5	1,157	15.4	443	8.5	816	9.9
Africa[d]	1,016	22.8	1,632	19.9	979	21.0	1,483	19.7	668	12.7	854	10.4
Asia[e]	151	3.4	1,050	12.4	41	0.8	144	1.9	22	0.4	72	0.8
Total developing	2,097	49.5	4,338	53.0	1,836	39.6	2,784	37.1	1,133	21.7	1,742	21.2
Total	4,439	100.0	8,178	100.0	4,646	100.0	7,500	100.0	5,219	100.0	8,180	100.9

[a]Excluding socialist countries, whose output (excluding China) in 1966 was for mines: 944,000 tons; smelt-ers: 956,000 tons; and refineries: 1,129,000 tons (17 percent of total world refined copper).

[b]Estimated.

[c]Includes Japan, Australia, South Africa.

[d]Excluding South Africa.

[e]Excluding Japan, and Australia (including Bougainville, New Guinea).

Sources: Calculated from International Wrought Copper Council, "Survey of Free World Increases in Cop-per Mines, Smelter, and Refinery Capacities, 1971–1977," and World Bureau of Metal Statistics, November 1967.

9

TABLE 1.4

Shares of World Copper Output of Biggest Copper Companies

Companies	1948 Short Tons	1948 Per-cent	1960 Short Tons	1960 Per-cent	1969 Short Tons	1969 Per-cent
U.S.-Latin American-based:						
Kennecott	514	30.8	571	23.7	699	24.2
Anaconda	362	21.7	476	19.8	597	20.2
Phelps Dodge	247	14.4	476	9.7	284	9.8
International Nickel	118	7.1	155	6.4	110	2.8
African-based:						
Roan American Metal Climax Group	134	8.0	241	10.4	368	12.7
Anglo-American Group	118	7.1	392	16.3	426	14.7
Union Miniere	171	10.3	331	13.8	399	13.8
Total: big seven	1,664	100.0	2,400	100.0	2,883	100.0
Percent of world production	—	70.0	—	60.0	—	54.0

Source: A. Geddicks, "International Production and the Case of Chile," Natural Resources and National Welfare: The Case of Copper Mining (newsletter), no. 2 (March 1973).

dicated by overall new trends, which seem to reflect the uneven development of capitalist economies and firms. Particularly in times of prosperity, new ones may expand their output in part at the expense of older ones.[2]

Copper-mining and processing has been relatively profitable for most of the big companies. Even after the fall of copper prices in 1970, Anglo-American reported to its shareholders that copper, while making up only 6 percent of its investments in 1971—after Zambia's government purchased 51 percent of Zambian mines—provided 21 percent of its group income.[3]

ALTERNATIVE EXPLANATIONS AND
SOLUTIONS OFFERED

Broadly speaking, two major sets of theories provide sources of hypotheses to explain and suggest solutions to the problems confronting less developed countries like the copper-exporting nations, whose economies are, to a large extent, dependent on mineral exports. These sets of theories are neither monolithic nor mutually exclusive. They do, however, offer significantly different kinds of explanations of the problems. Furthermore, the adoption of either—whether by implicit assumption or explicit declaration—is likely to lead, over time, to the implementation of qualitatively different development strategies.

One set of theories is essentially founded on the assumption that developing countries have no alternative but to create a "hospitable climate" for foreign private investment in their mines. Its proponents explain the problem as essentially one of lack of capital and highly skilled manpower. Underdevelopment is perceived as the consequence of a vicious cycle of poverty: People in underdeveloped nations have no capital, so they cannot invest in productive activities to generate savings, so they cannot accumulate capital to invest. It is often added that this cycle is difficult to break because of the lack of skilled high-level manpower and/or the inhibiting influence of traditional attitudes and institutions. Therefore, foreign investment must be attracted as the primary means of breaking into this vicious cycle and creating the conditions for "takeoff."

The investment of foreign funds in the "modern" export enclave of the economies of underdeveloped countries is expected to provide an "engine of growth," setting off a multiplier effect that will, over time, lead to development throughout the economy as the rest of the population adopts the required modern skills and attitudes. The mines will provide jobs and contribute to demand for raw materials and parts that may be supplied by the expansion of domestically owned industry and agriculture. Domestic tax revenues obtained from foreign mine investments will enable governments to finance economic and social infrastructure. These conditions will attract private capital—domestic as well as foreign—to build new industries and expand agricultural output. As productive activities spread throughout the economy, savings will be generated. Given education to create appropriate skills and entrepreneurial attitudes among the local population, "takeoff" will be assured.

The debate among proponents of this approach revolves around the question, not of whether this is the best way to stimulate economic development but of how to attract foreign investors. The various devices proposed include tax concessions: guaranteed remittances of profits of foreign firms and salaries of foreign skilled employees; tariff protection for "infant" industries; and, more recently, government participation in the ownership of the mines to ensure government support for company policies.[4] Proponents of this approach may argue as to the potential advantages and disadvantages of these and other specific proposals for creating a "hospitable investment climate." But the central objective is to encourage foreign mining companies to invest in expanding mineral exports.

The other set of theories, while agreeing that lack of capital and skilled manpower constitutes a serious economic constraint, rejects the notion that private foreign investment will automatically provide an "engine of development." This rejection is founded on a fundamentally different explanation of the problems of underdevelopment. The proponents of this alternative hold that the opening of mines in Third World countries, whether under the umbrella of colonial rule or in politically independent countries, led to the domination of so-called modern export enclaves by giant foreign corporations, which geared them to the profitable export of crude minerals and the import of manufactured goods. The local population was drawn in primarily as a source of cheap labor. An entire set of externally dependent institutions and working rules was shaped to perpetuate this enclave type of growth, controlling the banking and financial system, and export-import and wholesale trade, as well as basic industries. Domestic private entrepreneurs, insofar as they emerged at all, could not muster enough capital to compete effectively with the dominant foreign firms. Instead, they have tended to become associated with them in a variety of relationships through personal and financial links.

Whether the state machinery is controlled by a colonial government or local classes benefiting from the multinationals' activities, the resulting economies of most Third World countries remain, like those of the four major copper-exporting nations, basically similar. Outside the narrow "modern" enclave, centered on export-import trade, productive employment opportunities are limited. Traditional agricultural and handicrafts pursuits have tended to be undermined by importation of manufactured goods. This process has been accelerated more recently by the establishment of relatively capital-intensive, import-substitution projects utilizing imported machinery, parts, and materials—a change in the form, which leaves the national economy as dependent on imports as before.

The entire institutional structure, proponents of this second set of theories maintain, has systematically contributed to excluding the peasants and wage-earners from modern productive activities except as members of a low-cost labor force. The drift of tens of thousands of "fugitives" from the resulting rural poverty builds up a vast labor reserve of urban unemployed, which serves to depress the wages of those lucky enough to obtain jobs.

A skewed income distribution inevitably accompanies this pattern of development. The major share of investable surpluses produced in the export enclave, in the form of profits, interest, and high salaries, is funneled into the pockets of the classes associated with ownership of the means of production—domestic businessmen, managers of foreign firm interests, top-level civil servants, large farm owners, and wholesale traders—or drained away to the home countries of the foreign firms.

This second explanation of underdevelopment leads to proposals for reorganization of the entire set of institutions shaping the national political economy of mineral-based underdeveloped economies. Political institutions must be rebuilt to ensure that the working classes, wage earners, and peasants, are increasingly involved in crucial government decisions. The government should, as quickly as possible, take control over the "commanding heights"—basic industries, banks and the entire financial system, and export-import and internal wholesale trade—the first step in a planned transition (which might take 10, 20, or more years) to the complete socialization of the means of production. Investable surpluses, previously drained out of the country or enjoyed by a small domestic class, would then be directed, in the context of carefully formulated long-term physical and financial plans, to the creation of new productive and agricultural activities in all sectors and regions of the nation. Practically oriented educational programs would be designed to upgrade the skills of workers at all levels to fulfill manpower plans directly related to implementing the overall industrial and agricultural development strategy. Control of external trade would ensure that imports of consumer luxuries would be replaced by the import of capital goods and equipment. As industry expanded, more and more processed goods would be sold abroad, contributing to increased ability to buy imports required for expanding development. The resulting expansion of productive employment opportunities would enable the mass of low-paid workers and peasants, over time, to expand their effective demand for the goods produced by the growing productive sectors.

The role of multinational firms in this context is perceived as being limited to providing essential technologies, capital equipment, and, perhaps for a short period, some high-level manpower. They would be permitted to operate only within the framework of an economy in which the state, representing the workers and peasants, firmly controlled the "commanding heights." Even this role would be encouraged as long as—and only if—they could be shown to contribute directly to implementation of national long-term plans. Continual vigilance would be required, proponents of this approach argue, for the companies would be expected to use every possible means—political as well as economic—to return to the status quo ante in order to maximize their own profits without interference.

THEORY AND PRACTICE IN THE COPPER-BASED ECONOMIES

A wide range of variations and permutations of these theories has emerged as they have been used implicitly or explicitly as guides to day-to-day activity. Government policies have often been formulated in a rather ad hoc way, without rigorous analysis of the problems they are supposed to solve or adequate evaluation of the consequences. They have often emerged as the outcome of political struggle between conflicting interests, both between the governments and the companies and within the countries between groups benefiting from the status quo and those seeking to change it.

This is illustrated by the case of the four copper-exporting countries. Each has devised quite different development strategies in the face of rather similar problems over the 20 years since the mid-1950s. Chile's government first sought to attract further mine investment through tax incentives in the 1950s. It then took 51 percent ownership of the mines in the late 1960s. Finally, a united-front government nationalized the mines outright. In mid-1973, following a coup, the new military government declared its aim of paying compensation to the mines' former owners while seeking to attract new foreign investment.

In Peru, where the copper mines have not constituted such an important feature of the economy as in the other three countries, the government, despite many political changes, has tended to continue to pursue policies of trying to attract further foreign investment in the mines. In recent years it nationalized big foreign-owned sugar estates and attempted to formulate a strategy designed to foster growth in other sectors of the economy.

After attaining independence in 1960, the Zaire (then Congo) government experienced a bloody civil war in which a separatist movement sought, with foreign assistance, to break the Copper Belt away from the newly created nation. The civil war interrupted but did not halt the continued expansion of mine output. In 1967 the government purchased 100 percent of the ownership of the existing mines, while leaving the management in the hands of Union Miniere, the company that had originally developed them. It also invited other consortiums to open new mines in which it retained only a minority interest.

Zambia's government, following attainment of independence in 1964, initially confined itself to increasing the tax revenues obtained from the mining companies and demanding that they Zambianize middle-level and management posts. In 1969 it reached agreements with the Anglo-American Corporation and Roan Selection Trust (American Metal Climax—Amax) to purchase 51 percent of the shares of the mines, leaving the management, investment, and marketing decisions in the hands of the companies, which retained the remaining 49 percent of the shares. Tax incentives were introduced to induce the firms to expand their investments. Simultaneously, efforts were made to attract other foreign investments, both from capitalist and socialist countries, to expand mining output. In 1973, the government refinanced its purchase of 51 percent of the mines with the announced aim of increasing its ability to exercise tax and exchange control measures to reduce the continued outflow of profits and interest and to acquire greater control of management and marketing decisions.

The Copper Project was designed to examine the consequences of these policies for these four countries over the 20 years since the mid-1950s in an effort to contribute to a systematic evaluation of the advantages and disadvantages of alternatives suggested by the two sets of theories outlined above.

THE RESEARCH PROGRAM

From the outset, it was agreed that the Copper Project should primarily involve research by university and government personnel in the participating countries, rather than outside "experts." It was hoped that the primary support for the project would come from the governments and research institutions of the participating countries.

The methods and levels of work conducted during the two-year project varied from one country to another. Already existing research resources were drawn upon wherever possible. These were relatively developed in the context of the widespread growth of higher education in the Latin American countries, which had enjoyed political independence for 150 years. In Chile, Ceplan, located at a leading university in Santiago, had already initiated extensive research on a range of issues relating to Chilean development and had sponsored a conference on copper in Chile in early 1973. In Peru, Desco, an independent research institution in Lima, was the center of ongoing research on many development issues.

The two African copper-exporting countries had attained political independence only during the period under study. Higher educational institutions for Africans had been neglected by the colonial administrations. The University of Zambia, which now has about 2,800 students, was established only in 1966. Research, insofar as it was developed at all, had in the past been dominated by colonial experts. The institutionalization of policy-oriented research directed to examining the consequences of strategies for attaining greater national benefit from the mines had to start largely from scratch.[5] Researchers from several disciplines at the University of Zambia began to meet biweekly with government personnel to examine the consequences of the rapidly changing government policies.

In Zaire, Mulumba Lukoji worked with an interdisciplinary group of colleagues to develop the research at the University of Zaire in Kinshasha and Lubumbashi. With university and government support, they organized two conferences in the two-year interval before the Lusaka conference. The proceedings were published.[6]

Researchers from Papua New Guinea and Botswana joined the project later. Both countries had become the scene of vast new mining projects in the last decade. Botswana attained independence in 1966; Papua New Guinea had not yet attained that status. Their universities had only recently been established. As a result, although the participants from both countries consulted their colleagues about their papers, extensive team research had not yet been initiated.

The project's newsletter was a significant stimulus to maintaining the initial impetus of the project, keeping the participants informed of their colleagues' progress.[7] Nevertheless, the difficulties of communication between the participating countries,* plus the involvement of new local researchers, explains, at least in part, why the chapters from the different countries do not all follow a common research design, as originally anticipated. In particular, Chile and Peru concentrated their analyses on the consequences of government policies for capturing the investable surpluses produced by the mines, drawing on a wealth of detailed historical and analytical data.** The participants

---

*Letters, if they arrived at all, often took two or three weeks going in one direction, and many never reached their destinations.

**Ffrench-Davis emphasizes that even in Chile the statistical data available over the last 20 years was inadequate and offers useful suggestions for its improvement (see Chapter 6).

from Zambia and Zaire, although having less detailed information at their command, * broadened the coverage of their chapters in an attempt to assess the consequences of overall development strategies. The Papua New Guinea and Botswana chapters examine the overall impact of the newly established copper mines on national development. These new copper-exporting nations confront different problems than the older ones, although they may benefit from the experience of the latter.

The Lusaka conference centered primarily around four issues: (1) the formulation of criteria for evaluation of development and the extent to which research of this type must be based on an analysis of the political stance of the governments involved, an issue that came up repeatedly in a variety of forms; (2) the possibility of forecasting world copper prices, which inevitably affect the incomes and development possibilities of copper-based economies; (3) the difficulties confronting governments attempting to capture the investable surpluses produced by the copper mines, as well as to increase national control over their operations; and (4) the possibilities of integrating the mines more effectively into national industrial growth, while investing surpluses captured from them in other industries to achieve a more balanced economy.

The Lusaka conference participants expressed their recognition that there were several issues with which they were not dealing but which were nevertheless important. One of these was the comparative level of working and living conditions of the mine workers, as well as the status of their trade union organizations. Another was the chronic international inflationary rate of 10 percent or more a year, which affects the copper-exporting countries in many and contradictory ways. Further research, it was agreed, should be directed to these as well as other issues identified in the course of the discussions.

This Introduction seeks only to summarize the main arguments on the four issues around which most of the debate centered.

THE CONFERENCE SUMMARIZED CRITERIA
FOR EVALUATING DEVELOPMENT STRATEGIES
IN COPPER-BASED COUNTRIES

The conference participants agreed that the basic criteria for evaluating development strategies should be the extent to which they

---

*In Zambia, at least, the lack of legislation requiring the companies to publish critical data, even though the government holds 51 percent of the shares of major sectors of the economy, constitutes a serious hindrance to the development of such detailed analyses.

contribute to the spread of productive employment opportunities and increased living standards for the masses of the people in the framework of more nationally integrated, self-reliant economies. Several participants argued that it was impossible to consider specific development strategies without examining the governments' political ideologies. After extensive debate, it was agreed that, while political ideologies undoubtedly do affect particular strategies adopted, the conference should not adopt positions on the governments' political stance. Given the nature of the institutionalization of the research, involving government and university personnel, * this was impossible if the research was to continue. The Copper Project should, instead, examine the explanations of the problems of underdevelopment underlying particular strategies adopted to see to what extent they are warranted by the data and look at the facts showing the consequences of efforts to implement them in light of the agreed criteria for evaluating development.

State intervention, conference participants repeatedly stressed, is not of itself "progressive." Rather, it must be evaluated in terms of its actual effect on the working and living conditions of the people of the country. Participants emphasized, for example, that the declared "will" of the government or its international image as "progressive" or "regressive" could not provide a yardstick for evaluating its particular policies. It was considered very necessary to evaluate carefully the effects of the changes of policies relating to the copper mines over the last two decades in terms of their impact on the lives of the people. More specific criteria mentioned included the contribution to permanent productive employment opportunities: autonomous, self-reliant development; balance of payments stability; and the structural homogeneity of industrial development.

Some participants objected that, in the past, the World Bank's insistence on the "credit-worthiness" of particular countries and projects had been applied with a political bias. Continuation of this policy would undoubtedly affect new proposals for financing copper mines, perhaps to the detriment of the national welfare of the countries involved. Among examples cited were the refusal to lend funds to the Chilean government of Salvador Allende; failure to provide finance for the Tazara Railway as requested by the Zambian government to end its dependence on Southern Africa; reluctance to finance the opening of Botswana coal mines to provide an essential power source for the new mines, instead of importing coal from South Africa. Reichelt and Kenji Takeuchi maintained that these issues had been decided on

---

*In Africa and Papua New Guinea, the universities are government-financed.

purely economic grounds. Those disagreeing held that the bank's narrow interpretation of "credit-worthiness" neglected the criteria of
self-reliance adopted by the conference and involved political judg-
ments at variance with those of the governments involved.

While the implications of the specific issues debated were not
fully resolved, the conference participants did agree on the necessity
of including, in continuing research, an analysis of the various domestic and international interests that espouse alternative development
strategies. Such analyses are essential to explain why particular
strategies are adopted, an important aspect of understanding the continuing problems of poverty despite the wealth produced by the copper-
based economies.

THE COPPER MARKET: SUPPLY,
DEMAND, PRICES, AND STRUCTURE

A major problem confronting the copper-exporting nations is the
sharp fluctuation of the prices for which they sell their copper on the
world market, which seriously impedes efforts to achieve effective
planning. Radetzki argues that improving technology and the opening
of new copper sources will ensure an adequate supply of copper in the
foreseeable future at real prices that are unlikely to increase.* Takeuchi, working on World Bank commodity studies, concludes that by
1980 there is likely to be an oversupply of copper and falling prices,
due to the extensive exploration and development of new mines now
taking place. Jelenc maintains that world prices will in the long run
remain at the historic average ratio in relation to costs.

The conference participants agreed on the necessity for concerted action by the copper-exporting nations to stabilize the copper
price. The speculative features aggravating the short-run fluctuations
of the London Metal Exchange (LME) price, to which most of the copper-exporting countries peg their copper sales, render it entirely unsatisfactory. It was agreed that there was a need to know more about
the factors affecting the LME price-setting mechanism, in particular
the role of multinational mining corporations, which are "ring dealers"
on the exchange.

Jelenc's essay outlines a range of possible techniques for stabilizing prices. To the extent that the producing countries market their
own exports, and work in concert through the Intergovernmental Coun-

---

*M. Radetzki, Chief Economist of CIPEC, contributed a paper
that is to be published under the title, "Does Mineral Resources Exhaustion Cause a Threat to Material Progress," in World Development.

cil of Copper Exporting Countries (CIPEC), the possibilities of stabilizing the world price appear greater. Luis Pasara's essay suggests some techniques for state marketing. Lack of more detailed discussion of this issue reflected, in part, the fact that the CIPEC conference of government representatives, who had met in Lusaka the previous week, had announced an agreement to attempt to control prices and output, but the details were not yet available.

The question was never fully resolved as to the probable effect on world prices of growing governments' efforts to expand their nation's output as rapidly as possible. Some governments appeared willing to risk serious reduction in current revenues in order to stimulate investment in expanded copper output in hopes of obtaining greater returns at a future date. A number of participants felt that if each country attempted to maximize its revenues by expanding output, the combined effect would inevitably depress the world price in real terms as a result of oversupply in relation to slowly growing demand.* This seemed to be supported by Takeuchi's forecast of oversupply by 1980. The issue of appropriate levels of output for each country must be tackled seriously in the context of any attempt to stabilize world copper prices.

Mezger's essay seeks to identify the conflicting interests of multinational corporations in the framework of the structure of marketing and processing copper in Europe. Some of the conference participants felt that her conclusion that multinational corporations could continue to dominate copper exporting through control of technology was too pessimistic. It was agreed that to break away from the international system dominated by the multinational companies and/or to take advantage of the potential contradictions between them, the copper-exporting countries should formulate a joint strategy to produce their own technology. If the countries could devote 1 percent of their copper revenues to research and development, they could, over time, produce their own mine inputs, as well as establish their own refining and fabricating facilities. This would be easier if they could increase their control of copper processing and marketing.

Many of the chapters below illustrate the extent to which multinationals had, in the past, dominated the copper-exporting nations. Cabieses-Berera and Brundenius describe how the U.S. firm of Cerro de Pasco expanded its domination of the Peruvian economy after World War II, forging close links with the national bourgeoisie and smaller Peruvian enterprises. Their essays, as well as that of

---

*With the fall in the 1974 world copper price, the CIPEC members did agree to reduce their exports in an effort to obtain a more favorable price.

Ffrench-Davis of Chile, detail means by which the companies siphoned surpluses out of the Latin American economies, both openly and covertly. Brundenius points out that the introduction of new capital-intensive technologies had already led to stagnation of employment and real wages in Peru. The limited employment impact of new technologies, especially in open-pit mines, is vividly illustrated in the case of Papua New Guinea, where total employment is about 5,000 local workers.

The essays on countries discuss several aspects of the possibility of formulating strategies to reduce their nations' dependence on the giant multinationals, which, despite occasional conflict between them, control the U.S. and European copper market. The gist of these discussions is summarized below.

CAPTURING INVESTABLE SURPLUSES

Much of the time of the conference was devoted to analyzing the consequences of various government policies relating to the generation and capture of investable surpluses produced by the mines during the 20-year period under review.

Initial Negotiations

The evidence from all the countries suggests that the multinational corporations initially developed mines under conditions in which the governments for one reason or another granted them the right to take away a major share of the investable surpluses produced. Garcia-Sayan shows how the Peruvian government's post-World War II mining code opened the door to a U.S.-dominated consortium to establish the vast, profitable Toquepala mine in Southern Peru. Ffrench-Davis shows that, despite the Chilean government's effort to create an "attractive investment climate" in the 1950s, the giant U.S. mining companies expanded their output in other areas of the globe, in the context of their worldwide profit-maximizing considerations.

In Africa, Zambia and Zaire were colonies during the first postwar decade. The South Africa-based conglomerate, Anglo-American, and Roan Selection Trust, a subsidiary of the U.S. firm American Metal Climax (Amax), which owned the Zambian mines, and Union Miniere du Haut Katanga (UMHK), which owned the mines in the then Congo, developed them with the full support and encouragement of the colonial governments. In these conditions these companies expanded their output more rapidly in the postwar period than did the big U.S. copper companies in Latin America.

The most recently developed mines, those in Botswana and Papua New Guinea, were established by mining companies in negotiations in which the governments were hardly in a position to bargain successfully. Botswana, economically dominated by South Africa, despite its political independence, is not really a bargaining equal for the Anglo-American-Amax consortium in developing its mines. Papua New Guinea was not yet politically independent of Australia—whose authorities negotiated Rio Tinto Zinc's* mutual favorable agreement with Papua New Guinea. Despite extremely profitable operations, especially in the case of Papua New Guinea, where Rio Tinto Zinc recovered almost half its initial investment in one year, the countries received a very small share of the surpluses generated. The U.S. Kennecott Company, the Anglo-American Company, and Japanese interests also have begun to negotiate with the Papua New Guinea government to develop other concessions.

New Arrangements

By the end of the 1960s, most of the copper-exporting countries had begun to attempt to capture a greater share of the investable surpluses produced by their mines. Chile's government first took 51 percent of the shares of the two big U.S. firms, Anaconda and Kennecott, which produced most of its copper. Then, following the election of Allende's coalition socialist government, the mines were nationalized. Ffrench-Davis's essay describes the 20 years of experience that led to that nationally popular decision. In the discussion, it was suggested that the preceding efforts to control the companies through 51 percent ownership and increased attempts to take over marketing, while clearly not achieving the desired degree of national control or increased government shares in the investable surpluses, may have contributed to accumulation of essential Chilean skills and knowledge about production and marketing.

In Peru, both Pasara and Cabieses-Berera explained, the legislation adopted in 1968 by the newly established military government began a process of restoring national control of the mines to the state. Pasara analyzed the post-1968 attempts to expand mining output, including the efforts to increase the percent of copper refined in the country and the difficulties impeding the government's attempts to increase control of marketing.

---

*Rio Tinto Zinc also has significant investments in South Africa and Namibia.

Cabieses-Berera examines the details of Cerro de Pasco's operations as the background of the prolonged company-government negotiations that led to complete government ownership of the underground Cerro mines at the end of 1973, following agreement on the payment of about $76 million compensation. He suggests the company was eager to sell the mine, for it was making far more profit on its shares in the newer open-pit mine in southern Peru.

Mulumba Lukoji explains in Chapter 15 how the Zairois government has placed the holdings it purchased from Union Miniere in 1967 under a government firm, Gecamines. A Union Miniere affiliate, SMG, handled the marketing and management for fees totaling 6 percent of sales. A new investment code was passed, described by Katanga Mukumad Yamutumba, under which the government granted new concessions to two new multinational consortiums to open up new mining operations. As Mulumba Lukoji points out, one of these consortiums, Sodimiza, consists of Japanese firms that ship home their mines' output in smelted, unrefined form. The second, SMTF, includes American and Japanese interests as well as Charter Consolidated, an Anglo-American associate. The discussion at the conference suggested that it would, if possible, be useful to obtain more information as to the details of the actual operation of these arrangements.

At the end of 1969, Zambia's government took over 51 percent ownership of the two big mining companies, Anglo-American and RST-Amax, in an effort to increase its control of the mines' operations as well as to foster expanded output.

Ushewokunze's and Simwinga's essays argue that the 51 percent takeover agreement did not provide the Zambian government with a significant degree of control over the critical decisions relating to the mines. The compensation payments,* together with tax and other concessions, contributed to a sharp fall in government revenue. Data presented by Raj Sharma show a sharper fall in Zambia's government revenue in the years of lower copper price, 1971-72, than in Zaire's.

The discussion emphasized that the benefits to be attained from government ownership of shares, even a majority, depend to a considerable extent on the kind of agreement reached. The mere fact of government ownership, of itself, even of as much as 100 percent of the shares, may not eliminate dependence on and losses of surpluses to multinationals, to the extent that they continue to handle management, processing, technology, and marketing decisions and receive a major share of surpluses as compensation. Ushewokunze notes

---

*Zambia was to pay $40 million annually for its share until 1978 and about $25 million a year more from 1978 to 1982.

that the Zambian government's efforts to achieve a greater degree of control of marketing and management of the mines by paying off in full the cost of its 51 percent of shares of ownership could be thwarted by the initial agreement itself, which remains binding as to these aspects until 1980. Negotiations were in progress on this issue at the time of the conference.

The Botswana and Papua New Guinea governments were reported to have rejected what they alleged to be the "symbolic" notion of majority ownership. Botswana's government, as Silitshena's paper shows, has acquired only 15 percent of the shares of the Amax-Anglo-American consortium, which is developing its Selebi-Pikwe deposits. Papua New Guinea has recently initiated new negotiations in an effort to increase its share of the investable surpluses produced by the Rio Tinto Zinc's open-pit mine but does not envisage attempting to acquire a majority of the shares of the company.

The appropriate relationship to be established between the government companies holding the state's shares of the mines and the rest of the government requires further study. In Chapter 7, Ernesto Tironi explores the possibilities as well as the problems of horizontal rationalization of the operations of the two big mining complexes into the economy of Chile. Prior to nationalization these had been integrated into their multinational owners' international vertically integrated corporate structure. Tironi maintains that the mines should still be operated according to profit maximizing criteria, if and only if the government retains a strong planning authority and clear policy directives. This, he argues, will facilitate a degree of decentralization to achieve greater efficiency. In the discussion it was pointed out that the post-coup government in Chile, while retaining the nationalized companies under the aegis of the government corporation, the Corporacion del Cobre (Codelco), extended a decentralization process. Overall coordination of the mining sector in the context of national planning appears to have been rejected.

Some Lusaka conference participants asked if there was not a danger of creating a "juggernaut," a state within a state that might not be responsive to the needs of the nation. This might equally be true in the case of the autonomous parastatal sectors being established to manage the mines in the African economies, especially since the management is typically handled by foreign firms. This consideration appears to be involved in the issue of taxation affecting Gecamines mentioned by Mulumba Lukoji. If the Zairois government leaves the revenue with Gecamines, rather than taxing it, then Gecamines' management alone has the power of deciding on the appropriate investments to be made.

There was some discussion of measures introduced to increase worker participation in mine management, particularly in the Latin

American countries. The post-coup Chilean government insists that too many meetings and involvement of the workers in political activities reduces their efficiency. A number of conference participants argued that it is crucial to develop appropriate institutions for worker participation, even though this may take time. On another aspect, Luis Pasara notes a difficulty in the formula adopted for worker participation in the profits produced by the Peruvian mines. Given the large amounts of capital invested in the mines and the existing salary inequalities among the mine workers themselves, he suggests, the formula may actually aggravate the income inequalities among the working class of the nation. Other participants maintained that, nevertheless, it was an imaginative and original mechanism that perhaps with some modification might contribute to transforming the economy. It was agreed that these issues are worthy of considerably more research.

Multinationals' Device for Retaining Investable Surpluses

The Chilean and Peruvian chapters detail a variety of techniques by which the mining companies, over the past 20 years, shipped significant shares of investable surpluses out of the countries despite governmental efforts to attain greater control. Ffrench-Davis' and Brundenius' essays, in particular, focus on the way overpricing of imports of mining equipment and machinery, the introduction of large depletion and accelerated depreciation allowances, and the manipulation of exchange control devices enabled the firms to reduce the shares of investable surpluses accruing to the government. Taxes on incomes, rather than on exports, it was pointed out, tended to increase the possibilities of utilizing these devices, for book incomes may be adjusted within multinational corporate structures to reduce the apparent income produced in the countries. To the extent that a government can handle its own management and marketing and reduce its import requirements, it might be more likely to be able to reduce losses resulting from these devices.

If the governments could, through CIPEC, arrive at minimum conditions for investment and taxation of mines, the conference participants agreed, then their bargaining positions vis-a-vis the companies might well be enhanced in regard to all these questions. Reduction of the length of time of particular agreements with multinational consortiums would enhance the governments' opportunities to improve their status as their bargaining position improved. The establishment of long-term contracts, as in Botswana and Zambia, tended to hamper the governments' efforts to improve the conditions at a later date.

Localization of Staff

The evidence presented to the conference suggested that, while Chile and Peru had substantially sufficient local personnel to handle management, marketing, and technical posts, this was far from the case in other countries. The inadequacies of colonial education had bequeathed on the African states as well as Papua New Guinea a shortage of highly qualified manpower that could be expected to persist for some years.

Zaire, having been independent the longest, appeared to have achieved the best status in this regard among the African states. Mulumba Lukoji reports that about 75 percent of the administrative personnel of the mines are Zairois, although three-fourths of the 2,000 or so technical personnel are still expatriates. In Zambia, George Simwinga shows that Zambianization has been proceeding at a slow pace, with an apparent reluctance to fill either technical or administrative posts beyond a small number of hand-picked individuals. In Botswana and Papua New Guinea, the lack of highly skilled local cadre is still more serious.

Shortages of skilled manpower have two consequences. First, efforts to assume national control of management may need to be postponed until more local cadres are created. The existence of trained local personnel does not, of itself, however appear to ensure national-oriented, as opposed to company, control. The Chilean conference participants noted that some of the top-level Chilean technical personnel had accepted the inducements of the international copper firms to leave Chile for company posts in other countries when the mines were nationalized.

The second aspect of this problem is the high cost of expatriate personnel. In Zambia, about a third of the mines' wage and salary bill is paid to the expatriates who constitute about 10 percent of the total labor force. This excludes gratuities, housing subsidies, and other perquisites, which, added together, probably double their cost. In the highly capital-intensive mines of Papua New Guinea, Zorn shows that the total salary bill paid to the expatriate staff, excluding fringe benefits, is today more than double the total wages paid to local personnel. In 1980, it will still be a third more. Only by 1990 is it expected that the wages and salaries paid to expatriate staff will drop to a third of the total wage and salary bill!

These two aspects emphasize the necessity of introducing training programs for local personnel as a vital feature of any program to develop mines in any country. Meanwhile, it was suggested that perhaps the Latin American countries, with many more years of political independence and hence more high-level manpower, might well offer technical assistance to the countries for which lack of skills remains a major factor fostering continued dependence on multinational corporations.

Small and Medium-Sized Mines

In the last 15 years, Chile's government has implemented measures that fostered the expansion of small mine production from 10 percent of exports in 1960 to 21 percent in 1970. Ernesto Tironi in Chapter 9 examines this experience in some detail, looking at the system for marketing and processing the output of these mines, as well as taxation of the sector. He points out that in fact five of the "small firms" are in reality medium-sized, producing almost half of the sector's output. Nevertheless, the employment per unit of investment is considerably greater than in the large mines, and the degree of Chilean—as opposed to foreign—ownership appears significantly greater, although the medium-sized mines did have important foreign participation. These apparent benefits may, however, have been offset at least in part by the major loss of investable surpluses produced due to excessive tax concessions granted to domestic and foreign investors, whose returns at times exceeded 100 percent a year.

In the discussion, it was questioned whether this experience could be duplicated elsewhere, or was dependent on the particular character of the Chilean deposits. While more information would be needed to answer this question, it was emphasized that, unless government created the necessary institutional structure to foster small mine production, it was unlikely to expand even if the nature of the deposits was favorable. On the other hand, careful consideration should be given to tax and other concessions granted to small and medium-sized mine firms if their contribution in terms of employment, foreign exchange, and other aspects is to outweigh the costs to government.

RESTRUCTURING MINERAL-BASED ECONOMIES

Most Lusaka conference participants agreed that it was essential to perceive the mines as more than a "cash register" from which governments may hope to extract funds for expanded social and economic services. The mines need to contribute to the establishment of crucial productive projects to spread productive employment activity to all sectors of industry and agriculture.

The Negative Impact of Externally
Dependent Mining Enclaves

In all the countries, the mines were initially imposed on the preexisting economies essentially as externally dependent enclaves. Their main linkages in the form of trade, technology, and finance were with the metropolitan-based multinational corporations. In all cases, although the detailed sets of institutions differed, the remaining sectors of the economy served primarily to provide a vast reservoir of low-cost labor.

In Zambia, the creation of a handful of capitalist-settler estates on lands taken from the Africans along the line of rail, combined with a poll tax and a discriminatory marketing structure, created conditions forcing a majority of the male inhabitants of rural villages to migrate to the mines and estates of Southern Africa in search of paid employment. The resulting drain of manpower led to the underdevelopment of vast rural areas, previously at least capable of supporting their populations. The ensuing deterioration of living conditions and rural stagnation fostered a steady urban drift. In the decade after independence, when colonial government restrictions on migration have been lifted, the number of urban inhabitants jumped from 25 to 40 percent of the total population. It is estimated that some 20 percent of these are unemployed. Over a third live in squalid shanty compounds lacking water, sewerage, electricity, and other social amenities.

In Zaire, the process differed in that the mine companies relied on African peasants to produce food crops for the mine workers. Large-scale capitalist plantation agriculture was encouraged primarily to produce export crops. At first, the colonial regime deliberately prevented the miners' wives from living in the mine compounds, hoping to hold wages down by letting them provide foodstuffs for themselves and their children in the rural areas. Later the miners' families were permitted to live with them, in an effort to reduce labor turnover. Despite the fact that African peasants were permitted to produce crops for the miners, however, the low prices paid and the siphoning off of significant shares of the resulting surpluses by middlemen and taxes hindered the emergence of an extensive prosperous African peasantry. When the local peasants could not produce enough staple food crops to meet the mines' needs, supplies were imported from estates in the then Rhodesias (now Zambia and Rhodesia).

In Peru and Chile, despite the existence of patterns of landholding very different from those in Africa, the imposition of externally dependent large-scale mining sectors and the emergence of associated activities also fostered extensive rural stagnation. The establishment of a significant agricultural export sector in Peru, as in Zaire, really constituted an expansion of the export enclave, rather than fostering a more integrated national economy. Urban migration has already gone much further in both Chile and Peru than in Africa. Over half the populations of the former countries now live in a handful of cities. A very interesting study could be made of the similarities and differences in the sets of institutions that had, in the historically differing settings of these two continents, bred what appears, on the surface at least, to be significantly similar results.

Several participants questioned the realism of the Papua New Guinea government's proposal to permit mining to continue to expand as an enclave in the hope that it would provide funds for social and economic infrastructure for an essentially rural population. Despite the capital intensity and limited employment creation of Rio Tinto Zinc open-pit mines, it was felt that the imposition of a highly complex, modern mining sector would set in motion a series of events that would inevitably gradually erode the balanced rural economy allegedly still existing.* The continued introduction of large-scale cash-crop farming, whether for local or export consumption and whether domestically or foreign owned, inevitably forces marginal peasants out of business, contributing to the growth of a landless labor reserve. The relatively equitable distribution of wealth said to still exist would be rapidly undermined by the injection of large sums through the creation of an elite around the mines and urban service sectors. In Botswana, where Silitshena reports that about 50,000 laborers migrate annually to South African mines and plantations, this process is clearly well under way. It was argued that unless governments of countries like Papua New Guinea and Botswana planned effectively for the creation of an integrated, balanced development their economies would soon increasingly exhibit characteristics typical of the four Latin American and African nations whose economies had been for so much longer geared to the export of copper.

A considerable amount of discussion centered on the ingredients of a long-term strategy, utilizing investable surpluses produced by the mines, that might contribute to more balanced, self-reliant economic development. The discussion dwelt on two main aspects. First, as all participants emphasized, government policies need to stress the creation of backward and forward linkages from the mining sector to the rest of the economy. This could contribute to the creation of a basic industrial sector to reduce dependence on foreign technology and increase domestic employment and revenue. Second, such

---

*Whether such a balanced rural economy really still does exist seems open to considerable question, given the long-standing, Australian-dominated plantation economy employing thousands of wage-earners at extremely low wages, as described in Chapter 24.

a strategy should incorporate specific institutions to implement balanced industrial and agricultural activities based on the employment of local resources and designed to spread productivity and increase living standards in all sectors of the economy.

The Potential for Backward and Forward Linkages

A number of specific suggestions were detailed for linking the mining sectors more directly to stimulation of activities in the rest of the copper-based economies. These included the establishment of plants to produce essential mine inputs, beginning with power, construction materials, and simple tools, and gradually expanding to include explosives and more complex equipment and machinery. At the same time, the copper-exporting countries could increase the value of copper exported if they established refineries and initiated production of an increased range of fabricated copper and copper-alloyed products.

There are obvious economies of scale to be achieved in industries producing inputs for the mining industry and for further fabrication of mine output. Since independence, both Zambia and Zaire, separately, have been in the process of expanding the range of inputs they produce. If they could plan these jointly, they could benefit from potential economies of scale and perhaps establish additional linkages that neither country, alone, could afford. The explosives industry, for example, might contribute to making feasible a range of chemical projects. If Botswana, too, could buy such inputs from Zambia and/or Zaire and perhaps specialize in the production of some of them in cooperation with its northern neighbors, it could reduce its dependence on South African sources of supply.

At the other end of the copper production sequence, it was noted that the new mine complexes established in Botswana and Papua New Guinea did not include refineries. Even Peru as yet refines only about a fourth of its copper exports. The more copper a country exports in a refined state, however, the greater the domestic value added. Moreover, copper exported in an unrefined stage may include unidentifiable valuable minerals that the country may lose without even knowing of their existence.

Zaire is now planning to expand its refinery capacity to process all of its output domestically by 1980. Mutual benefit might be derived from planning this with Zambia, which has, historically, refined a higher proportion of its output than any of the other CIPEC countries. Zambia, on the other hand, is expanding its sales of wire and cable through Zamefa, a government plant managed by the U.S. firm Phelps

Dodge, which sells its own imported products to supplement those
produced in Zambia. As a result, Zambia has ended imports of these
products from Zaire, imports that previously constituted an important
market permitting the establishment of a Zairois wire and cable factory. The possibility of attaining greater economies of scale by joint
planning of the refining and fabricating processes ought to be considered. These possibilities could be still further enhanced by including
Botswana, which, at present, is compelled to ship its unrefined ores
to an Amax-owned refinery in the United States, and thence to Metal-
gesellschaft in Germany. It would seem preferable from an overall
African viewpoint to process Botswana ores in a Zaire-Zambia refinery and fabricating complex, but Amax refused to permit this. The
construction of the infrastructure needed to make this possible would
also contribute to reducing Botswana's dependence on Southern Africa,
opening the door for further integration of its industrial and agricultural sectors with that of its northern neighbors.

The initiation of the joint planning required to expand regional
copper-based linkages, the conference participants concluded, would
best be planned in the context of cooperation for the development of a
more extended regional industrial complex. Fabricated copper is
sold primarily to factories that process copper into a range of products to meet the needs of final consumers. The copper-exporting
countries would find difficulty in selling their processed output in
competition with the already established copper-based consuming industries of Europe and the United States—markets that are protected
by tariffs and the interests of multinational firms. Increased copper
fabrication in the copper-producing countries requires the expansion
of the copper market in the developing countries. This in turn necessitates extensive industrial growth geared to the economic and social
transformation of the entire region.

Planned Reinvestment of Surpluses

The Lusaka conference participants stressed the necessity of
careful investment of the surpluses captured from the mines in the
context of a long-term industrial strategy. Ann Seidman in Chapter
22 showed that the adoption of an import substitution policy for industrial development in Zambia was based on the notion that government
should concentrate its augmented expenditures on expanding administrative capacity and social and economic infrastructure. This, in
turn, assumes that private firms, mainly foreign, must be attracted
to provide the skills, technology, and capital to expand industrial sectors. The Zambian government's holding company for industry, Indeco, functioning autonomously, is expected to work together with its

foreign "partners" in response to short-term profit maximization possibilities to establish industries to produce goods formerly imported.
The consequence of this approach has been that the industrial sector,
tripling in size since independence, has tended increasingly to produce luxury items demanded by the higher-income groups; to process
imported parts and materials; to be capital-intensive, using imported
complex machinery and equipment that employs relatively few workers
per unit of investment; and to be concentrated in the relatively developed strip of the nation lying alongside the line-of-rail leading south
from the Copper Belt. The failure of industry to spread increasingly
productive employment opportunities into rural areas has tended to
increase, rather than reduce, external dependence. At the same
time, the expansion of infrastructure and administration has burdened
the government with heavy current operating expenditures, which have
to be met, when world copper prices and tax revenues fall, by borrowing funds. Participants emphasized the crucial role that wrong foreign trade policies had played in promoting the production of luxury
and capital-intensive goods. Attention was drawn to the risk that African countries might be repeating the same error committed by the
Latin American countries in past decades. It was asserted that there
was a wide field for a much more efficient industrialization in labor-
intensive industries producing goods for the use of low-income people.
Chapters 7, 9, 17, and 22 shed light on different aspects of these possibilities.

In the discussion in Chapter 22, as well as in Luabeya Kabeya's
analysis of Zaire's expanding industrial sector (Chapter 17) it was
suggested that a long-term industrial strategy should be formulated
to ensure the construction of factories to produce items designed to
meet the productive requirements of the economy and to increase the
levels of living of the majority of the people in the low-income sectors.
Few people, it was pointed out, can afford to buy imported items
like television sets or private autos. What they need is a range of
simple farm tools, sturdy public transport that can travel into rural
areas to collect agricultural produce, improved construction materials, and small factories to process their output for use in the resulting expanding local consumer markets. At the same time, perspectives for the establishment of basic iron and steel, chemicals,
engineering, and associated industries should be worked out to permit increased production of technologically appropriate tools, machinery, and equipment for all sectors, including the mines. Over
time, this would help to reduce dependence on the technologically
sophisticated imports requiring heavy capital expenditures and foreign skilled operatives supplied by the multinational corporations.

A debate emerged in the discussion of Chapters 23 and 24 on
Botswana and Papua New Guinea over whether formulation and imple-

mentation of such a longer-term development strategy should even be
attempted in these more underdeveloped, predominantly agricultural
economies. Consensus seemed to emerge on the inevitability of some
form of industrialization. Already, in both countries, the urban
drift has begun. In both, a limited range of industries has begun to
be established, but they could not provide adequate employment nor
were they adequately linked to the stimulation of agricultural productivity in rural areas where the majority of the populations still live.
It was suggested that it was important to begin planning in these countries as soon as possible to ensure that revenues obtained from the
mines contribute to more balanced industrial and agricultural growth
than that which characterizes the more advanced mineral-based CIPEC
countries.

The analysis of the Zambian experience suggests the danger of
a long-term development strategy relying on competitive "market
forces" to ensure efficiency or appropriate resource allocation patterns. In the first place, the typical Third World market cannot consume the output of more than one or a few plants, even in consumer
goods industries. Second, private, or even parastatal firms, seeking
to maximize their profits, may formulate investment or price policies that contradict national requirements. Institutional changes are
required, especially in the area of investment in and management of
basic industries, import trade, and banking.

Participants in the Lusaka Conference from the Economic Commission for Africa emphasized further that the small economic size
of individual African nations necessitated implementation of a regional industrial strategy based on joint infrastructural and industrial
projects. These might initially be based on cooperation to produce
inputs for copper mining and to fabricate copper outputs, as outlined
above, but it should be extended to increase specialization to attain
economies of scale in all industrial sectors. Both Zambia and Zaire,
it was pointed out, are attempting to build iron and steel industries,
as in fact are the neigboring countries of Tanzania, Uganda, and Kenya.
The existing narrow market and heavy capital investment required
have initially restricted Zaire to establishing a plant to use imported
scrap rather than develop its own inland iron deposits. It would make
far more sense for each country to specialize within the context of a
regional plan. Zaire, for example, could expand its iron and steel
project, while Zambia might focus on development of a petrochemicals industry, built around the existing petroleum refinery on the Copper Belt, to serve the needs of both countries. These possibilities
should be examined carefully in the context of an overall regional
plan.

(The conference participants did not attempt to spell out the details of a regional industrial strategy. The Zairois participants indi-

cated that they would arrange a joint conference with participants
from Zambia and Botswana in the near future to explore the possibilities in greater depth. The Economic Commission for African participants agreed to assist in arranging such a meeting.)

UNFINISHED BUSINESS: THE FORMULATION OF PROPOSITIONS OF RELIABLE KNOWLEDGE

A final session of the conference was devoted to consideration
of a series of propositions presented by Robert Seidman (Chapter 25),
which attempted to draw out of the other essays and discussions a
set of logically connected explanations of the basic problem that
seemed to confront all the copper-based economies: the persistence
of poverty and growing unemployment among the majority of workers
and peasants of all the countries, despite the relatively great wealth
generated annually by the predominant mining sectors. It was agreed
that the conference had not yet provided evidence to substantiate all,
or even most, of the explanations suggested. Yet enough information
and data had been advanced relating to the last 20 years of experience
of the major copper-exporting countries to indicate that they should
stand as hypotheses that ought to be tested. If, over time, they
should be proven valid, they would then constitute the kind of reliable
knowledge that might lead to more useful policies for solving the problems, not only of copper-based economies, but all mineral-rich countries—perhaps even, as one participant suggested, for all monoculture
export-dependent nations.

The Lusaka conference participants expressed their deep appreciation to the Government of Zambia for paying local costs and to
UNIDO (United Nations Industrial Development Organization) for financing the travel of those participants who could not otherwise have at-
tended the first conference.

NOTES

1. Federal Report on the Copper Industry, 1947 (Washington,
D.C.: Government Printing Office, 1947).

2. Compare Theodore H. Moran, "The Multinational Corporation and the Politics of Development: The Case of Copper in Chile,
1945-1970" (unpublished Ph.D. dissertation, the Department of Government, Harvard, 1970).

3. Anglo-American Corporation, Annual Report, 1972.

4. A.Geddicks, "International Production and the Case of Chile,"
newsletter: Natural Resources and National Welfare: The Case of Cop-

per Mining (Madison: University of Wisconsin, mimeo.), no. 2,
March 1973.

  5. This is not to say that very important books and articles
had not been written dealing with these issues—for example, C. Harvey and N. Bostock, Economic Independence and Zambian Copper
(New York: Praeger Publishers, 1972); and M. Burroway, The Color
of Class (Lusaka: Institute of African Studies, 1973).

  6. Industry Miniere of Developpement in Zaire (Kinshasa:
Presse du Universite du Zaire, 1974). Some of the papers presented
at Lusaka were chapters translated by Neva Seidman, who received
the thanks of the conference participants for her typing and general
assistance in making the conference a success.

  7. Much credit for the success of the Lusaka conference should
go to Nora Hamilton, then a graduate student at the University of Wisconsin, who edited this newsletter in her spare time.

THE COPPER MARKET:
DEMAND, SUPPLY, PRICES,
AND STRUCTURE

2

COPPER PRODUCERS AND
THE WORLD MARKET:
A CASE STUDY
D. Jelenc

Copper receipts, cost of inputs, cost of development, dividends, and tax are interrelated variables. Taxation and dividends depend on copper receipts, cost of sales and capital expenditures, and other appropriations.

The price of copper largely determines the copper receipts, whereas the cost of sales and capital expenditures depends on imported inflated prices of materials, equipment, machineries, and services. The operating profit before taxes is largely the function of prices on world market.

The depressed prices of copper and inflated prices of inputs threatened, in 1971-72, to close the gap that had formerly been sufficiently open to accommodate fair tax revenues and dividends. Though the copper price increased in 1973 the cost of sales continued to rise too. To keep the gap between the copper receipts and sales cost plus capital expenditures open enough, the copper-producing countries will be tempted to put a squeeze on industrial countries by raising the price of copper. Though the ratio of the price of copper to the cost of sales might remain constant over a longer period, the relative increase of capital expenditures may reduce the tax and the dividends.

Capital expenditures are expanding at a fast pace. Less than half of the capital expenditures are invested in net increases of mines' capacities. The larger part is necessary to counteract depletion of deposits and the decreasing content of the ores, and to replace obsolete plants.

There are two possible alternative outcomes: to slow down copper mine expansion and/or reduce taxation and dividends or attempt to get a fair and stable price for the copper in the world market.

FIGURE 2.1

Copper Mines Equation

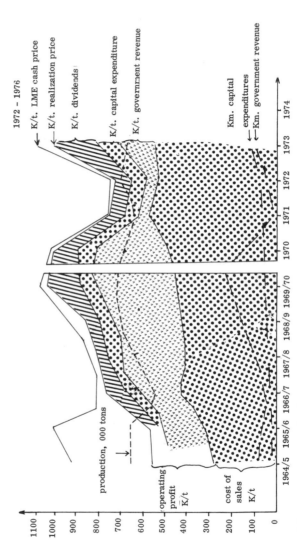

Source: Data on file at Ministry of Planning and Finance, Lusaka, Zambia.

40

COPPER MINES EQUATION*

The defense of the reasonable copper receipts and of low cost is of paramount importance for countries where copper is responsible for the majority of exports, government receipts, and even gross national product (GNP). The importance of copper receipts in relation to the other factors may be derived from the equation:

$$R - C - I = T + D \ (1)$$

where

R = copper receipts
C = cost of sales
I = capital expenditures
T = tax
D = dividends

Its graphic presentation is shown by Figure 2.1.

The variables of the equation will be examined one by one in relation to Zambia in the context of the world market. The aim is to assess the degree that national level of decision-making can, in present institutional conditions, influence them to obtain optimal benefits.

COPPER RECEIPTS

The most important factors determining copper receipts are the level of physical production and the price.

The physical production of copper does not display dramatic changes for two reasons: first, the limitation of proved ore reserves, which correspond to the criteria of profitability used by multinationals, and, second, the limited capital directed to the development. As a result of both factors, production is stagnant for many years. Most capital expenditures are aimed at replacing the production of the depleted deposits and the obsolete plants. The most important factor affecting receipts, therefore, is the realization price of copper that largely depends on London Metal Exchange quotations. Therefore this paragraph volens nolens deals mainly with the behavior of the copper price and methods forecasting and controlling it.

---

*The use of "equation" is to show the interlinkage of variables. It does not, however, imply regularities of world market. On the contrary, these irregularities are responsible for the uncertainties of income from copper.

METHODS OF COPPER PRICE FORECASTING

In the methods of copper price forecasting, two components are clearly distinguished: more regular long term and the irregular short term. The long-term component of gradual increases of prices in real terms is guided by the gradual deterioration of mining conditions because of the poorer content of copper in the ores and more complex nature of copper in some part of the deposits. This effect has exercised a constant pressure via higher cost, causing constant copper price increases during the last few decades.

It was observed by A. E. Notman among others in 1933 that in the long run the ratio between average world mining cost and copper price is constant. Therefore in the long run, forecasts of mining cost also determine the copper price projections.

A more sophisticated projection of cost is given by T. J. O'Neil using a computer program considering each cost element separately.[1]

The short-term fluctuations are a result of factors in which the changes of sales costs in real terms are less important. For copper producers they are exogenous factors depending on LME "ruling," reflecting certain market anatomy and conditions. The LME and similar institutions do not base their quotations on value of work contained in the refined copper and on the socially reasonable surplus required by the producing countries. They exploit the fact that the producers are separate entities that at present do not efficiently control their production in order to meet the consumer's demand. The real interests of the producers and consumers are neglected. The LME quotations are in addition largely speculative.

Some models have been developed to forecast short-term copper prices in spite of this. The only models used are econometric, based, in essence, on demand/supply projections. The supply is determined by assessment of copper mines expansion, projected by project. However, supply is commonly overestimated in such studies.

Demand is assessed by considering commodities containing a substantial proportion of copper, using the following formula:

$$C_1 = c_1 \times a \times p \times F \ (2)$$

where

C_1 = forecasted copper consumption of first commodity
c_1 = present copper consumption of first commodity
a = ratio of production, forecasted/present
p = forecasted penetration ratio of copper
F = ratio of prices of substitution materials forecasted/present

Based on analogy, using formula 2, C_2, C_3 . . . C_n will then be forecasted. Substitution becomes an important phenomenon if the prices of copper are high compared with the prices of substitution materials.

The increases or decreases of prices of substitution materials are taken into account in the formula (2) by the factor F of the same equation. For example, the aluminum price* 99.5 percent CIF (cost, insurance, and freight) Europe, has risen steadily over the last few years from $4.21 per metric ton in 1971, to $1,017 in 1974.

The total copper consumption is calculated from individual results obtained by the formula (2) for single commodities:

$$C_{Total} = C_1 \cdots + C_n \quad (3)$$

The demand projections are prepared by multinationals, various national/international associations, and fabricators. They are not normally published. For the governments of copper-exporting countries, it seems to be a matter of high priority to join efforts in obtaining and improving available information on demand projections.

Various estimates may also be obtained that take into consideration the secondary copper supply.

Nevertheless, as the assumptions to be used in demand and also in supply forecasts are vague, in the long run the resulting price forecasts are uncertain.

METHODS OF PRICE CONTROL

It would be possible to reduce the fluctuation of copper receipts by strengthening the bargaining position of copper-producing countries in the world market. This would also reduce the hazards in short-term planning of receipts of copper. Though a group to promote the position of copper in the world market is in existence (Intergovernmental Council of Copper Exporting Countries, in French abbreviation, CIPEC), its influence is at present insignificant. It could, however, be more effective in the future in stabilizing copper receipts. **

---

*The industry determines the value of aluminum vis-a-vis copper by comparing the price of "physical properties" purchased.

**CIPEC conference of ministers, 5th session, Lusaka, June 24–26, 1974, stipulated the following:

The right of countries to freely develop their economies and, in particular, adopt sovereign de-

One possibility would be to average the LME prices. The expected consequence would be the dual price of copper on the international market and would still exhibit short-term fluctuations.

Producers' prices have been enforced by the mining industry in the past. They also led to the pronounced dual pricing on the international market of copper, as shown in Figure 2.2.

Buffer funds may be used to subsidize buyers when the price is high and vice-versa.

Buyers' price is the prevailing system used recently in copper trade of CIPEC countries. The buyer selects within a limited period, copper LME quotation, fixed dates or monthly averages, prior or subsequent to shipping.*

Last but not least would be the policy of price flooring with or without a reference price and with or without ceiling. This pricing policy leads also to dual prices in the periods when the LME quotations are higher than the ceiling. Use of this method would require the formation of buffer stocks by the producers.

It seems that the above methods are cheap in cost of implementation for producer countries but have the disadvantage of introducing dual-pricing systems. This could lead to cancellations of quotas or contracts and require stockpiling of certain quantities of copper within the stores of producers, at least for limited periods.

---

cisions in respect of the ownership, method of exploitation, and procedures for marketing their natural resources was universally recognised in various international bodies. . . . Nevertheless, its exercise by the developing countries which export raw materials has posed many problems and has given rise to acts of reprisal, in the face of which the international community lacks adequate means of reacting effectively.

The member countries of CIPEC, besides being large-scale producers of copper, are nations whose development possibilities are almost entirely conditioned by the sovereign exploitation and marketing of this natural resource. Hence it is imperatively necessary for them to take concerted action in the face of whatever measures of economic or political coercion may be adopted against any member country and which affect this right.

*In Zambia, more than 50 percent of copper was sold in 1973/74 by averaging, 40 percent by fixing, and less than 9 percent by averaging/fixing options and other terms.

FIGURE 2.2

Copper Prices, Monthly Averages

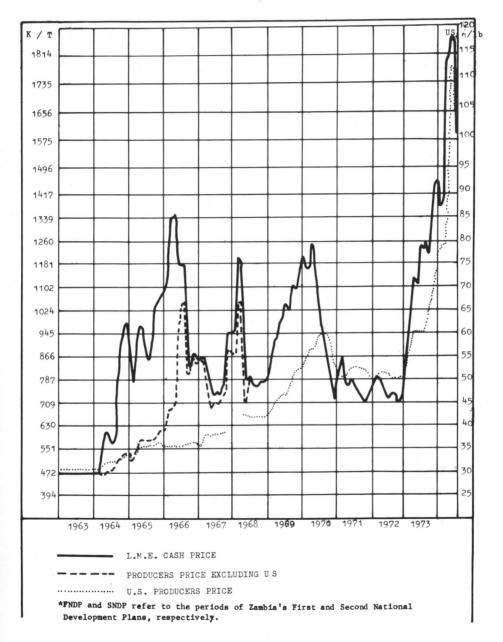

L.M.E. CASH PRICE

PRODUCERS PRICE EXCLUDING U S

U.S. PRODUCERS PRICE

***FNDP and SNDP refer to the periods of Zambia's First and Second National
Development Plans, respectively.**

<u>Source</u>: Ministry of Planning and Finance, Lusaka, Zambia.

All the above pricing methods have been suggested to <u>replace
LME quotations</u> or at least to reduce the impact of their fluctuations.

The <u>second group</u> of possible pricing schemes seems to be more
fitting in the economy of free competition. They mainly originated
from fear of substitution of copper by other cheaper materials, for
example, aluminum, plastics, and sometimes even more costly materials with advantageous physical properties. They seek to control the
volatility of LME prices by intervening in the supply and demand relation.

<u>Copper buffer stocks</u> would intervene on LME when supplies are
depleted and buy copper when the stocks are substantial and depress-
ing the price. The establishment of copper buffer stocks would, however, be an expensive venture.

<u>Export and production quotas</u> could be used to reduce the supply
when the demand is low. The disadvantage of this method is the adverse effect on development of copper-producing countries.

The establishment of an <u>international marketing</u> organization
has been suggested that would itself manage a buffer stock. It would
need to be spread all over the world and is unlikely to be implemented
because of the high cost involved.

Evaluation of a proposed pricing policy must be made in terms
of a quantitative analysis regarding its impact on capital cost and cash
returns.

The selection of the alternative could be suggested by institutions such as CIPEC. From press reports, it could be concluded that,
during the depression of prices of copper in 1972, the actual policy
supported by the CIPEC Permanent Secretariat as the most favorable
was the building up of stocks of copper and producers' stores. Other
alternatives appeared more expensive in cost required for their implementation and would have generated fewer benefits.

Since then, the situation has changed. After the oil crisis, some
economists expected that other primary producers might combine to
put a squeeze on industrial countries by deciding to cut production or
raise prices in OPEC-like agreements. When the demand for raw
materials rises again, this could indeed tempt groups of primary producers.

In any case, the above-mentioned proposals could reduce the
fluctuations of copper receipts. Complete elimination could be achieved
only if producers and consumers undertook joint action to defend the
price of copper.

Zambia's establishment of its own Metal Marketing Corporation
may promote this integration process. When this corporation is in
full operation, it will generate two benefits. First, it will give Zambia
greater bargaining power with the other nations in the implementation
of the pricing policies, and, second, it will facilitate the promotion of

a Zambian copper manufacturing industry to export finalized commodities.

COST OF SALES

The cost of sales is the cost incurred in a unit of time, normally a year.* The measure of the success of any mining is the size of the difference between realization price and cost of sales in which are normally included the wages, materials, replacement services, and general expenses. Projecting sales costs is a very important procedure not only for government planning but also to establish the profitability of the mines. The development of cost data is a difficult task. Mining sales cost is a result of a great variety of variables, including changing working conditions and mining practices, and changes in mineralization and metallurgy.

In the last years, with the decline of purchasing power of money, the cost in real terms has a meaning other than in money terms. With this in mind, the cost of sales—the second variable of our equation—displays two distinct components of changes:

The first is that noted over longer periods immanent to the secular changes in productivity. Since 1929-31, the technology has not been able to cope with the deterioration of grades, increased complexity of ores, and recently, also, other factors like environment control. Therefore there is an overall trend of increased cost of sales of copper, which also determines the long-term regular increase of prices since then.

The second component of cost of sales increases, which has had a predominant role since the monetary crisis developed, is an overall race of escalation of prices in the international markets. Zambia's economy, based on copper, is heavily dependent on inputs from imports. The imported inflation has had a predominant influence in the formation of the cost of sales of copper mines, which rose sharply by 13.7 percent from 1971 to 1973.

The only impact of changed technology on mining cost in this period was due to the use of sand filling in the rehabilitation of the Mufulira mines. The resulting permanent increase in cost of sales of this mine in 1971 had an inevitable impact on average sales costs of Zambian copper.

---

*In Zambia, the Nchanga Consolidated Copper Mines' (NCCM) fiscal year ends in June and Roan Copper Mines' (RCM) ends in March. Therefore the calendar annual cost is summarized from quarterly financial reports. In addition, the cost of sales is not identical with the production cost.

Cost of sales are affected most by the nature of the deposit and the system of extraction used. The open-pit mining in copper permits the use of the development of high-productivity mining. Only half of Zambia's production is obtained from open pits. The main cost element determining cost of sales are labor, materials, power, and others (replacement, general, services, and so on).

The escalation of cost due to inflation may be discussed in terms of the individual elements:

Salaries and wages in 1973 were 118 if 1971 was 100. This is a result of a new three-year agreement negotiated in 1973 for Mineworkers' Union of Zambia employees. These increases are related to the increases of consumer price, which, compared to 100 in 1969, reached 126 in 1973. (Wages constitute about 36 percent of total sales cost.)

In material, about 75 percent is imported. The import price reached index 123.7 in 1973, given a 1971 value of 100. As the domestic materials are based on imported manufacturing industry inputs, their contribution in cost increases is not less pronounced with the index 122.6 in 1973 considering the same comparative basis.

The cost of power displays marked differences as the metric ton of coal cost is now 250 percent more than 10 years ago, the cost of mineral oil products 300 percent more. Electricity, for which costs have increased only slightly, is the exception. Fortunately, over 70 percent of the total power supply in Zambian Copper mines is provided by electricity.

Other costs encompass the following:

The replacement, which, in view of the more sophisticated technology introduced in the mines, besides the inflated prices for such technology, is becoming increasingly important. At present, an accelerated depreciation is used that is identical with the annual cost of the equipment, machineries, construction, and services required to maintain the mines. The size of plants is also becoming larger because of continuous deterioration of mining conditions. In view of limitations regarding the availability of mineral resources in nature, the supply cannot be expanded ad libitum without expanding copper mining into deposits that have subeconomic contents of copper in the ores. It is obvious that the present rich deposits will be exhausted in a few years, if world consumption continues at the same rate as, for example, that of the United States today.

The cost of services is increasing on a large scale. In the past, these included fees paid for management services of Anglo-American Corporation and the Roan Selection Trust based on operating profit and sales.

The general expenses include the cost of transportation, which, for the copper exports, has remained pretty stable. The financial cost of keeping the copper in the pipeline is different because of changed distances of the export routes.

Other costs combined show marked increases from a base of 100 in 1970 to about 105 in 1973.

DEVELOPMENT COST

The third variable of the equation is the annual capital expenditure for development. The extent of capital expenditure is a policy parameter, controlled mainly by management and boards of directors. However, this cost depends on the cost of capital goods overseas and the cost of mine development. The capital goods cost is subjected to fast increases. Even the special engineering services required in mine development, such as shaft-building, grouting, and so on, are subject to prices formed by foreign countries' economic policies. Engineering services are often a monopoly of industrialized countries.

The determination to invest in new capacities is largely a function of the investment climate formed by general economic conditions and legislation, as well as the availability of attractive proved ore reserves. The role of the Zambian Mining, Tax, and Investment Acts have been important factors encouraging new mining ventures and will be discussed more in detail below.

In the present policy of the companies, development cost is also increasing in current money terms because of the inflation of capital goods and prices. It is financed from operating profit. The amount financed by loans has been negligible, but their role in the development of infrastructure has been much more pronounced—largely through the government capital expenditures. The foreign loan component of electricity supply development, for example, has been substantial, amounting to over 50 percent.

The internal composition of the development cost is shown by the following equation:

$$I = cf \times t \times K^2/P \ (4)$$

where

    cf = capital output coefficient (kwachas per ton of capacity)
    t = growth rate of capacity (percent)
    K = production (tons)
    P = production cost (kwachas)

The capital cost for the net expansion of capacity is only that spent for this purpose in any stage of the copper extraction process, at any of the following intermediate capacities:

1. Copper mine production: Mining production is usually assumed identical with the capacity of concentrators. Strictly speaking, however, it is composed further of individual capacities of stopes in the underground, open-pit mines, tailing, disposal reclaiming, milling, and benefication facilities.

2. Copper extraction: Smelter production capacity is in turn composed of individual capacities such as verberatory furnaces (producing matte), converters (producing blister), and so on.

3. Leaching capacity is often complementary to smelters, mainly for the acid-soluble ores (oxides, silicates) producing cathodes.

4. Other processes complementary to 2 and 3 for refractory ores, which are not suitable for smelting and standard leaching.

5. Copper refining: Electrolytic refining in tank houses producing cathodes.

6. Fire refining by furnaces producing wire bars or ingots.

7. Copper casting: Casting of shapes, most commonly wire-bars but also rods as required.

Appropriations of surpluses for development are very large compared with the production gains achieved. The part of capital expenditure used for expansion of productive capacities is only a fraction of the total amount. The part for net expansion could be derived from the total only by contemplating modifications on production* and capital expenditures** sides.

Moreover, there is a time lag effect on the investments that makes comparison with production gains impossible, unless considering project-by-project assessment. Today, also due to inflation and lower metal content in ores, much larger capital expenditures are needed in order to expand the copper mines to create the same amount of capacity as 5 or 10 years ago.†

The case of Zambian copper mines proves that, taking all these factors into consideration, the majority of the capital expenditures are needed in order to maintain the level of mines capacities due to the depletion, reconstruction, and the development of import substitution projects.

---

*(1) Capacity/production ratio; (2) decrease or increase of ore grades slightly influencing the concentrating capacity; (3) quality of fuels used in pyrometallurgy.

**(1) Mine development necessary to maintain the level of productive capacity; (2) the capital output coefficient for different projects.

†The capital coefficient growth indexes in 1973 (1965 = 100) per ton of expanded capacity were estimated for world mining conditions as 192 for new mines and 189 for old mines.

TAXATION

Taxation is the main variable determined by government policy decision. The benefits of the taxation for the government are, however, unpredictable and not controlled within the producer's country as it depends on the operating surplus and on capital expenditures. The taxation mechanism is controlled by the nation, but the results of taxation depend on factors outside of the national control.

Contradictory factors tend to influence policy decisions affecting taxation. Zambia's government, for example, recognizes the need to encourage the development of mineral resources, particularly those that would lead to diversification of production. This policy will take into account that the mineral resources represent an irreplaceable asset, which might bring increased benefits if developed when the country's industrial development may need them.

It would, however, be unrealistic not to embark now on development of mineral resources including the continuation of copper development in the circumstances of substantial lack of employment opportunities, the increasing costs of mining, compared to average world mining costs, the need for capital that can be provided from the mines, as well as the lack of general conditions for development that could be provided by the infrastructure created as an aspect of the expansion of the mining sector.

Until independence in October 1964, royalties on copper production in Zambia were paid to the British Africa Company on the basis of the price of copper irrespective of profitability of individual mines. Early in 1964 the major copper companies departed from the London Metal Exchange quotations and introduced producer prices, which were at the time much lower. The disparity between the LME price and the price of copper received by the Zambian mines on which the royalties had been based required revision of the tax policy.

After the Government of Zambia purchased the mining rights from the British South African Company in 1965, a mixture of direct and indirect taxes was enforced in 1966 to redirect to the government some of the windfall gains being enjoyed by the companies when the producer prices came close to LME copper price again. A royalty per ton of copper was paid at roughly 13.5 percent of the LME monthly average copper price minus U.S. $22.4 per long ton. In addition there was an export tax of 40 percent of the LME price minus U.S. $840 per long ton. The corporation tax was 45 percent of the remaining profit. The rates have been the same for all mines, although occasionally rebates and even refunds were granted to individual mines, because of large differences in mining costs of different mines.* The

---

*The cost of producing a ton of finished copper varied greatly between Zambia's mines. Mufulira mines cost of mining was 100 in

disadvantage of this system was that no depletion allowances and other fiscal facilities were granted to the new mines. The effective overall rate of taxation basically paralleled gross profits and the LME price.

A changed taxation policy was introduced in 1970 based mainly on mineral tax and income tax at a level that at the same world price level compared roughly with the superseded former royalty and export tax. Capital expenditures were entirely deducted. From the remaining operation surplus, 51 percent (the percentage is different for various metals) of profit is the mineral tax, and of remaining profit 45 percent is the income tax. The combined tax on the books was 73.05 percent of the operating profit after the appropriation for capital expenditure was deducted. This system makes no difference between the operating cost and capital expenditures. The last mentioned is totally deductible. As the capital expenditures were expanded as a result of this policy, and the average price of copper in 1970-72 was low, the tax receipts were reduced drastically, compared with the record year 1969.

All the systems, however, generate fluctuating government receipts because they are based on participation of taxation in operating surplus. The fluctuations display a time lag as the date of tax payment is different from the period of its generation.*

CONCLUSIONS

The international market controls the receipts (R) through the copper price quotations and the cost of sales (C) through the prices of imported commodities and services as well as indirectly the cost of locally produced commodities, in turn depending on imported inputs.

In the past, the management of multinational corporations determined the capital expenditures (I). As a majority of shareholders, the government may influence the capital expenditures now that its 51 percent share portion has been fully paid for. (The establishment of management of NCCM and RCM has been recently announced.)[2]

1967; the Chililabombwe (Bancroft) mines cost was nearly double, namely 188. Investment in high-cost mines was discouraged. The tax policy also led to inflation of break-even points and wastage of mineral resource reserves as the cutoff points of the ores have been increased correspondingly.

*The mineral tax is paid five to six weeks after the end of the month, during which it is incurred. The income tax is paid in two equal installments depending on company fiscal year, so there is a relatively large time lag between tax "generation" and "payment."

The government is the policy-maker regarding the taxation (T). The actual ability to exercise this power efficiently is minimized because of the "enclave" character of the taxation depending on the state of the international market. Dividends (D) are the residual of the equation.

Nonproducer countries are mainly responsible for the main variables, copper receipts, cost of sales, and capital expenditures. Even changes in one parameter against other parameters are possible only if the international context is fully taken into consideration.

Econometric methods are of some value for short-term planning of copper receipts. More reliable long-term planning of the receipts is based on forecasting of cost of sales that exhibit regular increasing trends since the year 1930, primarily due to the deterioration of mining conditions.

In the framework of the copper mining equation, almost all variables are determined by the world market. The planning policies can manipulate the growth of capital expenditures for mine expansion and taxation but only within limitations of the world market constraint.

The world market "parameters" are the relevant factor for national development. The management of the mines and national mineral policies in general should not be oriented only into "operational management" within one country but should be considered in the context of efforts to influence the international market as that is the only place to defend the real value of copper. Intensified international cooperation of the copper-producing countries is needed to increase their bargaining power vis-a-vis the world market in order to improve the terms of trade.

NOTES

1. T. J. O'Neil, "Estimating Minimum Copper Price Levels Through Production Cost Projections" (Tucson: University of Arizona, 1973).

2. More about formation of state-controlled companies in M. Bostock and C. Harvey, Economic Independence and Zambian Copper (New York: Praeger Publishers, 1972).

OTHER SOURCES

Battele Centre de Recherche de Geneve, report prepared for CIPEC, 5 vols., Paris, 1972.

CIPEC, various reports, Paris.

Conference of Ministers, 5th sess., Lusaka, June 24/26, 1974. Paris, 1974.

Dawis, J. E. "Mining Tax Variations in North America." Mining Engineering, September 1970.

Lovell, J. D. Copper Resources in 1970. New York: Mining Engineering, 1971.

MacKenzie, B. W., and C. G. Delbridge. "Changes in Taxation and the Profitability of Mine Development." Canadian Mining Journal, July 1971.

COPPER:
MARKET PROSPECTS FOR
1980 and 1985
Kenji Takeuchi

The world copper market is likely to be in a state of surplus in 1980, but the surplus may largely be eradicated by 1985. Based on the analysis of likely developments of demand and supply, the price of refined copper in London is projected to be 130-140 U.S. cents per pound in 1974 dollar terms in 1980 and about 200 cents per pound by 1985. These price levels are equivalent to about 80 and over 90 cents per pound in 1974 constant dollar terms respectively. (Inflation in "international prices" is assumed to be 8.7 percent a year in 1974-80 and 7 percent in 1980-85.)

DEMAND FOR COPPER

World* consumption of refined copper is projected to increase from 6.2 million tons (metric units) in 1972 to 8.6 million tons in 1980 or at 4.2 percent per annum. The long-term growth trend of consumption over the period 1950-73 is 4.2 percent per annum. World consumption in 1985 based on the same trend would be 10.5 million tons. The above projections of world refined copper consumption assume that, while economic growth in developed countries and non-oil-exporting countries might somewhat slow down in the coming decade, industrialization in oil-exporting countries would accelerate in the future. Copper may also benefit from the relatively more rapid increase in the prices of aluminum and petrochemicals; although cop-

---

*In this chapter, "world" excludes the centrally planned economies unless otherwise noted.

per will probably not regain markets previously lost to aluminum, copper's competitive position vis-a-vis aluminum and plastics has now improved as a result of the sharp increase in petroleum prices.

World demand for primary copper, including projected net exports to the centrally planned economies (CPEs), in 1980 and 1985 would reach 7.8 million tons by 1980 and 9.5 million tons by 1985. These projections assume (1) that the share of newly mined copper in world consumption of refined copper will average 86 percent in the future as it did in the past 10 years and (2) that net exports of copper to the CPEs will be 200,000 tons and 250,000 tons in 1980 and 1985 respectively. Net exports to these countries have been averaging 50,000 to 100,000 tons in recent years.

WORLD SUPPLY OF PRIMARY COPPER

A projection of world copper mine capacity is subject to a more than usual uncertainty at this time. Based on estimates made by industry sources and by the World Bank staff, world mine capacity in 1980 has been projected by region and by country as presented in Table 3.1. "Low" and "high" estimates are shown for major countries and regions for 1980. World mine capacity is projected to increase from 6.6 million tons at the beginning of 1974 (not shown in Table 3.1) to 8.1-9.6 million tons by 1980. No detailed projection by country or region is possible for 1985.

Latin American capacity depends on developments in Chile in the coming six years: Depending on the success of economic management, mine capacity could increase to 1,325,000 tons as planned, or only moderately to 1.1 million tons or show no significant increase by 1980. Chile presents the largest uncertain factor on the supply side of the world copper market for the next several years. The possible development of a large copper mine project, the Cerro Colardo, in Panama presents additional big uncertainty. For the present it seems unlikely that this mine will be in operation by 1980. Other uncertain possibilities in this region exist in Peru and Mexico.

As for other developing regions—such as Africa and Asia-Oceania—despite the relatively small differences between the "high" and "low" estimates given, uncertainty is not totally absent.

Among the developed countries, significant expansions are possible in the United States, Canada, Australia, and South Africa. However, recent trends in Canada and Australia toward reorientation of their mineral investment policies could have significant dampening effects on the pace of investments in copper-mining capacity in these countries.

TABLE 3.1

Copper: Projected World Capacity of Mine Production in 1980 as
Compared with Actual Production in 1972
(thousands of metric tons)

	Actual Produc-in 1972	Projected Capacity in 1980	
		Low	High
Developing countries			
Latin America			
Bolivia	8	10	20
Chile	717	1,100	1,325
Mexico	79	140	275
Peru	217	400	500
Panama	0	0	100
Other	9	10	20
Subtotal	1,030	1,660	2,240
Africa			
Rhodesia	32	35	45
Mauritania	15	20	25
Zaire	437	700	850
Zambia	718	900	970
Southwest Africa	22	55	130
Other	19	45	
Subtotal	1,243	1,755	2,020
Asia/Oceania			
Philippines	214	265	300
Iran	0	100	140
Indonesia	5	70	70
Papua New Guinea	124	175	270
Other	46	115	130
Subtotal	389	725	910
Developing countries total	2,662	4,140	5,170
Developed countries			
West Europe	297	445	500
Japan	112	50	100
United States	1,510	2,030	2,130
Canada	709	1,000	1,150
Australia	181	260	350
South Africa	162	200	240
Developed countries total	2,971	3,985	4,470
World (excluding CPEs)	5,633	8,125	9,640

Sources: Mine production in 1972: World Bureau of Metal Statistics, World Metal Statistics, December 1973; 1980 capacity projections: Estimates by IBRD Commodities and Export Projections Division; partly based on estimates by IBRD Regional Offices and Phelps Dodge Corporation, Statistical Section.

One of the problems for a rapid expansion of world copper mine capacity is the availability of capital. Capital investment required to establish a capacity to produce refined copper is estimated at around $4,000 (in 1974 constant terms) for each annual ton of capacity. Thus, if the "high" estimate of projected capacity—9.6 million tons per annum—is to be realized by 1980, $12 billion (in 1974 constant terms) would be required to build the additional 3-million-ton capacity needed. In view of the risks involved, however, it is unlikely that a total investment of such magnitude ($2 billion annually) would actually be forthcoming in this industry.

If world capacity is to increase by 0.41 million metric tons per annum in 1974-79 as in the 1968-73 period, capacity at the beginning of 1980 could stand at 9.07 million tons. Since projected demand for primary copper in 1980 is 7.8 million tons, the assumed capacity increases would create a surplus situation. Capacity increases at this rate can be achieved if there is an annual investment of $1.64 billion in terms of constant 1974 dollars. Actual investment in this industry would probably be slower than this, and yet the world copper market in 1980 is likely to be in a state of surplus with relatively weak price performance.

PRICE PROJECTION FOR 1980 AND 1985

The ratio of market price to the average cost of copper averaged 1.75 in the 1952-70 period. It appears that the cost of copper production has increased in "real" terms in the three years 1971-74, although it is difficult to determine by how much.* While the average cost of copper production will probably rise in real terms in the 1974-80 period, world copper market balance in 1980 is expected to be somewhat on the surplus side (as our earlier analysis tends to indicate) and the price-cost ratio in 1980 would probably be smaller than the historical average ratio of 1.75.

Assuming that the average cost of copper would rise by about 8 percent in real terms during the 1970 decade (from 32¢/lb. in 1970), the appropriate price-cost ratio to be assumed under the probable market conditions of surplus in 1980 is about 1.5-1.6.

As for the price of copper in 1985, it is assumed, on the one hand, that the cost in real terms would not rise materially in the 1980-85 period, and, on the other, that the copper market in 1985

---

*Reportedly, the extra cost of pollution controls for the existing smelting and refining facilities amounts to about 5 U.S. cents per pound of copper.

would be in near balance, warranting the application of a rate near the historical average price-cost ratio. The price in 1980 thus would be around 200 U.S. cents per pound in current dollar terms.

Above price projections assume that an effective OPEC-type arrangement will not be made or that, if such a cartel is formed, it would not be effective except to prevent substantial falls in prices in the face of massive surplus situations.

4

THE EUROPEAN COPPER INDUSTRY AND ITS IMPLICATIONS FOR THE COPPER-EXPORTING UNDERDEVELOPED COUNTRIES WITH SPECIAL REFERENCE TO CIPEC COUNTRIES

Dorothea Mezger

The present study is centered upon the relations between the European copper industry and national underdeveloped producers, especially in Chile, Peru, Zaire, and Zambia (CIPEC countries). The set of theses presented is preliminary and needs more discussion and further empirical verification.

THE EUROPEAN COPPER INDUSTRY IN THE INTERNATIONALIZATION PROCESS

As a case study of the problems of development and underdevelopment, this analysis is centered upon the internationalization of raw material economies, characterizing the development of international capitalist production.*

As a general rule, with some exceptions, copper is not extracted where labor is cheapest but in those areas where sufficiently large and rich copper deposits are found. This is true for mining, that over the time has become so capital intensive that wage costs have only a reduced impact on total costs. In processing of refined copper, wages have a major impact since manufacturing is less capi-

---

The editor wishes to express her appreciation to the Max Planck Institute for having made it possible for Ms. Mezger to make this contribution.

*The internationalization of the labor force as a new stage of capitalist production aiming at worldwide use of cheap labor plays a minor part in this context.

tal intensive, and substitution of copper by aluminum or other substitutes depends directly on the price of semifinished goods.[1]

The essential theses of this chapter are (1) that in future European and non-European copper corporations will seek to extract surplus value from underdeveloped countries by enforced development of technology, which will be sold in the form of licenses and so on and management instead of making profits through direct investment in production in underdeveloped countries; (2) that possibilities for underdeveloped national producers to organize a functioning cartel do exist, but that it will be rather ineffective; (3) that there are hardly any chances for national underdeveloped producers to raise price levels for copper substantially in the long run; (4) that the old dependency of copper-exporting countries will be replaced by a new type of dependency, by which the risks once assumed by private capital will be shifted onto the countries; and (5) that due to the tendency to rising copper prices, which are in the interest of both multinational corporations and underdeveloped copper-exporting countries, the related general augmentation of price levels may lead to a loss of real income of consumers. Moreover, increasing financing of the mining industry through public funds implies a higher rate of exploitation of non-capitalist classes, both developments tending to contribute to a deterioration of living conditions for large parts of the population of industrialized countries.

The analysis of the European copper industry will be carried out, examining the production process and the circulation process, as parts of the cycle of capital invested in this industry. The role of financial capital as fractions of industrial, trade, and banking capital as well as the role of the state and parastatal international organizations as agents of internationalization of the copper industry will be examined. An introductory section will demonstrate the share of European and CIPEC countries as well as some other countries in production, consumption, and international trade. Some information is given on technical details of copper production in order to facilitate understanding of the production process.

The European copper industry is defined in this study as the industry of the European Community (EC) member countries. It comprises corporations and institutions in mining, smelting, refining, and manufacturing. The Anglo-American corporations referred to are those with predominantly U.S., Canadian, South African, and British capital. The British copper industry is, as far as refining and manufacturing are concerned, included in the European copper industry, whereas British capital invested in mining is included in British-American multinational corporations. This fairly arbitrary definition will have to be revised as analysis of relations between both groups of industries proceeds.

PRODUCTION, CONSUMPTION, AND
TRADE IN COPPER

Copper Production of Important Producer
and Consumer Countries (Mainly
EC-CIPEC Countries)

The figures given in Table 4.1 comprise primary as well as secondary copper. Production comprises mining, smelting, refining, and manufacturing of copper. Whether copper is produced in the form of concentrates (ore) or as refined copper has far-reaching consequences for the industrial structure of the producing country that exports copper. (For example, if a country with its own mines exports its copper predominantly in refined form, it possesses an integrated industry, which is not the case if it exports predominantly in the form of concentrates.) Industrialized consumer countries have the choice of either producing copper from imported ores, producing refined copper from imported blister copper, or importing copper in refined form.

Table 4.1 shows the difference in attitudes of various European countries and of Japan with regard to the provision of industrial imports. West Germany especially is a relatively strong producer of copper from ore, following in this respect to some extent the outstanding example of Japan, with a high degree of smelter production from imported concentrates. In West Germany, there is also considerable production of refined copper from imported blister, to a slightly higher degree than in Belgium. Belgium has only an unimportant amount of smelter production yet is a strong producer of refined copper from blister, the bulk of which comes from Zaire. Great Britain has a relatively low share in refined production, importing nearly all copper in refined form. The reasons for this are historical ones. Due to transport difficulties in Zambia from which Britain imported great quantities, there was need to establish an integrated industry from mining to refining in order to minimize transportation costs. All other EC countries do not have a remarkable production of primary refined copper, not to speak of smelter production. It is evident that EC countries and Japan, as well as the United States, which is not included in Table 4.1, account for their overwhelming proportion of manufactured production, be it from copper previously smelted and refined within the country or directly from imported refined copper, whereas copper processing of the copper-producing CIPEC countries is negligible. To a great extent these countries have to rely on imports of manufactured items from industrialized countries.

TABLE 4.1

Copper Production of West European Copper Buying Countries and Producer Countries from Which They Buy
(thousands of tons)

Countries	1952	1972	1952	1972	1952	1972	1952	1972
Belgium–Luxembourg	—	b	12.0	13.0	211.8	314.2	85.8	195.2
Germany (F.G.R.)	b	b	42.4	159.4[a]	75.2	398.5	272.0	971.8
France	b	b	0.5[c]	—	23.9[e]	30.0[d]	277.5	546.7
Great Britain	b	b	18.1	—	228.9	180.7	742.5	758.9
Netherlands	—	—	—	—	84.0	80.0	102.8[e]	n.a.
Italy	—	—	—	—	—	—	204.4	442.0
Japan	53.6	111.9	49.3	724.2[g]	94.4	810.1	n.a.[i]	1,292.0
Chile	408.9[h]	720.5[h]	382.8	630.6	307.7	461.4	n.a.[i]	n.a.[i]
Peru	30.4[h]	230.0	20.5	176.2	20.5	39.2	n.a.[i]	n.a.[i]
Zambia	329.5	717.1	317.4	697.3	113.3	615.2	i	i
Zaire	205.7	437.3	201.2	427.0	109.5	280.0[j]	i	i
Canada	234.1[f]	708.8[f]	202.7	473.7	178.1	495.9	n.a.[i]	n.a.[i]
Australia	20.4[f]	180.5[f]	20.3	144.7	29.3	173.8	n.a.[i]	n.a.[i]
Mexico	n.a.	78.1[f]	n.a.	72.3	n.a.[i]	63.8	n.a.	n.a.
Republic of South Africa and Namibia	49.3	178.9	34.2[k]	193.9	11.4[k]	79.3	n.a.	n.a.
Philippines	13.2	213.7[f]	—	—	—	—	—	—

[a] Another 101.5 million tons of blister were produced from scrap.
[b] Unimportant.
[c] From ores; production of blister from scrap 6.5 million tons in 1972.
[d] From scrap and foreign blister.
[e] Semis only.
[f] Copper content of blister.
[g] Another 53.0 million tons of scrap were produced.
[h] Copper content.
[i] Negligible.
[j] Estimated
[k] Republic of South Africa only.

Notes: This data, drawn from different sources, is slightly different from that given in Table 4.2. Consumer countries are EC member countries and Japan.

Source: Metallstatistik, Metallgesellschaft, Ffm., 1951–60 and 1962–72.

COPPER CONSUMPTION OF IMPORTANT
CONSUMER COUNTRIES

The figure of total consumption of copper is obtained by the addition of refined copper, which is either primary or secondary copper, recycled from scrap and the direct use of scrap material.

Table 4.2 details the copper consumption of EC countries between 1960 and 1970. Between 1960 and 1970 the average annual growth rate of copper consumption has been rather slow. The share of European copper consumption declined in comparison with other regions, especially as compared to Asia, due to the expanding Japanese consumption. There are considerable differences in the growth rates of individual countries. Great Britain was the only one with a negative growth rate, losing first place to the F.G.R. as the biggest consumer of copper within the EC. Slightly over 60 percent of copper consumed within EC countries is primary copper produced from ore. Only France consumes as much as 75 percent in primary form.

The European industries recover as much copper as possible by recycling it, thus making use of their "mills above ground." An average proportion of 40 percent of secondary copper in total copper consumption may become an instrument to be used in price policies. Given the structure of the secondary copper-producing industry, which is fairly unconcentrated, however, this instrument is hardly effective as yet. It will become more effective to the extent to which concentration of capital in the copper secondary industries advances.

INTERNATIONAL TRADE IN COPPER OF
IMPORTANT COPPER PRODUCING AND
CONSUMING COUNTRIES

Final consumption of copper is generally in the form of refined copper. Copper can be imported as refined copper, as unrefined copper (blister), or as concentrate (ore), which has to be refined and smelted within the importing countries. Except for copper produced from scrap, nearly all copper has to be imported from foreign countries. There is a remarkable discrepancy between the different European countries with respect to their policy of copper provision in relation to the form under which copper is imported: refined copper, blister copper, or concentrates. A high quota of blister copper is an indication that parts of the production process are being transferred to the consumer countries, thus creating backward integration from manufacturing into refining and, in the case of imports of concentrates, into smelting. This hinders the construction of integrated copper industries in the countries with the mines. This policy is intensively

TABLE 4.2

Copper Consumption of EC Member Countries
(thousands of tons)

Country	1960	1970	Average Annual Growth Rate
Belgium-Luxembourg	124.6	210.0	+5.4
Federal German Republic (F.G.R.)	650.1	854.7	+2.8
France	339.1	459.0	+3.1
Great Britain	734.2	673.6	+0.9
Italy	250.0	443.0	+5.9
Netherlands	46.0	68.4	+4.0
Sweden	127.1	135.9	+0.7
EC total	2,271.0	2,844.6	+22.3

Note: Excluding Denmark and Ireland; including Sweden. Consumption is consumption by end users.

Source: Metallstatistik, Metallgesellschaft, Ffm., 1962-72, 1961-71.

followed by Japan and, to a lesser extent, by Belgium (imports of blister) and Germany (imports of blisters and concentrates). As a matter of fact, West Germany has increased its smelter capacities during the last years.[2]

EC countries as a whole import more than 50 percent of their total copper imports from CIPEC countries.[3] Analysis by country for 1972 shows the following:

1. Belgium imported the bulk of its copper, both refined and unrefined from Zaire, namely 70 percent of refined and approximately 60 percent of unrefined copper. Belgium does not import ores to any considerable extent.

2. The F.G.R. imported 394,000 tons of refined copper, 26 percent of it from Chile. Other important copper suppliers are Belgium, Zambia, the United States, and Canada. The F.G.R. also imported a considerable amount of unrefined copper, namely 117,900 tons, of which 26 percent came from Chile and over 40 percent from South Africa. The copper content of imported concentrates could be estimated to be 80,000 tons, the bulk of it coming from Australia.

3. Statistics of French copper imports make no distinction between unrefined and refined copper, yet France mainly imports refined copper. Main sources of imports are Belgium, Chile, and Zambia.

4. For Great Britain the same is true, importing only 53,100 tons in unrefined form, and nearly 400,000 tons as refined copper. Main sources are: Canada (27 percent), Chile (11.4 percent), and Zambia (30 percent). With 45 percent coming from Chile and Zambia, Great Britain is highly dependent on CIPEC countries.

5. Italy imports refined copper almost exclusively, mainly from Chile, Zambia, and Zaire.

6. The Netherlands have no refineries at all and depend completely on the import of refined copper. The bulk of it comes from Belgium, Zaire, and Chile.

The United States does not import large quantities of copper, relative to production and consumption, being the biggest producer and consumer in the world. Approximately the same amount of copper, imported to the United States, is reexported. The world's biggest copper importer is Japan, the predominant feature of its trade being the heavy reliance on ore imports. Over 55 percent of copper is imported in the form of ores or concentrates.

CIPEC countries have an important share in international copper trade. Taking the figures of 1971, it is evident that the real importance of these countries is not so much in production, where their share accounts only for 38 percent of world production, but in trade, a fact illustrated by Table 4.3.

TABLE 4.3

Share of CIPEC Countries in International Trade in Copper, 1971
(percent)

Type of Copper	1971
Ore, concentrate*	19
Blister	77
Refined copper	53
Total exports	53

*Copper content.

Source: "Copper in 1972," CIPEC, Annual Report (Paris, 1972), p. 76.

The importance of CIPEC countries in international trade is still greater than indicated by Table 4.3 for two reasons: (1) trade in refined copper is not classified under the country from which the ore comes, but rather where it is refined and (2) a considerable amount of copper consumed by the industrialized countries is secondary copper, previously imported from CIPEC countries. The share of CIPEC countries in EC country copper trade is shown in Table 4.4.

International Trade: Diversification of Supply Sources

Exploration in copper is mainly done by the multinational British-American companies, whereas the European companies do not seem to be very much inclined to open their own pits. This may be due to the fact that internationally, there is more security in long-term contract buying than in owning production in foreign countries. International jurisdiction in the recent past has decided in favor of fulfillment of long-term buying contracts, in cases where nationalization had taken place, as was the case in Chile, when Kennecott mines were nationalized.* But international jurisdiction cannot prevent nationalizations.

The risk for newcomers in a very close oligopoly in copper, especially in mining, is certainly the main reason for European reluctance to integrate backward to the mines. The rate of profits on capital invested is probably lower in view of the risks than necessary for a "sound investment" in underdeveloped countries.[4] Since the industrialized European countries are the biggest buyers of copper and account for more of the foreign trade in copper, they try to preserve their positions as buyers. This makes them especially sensitive to a nationalist producer cartel. European buyers are usually interested in diversifying their sources of supply to regions outside the CIPEC area, since they do not want to be confronted with a cartel of a few big nationalist producers. Hence the declining share of CIPEC countries in international trade with European countries.

---

*This refers to the last two decades. It does not refer to Union Miniere, which owned mines in Zaire.

TABLE 4.4

Share of CIPEC Countries in International Trade with EC Countries
(in thousands of tons and as percent of total copper imports)
(including Sweden; excluding Denmark and Ireland)

Import Countries	1952 (1,000 tons)	1952 (percent)	1972 (1,000 tons)	1972 (percent)
Belgium- Luxembourg	162.8*	95.7*	285.2	71.4
F.G.R.	21.4	26.8	295.5	39.3
France	53.8*	38.5*	142.9	37.4
Great Britain	254.4*	65.4*	201.0	44.8
Italy	31.8	49.7	186.4	65.0
Netherlands	—	—	18.2	43.4
Sweden	20.8*	47.0*	39.1	60.0
EC	545.0	60.6	1,110.4	52.1

*Including imports from Southern Rhodesia.

Notes: Table is based on import statistics. Copper content of ores is calculated on a base of approximately 20 percent. Amounts smaller than 0.1 ton are not included.

Source: Metallstatistik, Metallgesellschaft, Ffm. 1951-60 and 1962-72.

THE INTERNATIONALIZATION OF THE PRODUCTION PROCESS IN THE COPPER INDUSTRY

Some Technical Details on Copper Production

In order to understand the copper industry, especially the internationalization of the means of production, it is necessary to become acquainted to some extent with the technical procedure of copper production. In contrast to secondary copper, which is recovered from scrap, primary copper is produced from ore, which has to be mined, concentrated, smelted, refined, and fabricated (that is, manufactured). Secondary copper needs only to be refined in the majority of cases and is fabricated like any other primary copper (Figure 4.1 outlines the sequence of production and processing).

FIGURE 4.1

Production and Processing of Copper

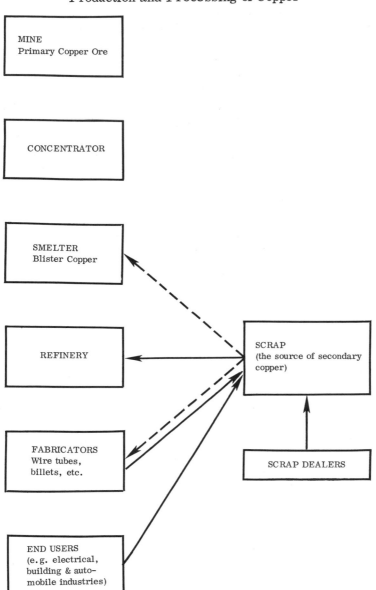

Source: Bankers Trust Company, "World Copper Prospects" (London, 1973), p. 27.

Mining is the most essential element in copper production. Highest profits in copper production accrue from mining, not from smelting, refining, or fabricating. Copper content of ores varies greatly, from about 0.4 percent to 5 percent and more. Over the last years, copper content of ore bodies showed a declining tendency. Copper is generally milled, and the product obtained is a concentrate with 20-30 percent of copper content, which is smelted into blister copper. Copper content of blister is between 98 and 99.5 percent. Most blister is either refined by electrolysis or by fire. Mining technology is influenced by the kind of ore body. Ores can be classified into two categories: sulphides, which account for about 90 percent of consumed copper, and oxides. An even more relevant distinction is between open-pit and underground mining. The majority of mines today are open-pit mines. In 1960, a total of 53 percent of primary copper came from underground.

Of the very large mining operations with ore capacities of 3 million tons, only a few are underground—for example, in Zambia. New developments in mining technology favor open-pit mining, allowing profitable exploitation of low-grade deposits, which to a large extent are to be found outside the traditional copper-exporting underdeveloped countries. There is evidence that the bulk of mining operations in mines opening up during the period up to 1977 will confirm the trend toward very large, low-grade open-pit operations.

Concentrating or milling is the second stage in copper production. In the early days of copper industry, copper ores were smelted directly whereas copper ores mined nowadays with much lower copper content have to be milled first, in order to get a copper concentrate of 20-30 percent. Sulphide ores are concentrated by a process of crushing, grinding, and floatation, whereas oxide ores are concentrated by way of leaching.

Smelting generally comprises three processes, in order to obtain a metal of reasonable purity: roasting (that is, oxidation of the concentrate); melting the obtained product in a reverberatory furnace or in a blast or electric furnace; and converting the "matte" in a converter. The product obtained is blister copper.

Copper refining is mainly done in two ways: (1) fire refining in a reverberatory furnace; and (2) electrolytical refining, which is of greater importance and gives a product of better quality in general. By this process, blister is melted into anodes and by electrolysis transferred into cathodes with a minimum copper content of 99.8 percent. While current copper technology is highly standardized and commercialized worldwide, there are new developments in mining, smelting, and refining, such as the continuous casting process. These may tend to restructure the copper industry leading to enforced concentration.

The Internationalized Process of
Production: The Basic Concept

Copper products, such as wirebars, cathodes, and many of the different semiproducts and castings are characterized by international standards. The standardized norms of, for example, wirebars and cathodes are the preconditions for trading them in such international stock exchanges as the LME. The production process for these products is completely internationalized, too, at least in mining, smelting, refining, and, to a lesser degree, copper processing.[5]

The Production Process: Copper Technology

Until recently, copper technology has not been very sophisticated and was easily accessible worldwide. Vulnerability of private multinational copper corporations seemed largely due to this, as well as to the fact that copper was to be found in relatively few areas. Consequently, the struggle of the copper corporations has led into the path of developing new technologies: (1) to exploit profitably new, low-grade deposits, outside the traditional underdeveloped copper-exporting countries—that is, CIPEC countries; (2) development of new, highly productive monopolized technologies often difficult to assimilate by underdeveloped countries, thus creating new forms of dependency as well as new possibilities of extracting surplus value taking the form of profits through the sale of licenses. The new technologies are developed by the multinational firms of the industrialized countries, mainly in the United States, Europe, and Japan. It is difficult to tell at this stage of research who is leading in this field. The United States has been leading so far, yet Europe and Japan are becoming increasingly competitive.

Worldwide competition in this industry presses the multinational corporations toward the development of technologies with increased productivity in order to make temporary extra profits. This forces the others in the same branch to adopt this technology or to develop their own technological process in order to maintain the profit rate. The use of continuous casting demonstrates the international character of these developments.

The process of continuous casting has been introduced for processing refined copper into different wires needed by the electro-industry, and thus it is relevant for more than half of all copper consumed.

Manufacturing of copper into wire was formerly done from the copper wirebar, the fabricator buying copper from the refinery. In continuous casting, refining and manufacturing are combined into one process, starting directly from the cathode and suppressing the stage of the wirebar. The cathode copper is melted directly into the required shapes. Since the product obtained is superior to the product obtained from the traditional wire bar, an extra premium of 30 percent over the current product is paid for semis obtained from continuous casting. This good quality is only obtained if excellent cathodes are used, so control of cathode production becomes compulsory, leading to the vertical integration of the production process, backward from manufacturing into refining and smelting. The German firm, Metallgesellschaft, states that during the next seven years cathode will become the base product for the production of semis, as continuous casting plants swell to between 70-80 percent of all copper rod mill capacity, thus making a strong vertical integration of copper production compulsory.[6]

New conventional plants cannot compete with continuous casting plants in terms of price per unit of output and general quality of products produced.

Since continuous casting plants require heavier capital investments per unit of output and operate at a larger scale, this will lead to concentration of capital in the copper industry. Only the very big corporations can afford the capital needed for an integrated production process. The continuous casting process has been developed by the American Southwire Corporation and is known under the name of Southwire rod cast (SRC) process. A series of prominent copper corporations all over the world have taken licenses, among them Norddeutsche Affinerie (NA) belonging to the Metallgesellschaft,[7] Selection Trust Ltd., Kennecott, and the British Industrial Chemical Corporation (BICC).[8]

On parallel lines in Europe a research group of Metallurgie Hoboken–Overpelt and Usine a Cuivre et a Zinc (Belgium) together with NA and Vereinigte Deutsche Metallwerke have been using a continuous casting process since 1965. NA and the other German firm left the research project and went Southwire. The first plant based on the Belgium continuous casting process, the Contirod process, came on stream in 1973. As Metal Bulletin states, it will be much harder now to sell SRC licenses, after the successful installment of this plant. Thus, Contirod and not SRC is going to be installed by one of the biggest copper corporations in the world, the American Asarco, at Amarillo, Texas. Hardware will be delivered by F. Krupp GmbH, which also installed Contirod in Hoboken. This in turn means that the margin of profits for SRC is shrinking through competition. At the same time, former technology is becoming obsolete, which means that capital invested in the development of this technology will become devalued. Similar developments are observed in continuous smelting. Mitsubishi for example developed a fast continuous smelting process. It produces copper from ore in only one process, through smelting and casting furnaces, eliminating conventional converters, and its labor requirements are about one-third of a conventional smelter of the same scale.

The Japanese copper industry, in collaboration with state agencies, is especially eager to minimize the risks of high import dependency by doing as much smelting and refining within Japan as possible. Thus it predominantly imports ores or concentrates. Due to the financial support of the state, it can offer very favorable prices for ores or concentrates (above the world market price) to underdeveloped countries, thus discouraging underdeveloped copper exporters, especially the Philippines, from developing integrated industries within their national boundaries and making them externally dependent for all further processing stages. This policy is to some extent also followed by Western Germany, whose NA recently largely increased its capacity for smelting and refining. The policy of developing new technologies, which are difficult to assimilate by underdeveloped countries for reasons that will be treated below, and the development of mining technologies that make possible the exploitation of low-grade deposits constitutes an effort of multinational corporations to control as much as possible of the production process, from the mine to the processing of refined copper. To gain control, it is necessary to reshape the production process, by shifting the barriers of entrance to that industry from mining to other stages of production that may be transferred to other areas considered politically safe. In copper, such safe areas are the industrialized countries and apparently even socialist countries.[9]

At the same time, rising costs per unit of copper produced in industrialized countries, due to antipollution measures and higher wage levels as well as rising transport costs, increasingly impede the transfer to developed countries. Hence multinational corporations may tend to produce only a "necessary" minimum of copper within their national boundaries, in order to ensure a minimum supply of copper in times of crisis. Since their aim is profit-making rather than production itself, multinational corporations may increasingly invest in development of copper technology at all stages, leaving production itself increasingly to underdeveloped copper-exporting countries that have to rely on the multinational corporations for technical expertise and management. As Mining Annual Review puts it, "The technical expertise largely rests with the major private sector corporations—and looking to the long term, this is their important strength."[10] Nevertheless, there is a tendency to develop and use advanced technology in industrial countries first. New technology, for example, the continuous casting process, seems difficult to assimilate by underdeveloped countries. At the present stage of this investigation, it is difficult to draw definite conclusions on this topic. Metallgesellschaft representatives have stated that the copper semis produced by continuous casting have to be produced near the market because otherwise its high quality would suffer from transportation

and because it has to be processed immediately after production.
If correct, this could be a key factor impeding the national underdeveloped producers from assimilating this process. Likewise, the
necessity of controlling the production of cathodes for quality reasons
might also favor smelting, which is particularly polluting, in industrialized countries. Pollution itself is not an insurmountable problem
in this context, as shown by the development of the Arbiter process of
Anaconda.

It could turn out that this problem is of limited duration until,
with some additional research financed by the license buyer (the underdeveloped copper-exporting countries), quality loss through transportation is overcome. Forced by international competition to expand
the market for licenses in order to maintain the profit rate, the big
corporations will move to the next level of technological development.
This would imply a time lag in the "transfer" of technology, while
higher wages and other costs in industrialized countries are compensated by higher productivity, thus creating optimal global conditions
for profitable reproduction of capital invested in copper technology.

In sum, since copper nationalizations and demands for higher
shares in profits by the nationals lessen the profit rate in production,
multinational corporations in copper have found a new method of securing profitability, extracting surplus value via sale of expensive
technology, creating thus at the same time a new form of dependency.
In many ways, this counterbalances, or more than outweighs, the
controls the underdeveloped copper-exporting countries have so far
gained via partial or total nationalizations of copper mines and organization into CIPEC.

It should be made clear that these "patterns" of copper corporations are in no case the outcome of a worldwide conspiracy against
poor underdeveloped countries. Rather they are the result of the inherent logic of profit-making and securing existing relations of production. The latter is another expression of the former with a long-
term perspective.

Any conflicts between national underdeveloped producers and
the private corporations, mostly the British and American ones, tend
to favor European and Japanese copper industries as sources of technology and management. While Europeans consider opening their own
pits in underdeveloped countries to be very risky, this is not the case
with development of technology. Mining technology for copper is in
many respects similar to mining technology in other base metals, so
that in this field the necessary experiences do exist. Large smelters
and refineries do exist, for example, in Western Germany (for example, Norddeutsche Affinerie). Copper and base metal technology has
already been developed by the French state-owned Bureau de Recherches Geologiques et Miniere, which is partly run on commercial lines,

and often cooperates with underdeveloped countries' governments in joint ventures. Investigation in this respect should be concentrated on the engineering societies of the big European copper and base metal corporations such as Lurgi, the engineering company of Metallgesellschaft, its spearhead of technological development.

It should be made clear at this point that technology is not the only weapon of multinational companies in this context. The opening up of new deposits and the organization of marketing and finance are others. Yet it is certainly the most powerful one, with regard to the technologically backward countries, like CIPEC countries, especially Zaire and Zambia, as well as the Philippines and Papua New Guinea.

INTERNATIONALIZATION OF THE PROCESS OF CIRCULATION: COMMERCIALIZATION OF COPPER

The Basic Concept

While there is considerable information on copper mining and production, there is hardly any with regard to the process of circulation of the copper merchandise. Nevertheless, it is especially in this field that European copper industry, together with the Japanese, has a strong position as an oligopolist buyer. The problem of marketing copper is of basic importance for structuring and restructuring of copper flows and related patterns of production.

For the purpose of the present study, the concepts of Christian Palloix concerning the crucial role of circulation of merchandise have been adopted, limiting their validity to the individual capital, in order to understand the strategies of multinational copper corporations in this field.[11]

The problems of commercialization, which form part of the circulation sphere, include price policies. It is through marketing and distribution of world copper production that copper companies are in a position to determine price levels in oligopolistic markets.

International Commercialization of Copper by Private Copper Companies

Since the big European and Japanese copper companies are typically refiners of ores and concentrates of mines that are not their own, attention has to be drawn to the terms by which they acquire the needed inputs. This is predominantly done via long-term supply contracts (15 to 20 years). Japanese firms are known for exporting technology (licenses) only, in return for long-term supply contracts, and not as is usually done for royalties and other payments.[12] Most probably, but less rigorously, the same policy is followed by European copper companies. Investigation in this respect is here concentrated on the big companies such as Union Miniere (Societe Generale de Belgique), Metallgesellschaft, and BICC/ICI.

In the market for refined copper, a strong oligopoly of suppliers is confronted with a far less concentrated buyers' market, the buyers being the fabricators of semifabricated copper and copper alloy products.

The bulk of refined copper is traded directly between producers and semifabricators, on one-year contracts. Marginal quantities are traded on the LME,* which has much greater importance than the physical volume of its operations suggests. Markets, and prices respectively, are producer-markets and prices. The European companies not only trade the copper refined in their own subsidiaries but they also handle marketing for other companies with limited marketing facilities, especially in the European market, where they have to face competition from the experienced Europeans. Thus, some quantities of Anglo-American Corporation copper are handled by Metallgesellschaft, the biggest copper trader in Europe and one of the biggest in the world. Belgium (that means Societe Generale) trades Zairois copper. European companies, as customs refiners, mostly do the marketing of the copper they smelt and refine for producers of concentrates or blister. Thus, Anglo-American decided to refine part of its copper with Metallgesellschaft by the continuous casting process, due to the necessity to produce and process it near the final consumers.

International Commercialization of Copper
by Underdeveloped Countries

Up to 1973, only Chile was able to handle her own marketing of all her copper via Codelco. In August 1973, the Zambian government announced its intention of establishing a national marketing organization, expressing Zambia's dissatisfaction with the marketing patterns of Anglo-American and Amax. Somewhat later, Minero Peru declared that it planned to assume the commercialization of all its copper in 1976, the amount commercialized by the end of 1973 being 75 per-

---

*The physical volume of copper traded at the London Metal Exchange is rather marginal as compared to total volume of copper traded. Yet LME prices are basic for determining prices for the bulk of copper traded internationally.

cent.[13] In February 1974, Zaire renegotiated an agreement with SGM
on a lowering of compensation for the nationalized copper mines, specifying among other items the establishment of a Zairois marketing organization. In the former 1967 agreement, it was set out that SGM
would carry on marketing of copper for another 25 years, being entitled to a commission of 4.5 percent of sales revenue.[14]

Multinational Private Companies and National
Producers of Underdeveloped Countries:
Possibilities of Organization of a
Price Cartel in Copper

The main reason for underdeveloped copper-exporting countries
to establish their own marketing organizations is that otherwise they
are unable to sell their product where they wish to sell it, on conditions that are not subordinate to the optimization policies of multinational corporations. The main disadvantage in setting up their own
copper-marketing arrangements is that the big producers of refined
copper have a worldwide network of marketing facilities and cooperate
in joint trade ventures. Concentration in the distribution network of
companies selling in industrialized countries is taking place quickly.
Penetration of new markets is often achieved through cross stockhold-
ing or acquiring capital stocks of trading companies in the area to be
penetrated. This is exemplified by the case of Nisso-Iwai, a Japanese company within the Mitsubishi Corporation, which took a 10 percent stake in Metallgesellschaft Ltd. (London). There are now five
major Japanese trading and/or producing companies in nonferrous
metals having a direct or indirect stake in the European market.[15]
The big producers often look for joint ventures with experienced buy-
er-traders: Anglo-Chemical Metals represents joint capital interests
in trade of Anglo-American and Mitsui. Anglo-American of South
Africa Ltd. and ICI (Great Britain) formed Acorga, a new company
marketing chemicals and technology developed by the two corporations in the metals recovery field.[16]
    It is evident that through the establishment of their own market-
ing organizations underdeveloped copper-exporting countries, at
least at the initial stage, can by no means compete with the worldwide
facilities of copper giants. Nevertheless, their own marketing facilities would be of considerable value, even if their only function was to
point out which quantities of copper should be commercialized by
which companies, thus benefiting from competition of the big corporations. The implementation of such a strategy would be more successful if undertaken jointly by CIPEC member countries so that in carefully selected areas their own marketing facilities could be estab-

lished, whereas in other areas commercialization would be done by
the corporations as agents of the national producers, eager to do the
profitable business. One of the conditions of this bargain could be
the training of nationals within those corporations, in order to increase
the number of experienced personnel. Concluding from the above, it
could be argued that there are fair chances for underdeveloped copper-
exporting countries to gain at least partial control of the commercialization of their own copper.

These policies may have important structural implications for
those private companies who traditionally used to produce copper in
the underdeveloped countries, for CIPEC countries themselves as
well as for the big trading and consuming corporations in Europe and
Japan. The establishment of their own marketing facilities originated
from a conflict of interest between national underdeveloped producers
and multinational corporations, which took out of these countries more
than the nationals considered to be justified. The impossibility of
immediately taking over the entire commercialization of copper exported thus benefits companies that traditionally did not have any mining and marketing interests in those countries. These are clearly
the big copper traders and consumers in Europe and Japan. Thus, it
may well work out that European firms as well as Japanese may increasingly become the intermediate agents of copper-exporting countries, thus increasing their importance in the worldwide competition
for raw materials.

Taking into account the possibility of Papua New Guinea, and
the Philippines joining the CIPEC agreement, these countries would
handle some 70 percent of copper traded internationally in the Western world (now 52 percent). It could be argued, especially if these
countries owned their own marketing facilities, even if rudimentary,
that this would be sufficient to control the price for copper in the
world. Apparently, it seems that there is no reason why copper-exporting underdeveloped countries should not be able to organize the
already existing CIPEC cartel along the same lines as OPEC has done.
Yet there are structural differences between oil and copper that will
make it very difficult, if not impossible, for CIPEC countries to have
the same effect concerning the price of copper. These structural differences are mainly to be found in the relatively easy substitution of
copper by aluminum, a process observed to take place continuously
and at a considerable pace.

In this context, it is important to know that all copper companies
of importance have their stake in aluminum. Thus, rising copper
prices will benefit those companies, increasing the possibilities for
profitable production of aluminum, a metal in which the many suppliers
of bauxite have no say, since bauxite is to be found in abundant quantities all over the globe and since the critical stage in production is

not in mining but in the production process itself, which is completely in the hands of the big companies of the industrialized countries.[17] Thus, an OPEC or CIPEC-like agreement in aluminum is impossible. So, whatever happens to copper prices, the industrialized countries, or, more precisely, their copper-producing industries, are always "on the sunny side of the street." The same is not true for underdeveloped copper-exporting countries.

Up to now, high copper prices have benefited the underdeveloped copper-exporting countries as well as copper companies. Yet the new copper technology will be an expensive burden for underdeveloped countries. They lack capital, which increasingly will have to be borrowed on the world market. Those countries will be pushed into high-cost production, pressing their governments to demand higher prices to obtain the necessary funds for development expenditures. This may speed up the rate of substitution, which may even lead to a drop in the demand for copper. On the other hand, high copper prices make new investments in both copper and aluminum more profitable, provided the copper produced can be commercialized.

In recent years, CIPEC countries have not been able to maintain their percentage in world production and world trade. This has to be interpreted as a successful strategy of private copper companies to foster production of copper elsewhere, outside the CIPEC area. Apart from the United States, this is mainly in Australia and Canada. Both countries are newcomers in copper with rich reserves. Under the present political conditions, it is not difficult to divine whose allies these countries would be. The same is true for Iran and South Africa, other important new centers of copper exploration and production.[18]

High copper prices favor both underdeveloped countries and copper-producing private companies, but for different reasons. In the case of raw-material-exporting countries, mineral resources shipped out cannot be replaced. Given an exponential growth rate of consumption of 4.1 percent a year, it is estimated that the known reserves of Peru will be exhausted in 60 years, Chile's in 48, Zaire's in 29, and Zambia's in 26 years. In view of these facts, it is a policy in the interest of the underdeveloped countries to preserve scarce resources. From this point of view, it is much more reasonable to produce less at a higher price.

Nevertheless, any augmentation of price levels by the nationalist producers will be limited by the possibilities of substituting other materials; developing alternative sources of supply; and backward integration of consumers to own ore bodies, outside CIPEC countries. This poses a danger for the CIPEC nations. As Theodore Moran puts it,

The nationalist producers, as they are seen as a
constant threat by the consumers, will most probably become suppliers of the last resort. Consumers try to buy from the integrated system of corporate producers as fast as these companies can
expand output, and they are constantly tempted to
integrate backward to production stage themselves.
As the international copper oligopoly becomes
[thus] more and more diluted, more and more of
the "regular" sales or long-term arrangements
will be covered between large corporate producers
and their major fabricators or industrial consumers, while the nationalists' share will be treated
as a spill-over market, subject to great fluctuations
in volume and price. Onto the nationalistic indepen-
dents will be shifted the burden of risk and instability for the international industry as a whole.[19]

Although CIPEC countries' share in international trade has
shown a declining tendency over the last 20 years, they still handle
more than 50 percent of all copper traded internationally. In view of
this, the suggestion that nationalist producers become "suppliers of
the last resort" is an exaggeration, yet with a grain of truth. Coun-
terstrategies by the nationalist producers aim at including more of
the underdeveloped copper-exporting countries such as the Philippines
and Papua New Guinea. European and Japanese copper consumers,
which are most threatened by any policy toward raising the price levels for copper, are following a policy of diversifying their sources of
supply, thus collaborating closely with the old private, integrated
corporate system. Research in and increased production of aluminum
and other substitutes for copper as well as the policies of developing
new technologies and, at least partly, more accentuated reliance on
concentrates (Japan, Federal Republic of Germany) and, as will be
shown in the following section, financial dependency aim at successfully preventing national underdeveloped producers from raising price
levels in the long run. This does not suggest that prices will not rise
in future, but not primarily because of CIPEC producers.

INTERNATIONALIZATION OF FRACTIONS OF
CAPITAL: THE ROLE OF FINANCIAL
CAPITAL IN THE COPPER INDUSTRY

The Basic Concept

The copper industry, especially mining, as outlined above, is
today highly capital-intensive. Large amounts of capital are needed

in order to open up new mines. Moreover, these investments need a
high percentage of long-term finance, given the fact that even after
exploration and proving of deposits, some five years are needed to
bring a mine on stream. Large amounts of total capital outlay have
to be invested in the material infrastructure, which to a considerable
extent does not have anything to do with mining itself. New developments in mining, especially development of technology and the related
low-grade deposit mining, have changed the pattern of financing in
mining, with the result that ever growing amounts of capital have to
be borrowed on the capital market. This implies that the copper industry is, to an increasing extent, becoming dominated by financial
capital.

The term financial capital used in the context of this study relates closely to the concept of Christian Palloix, who stresses its
functions in joining and dominating production and circulation of merchandise.[20] Within this "fusion," industrial capital as a fraction of
financial capital plays the decisive (determinant) role.[21]

The function of financial capital, as the most internationalized
expression of different (banking, industrial, commercial) capital
fractions, has to be seen in unifying the systems of production internationally.[22]

In fact, banking and industrial (mining and processing) as well
as commercial capital are the financiers of the big mining ventures,
whereby it seems that private financial capital, represented by private banking and mining capital, will be replaced to an ever increasing
degree by national and international public capital, as far as mining
in underdeveloped countries is concerned.

Patterns of Financing in Mining

Due to the limited information on finance in the copper industry
as a whole, the following details on patterns of finance are limited to
the mining industry alone.

The expanding demand for raw materials is one important reason for the industry's need for new capital, but even more significant
is the changing pattern of mine development, involving the exploitation of low-grade ore bodies. Moreover, the economies of scale pre-
clude a gradual mine-development program. "It is not generally possible with massive operations to open the property with a limited capital injection and then program subsequent development and expansion
of cash flow."[23]

In times when the exploited ore bodies were comparatively high
grade, the mining industry relied on relatively small equity issues in
order to establish a viable operation. Once self-financing, the subse-

quent expansion was based on cash flow, and thus, many of today's established mines were developed from comparatively small initial equity capital. This implied that only about 5 to 6 percent of the industry's total capital spending in mining had to be met by debt financing. This pattern is changing now. The massive economies of scale in low-grade mining required increased debt financing of around 30 percent of total capital outlay, given an unaltered pattern of raising finance within the industry itself. Yet, as the Mining Annual Review states, "in practice, with the continuing process of take-overs, and mergers, the general rationalization of its resources, the industry will probably be able to increase its proportion of the funding and the balance to be found is, therefore, likely to increase to between 20 and 25 percent of total annual capital requirements."[24] The Review explains that, in the past,

> the typical programme was for an ore-body with a potential refined metal output of 100,000 ton/year to be developed to, say, 15,000 tons on new equity and retained group funds. From that point, the operation would be self-financing and would expand as fast as cash flow allowed, up to its optimum size. Today, the trend is for all the pre-production planning and expenditure to be geared to, say, 75,000 tons initially on the basis of mixed funding —group retained earnings with possibly some new equity plus a major tranche [slice] of linked debt. The expansion from 75,000 tons through to 100,000 tons is again based on available cash flow after debt requirements have been met.[25]

It could be argued that one of the advantages of mining in underdeveloped countries is the possibility of finance on the traditional pattern, since relatively high-rate ore bodies are still unexploited. Yet private and financial mine capital is reluctant to invest in these countries, which are considered to be politically high risk.

The advancement in technology allowing now the profitable exploitation of low-grade deposits has led to a situation whereby there are more exploitable deposits of copper than funds to finance their development. Under the headline "When Will Copper Run out of Capital," Metal Bulletin writes:

> The problem of mining companies is not one of finding copper, but finding capital. As the remoteness of larger deposits increases, so the costs of providing production facilities as well as infrastructure and

the provision of social services increases tremen-
dously. This is not to say that there is a shortage
of credits worldwide in real terms, but the experience in the last ten years has created the im-
pression in the traditional financial institutions
that the mining industry might not be the safest
place to invest vast sums. Witnessing this is the
dramatic change in ownership of copper mines,
which has taken place in that period.[26]

The financial risk has thus contributed to the new exploration pattern
noted above, concentrating in the industrialized countries. The estimate is, that about 80 percent of all copper explorations are in developed countries, and 70 percent of that total is in the United States,
Australia, South Africa, and Canada.[27]

There will be a need in the long run to exploit fully and increasingly the copper reserves of underdeveloped countries. Also, the
Mining Annual Review argues that the investment risk increases in industrialized countries,[28] due to the uncertainties in the changing fiscal, monetary, and environmental climate, for example in Canada,
Australia, or the United States. Hence, there is a tendency for private financial and mining capital to invest in the industrialized countries, while a public financial infrastructure is built up internationally,
adapted to the purpose of financing base metals industry in underdeveloped countries. Pressure for public finance is exerted by copper-consuming firms, as well as by underdeveloped copper-exporting countries as a substitute to private equity capital in mining. Agencies
that could, in the future, assume these functions, are parastatal, national ones, as well as international ones, such as the World Bank,
which up to now have mainly financed infrastructural projects.

No underdeveloped country can today hope to afford its own mine
financing. The capital requirements of most modern mines (copper,
as well as other metals) and their attendant infrastructure is frequently above $100 million (the biggest require two or three times
that much), with an increased part coming from fixed-interest loans,
secured against prospective cash flow, assured by long-term preemp-
tive sales contracts. The necessary loan facilities are usually developed by an international consortium of bankers, who look for an assured cash flow, in order to guarantee repayments.

It is here that circulation of merchandise cuts into finance: Long-
term sales contracts have been associated in the past mostly with
smelter consortia. In the future, there might be a more accentuated
tendency toward developing purchasing groups (long-term contracts)
by semifabricators in the consumer countries, such as the EC countries and Japan. This would imply that the position of consumers in

international financing is emphasized, as a result of a policy of risk avoidance by the traditional private financiers of mining ventures. This is even more probable in view of the fact that in the industrialized countries, the big semifabricators and smelters are to an ever increasing degree integrated. Thus, there is a close liaison of the different fractions of capital "fused" into "financial capital," assuring production as well as circulation of the "merchandise" copper.

Conflicts Between Private Copper Corporations and National Copper-Exporting Countries' Governments

Although there is a tendency for private companies to withdraw from mining in underdeveloped countries, especially CIPEC countries, they still operate there, either as majority share-holders (Papua New Guinea) or minority share-holders, as in Zambia. In all cases, the participation of a national government offers advantages to the mining corporation. The state may be a source of finance in borrowing national and international funds, and, where it is the majority shareholder, it must guarantee for loans. Thus, while Amax and Anglo-American were repatriating nearly every dollar of profit accruing to their shares in the early 1970s, the Government of Zambia was forced to look for capital on the world market. As far as financing from national sources is concerned, this seems to be a direct contribution of the "development of underdevelopment."[29]

Far from being enabled to share the profits accruing to the multinational corporations, national governments find themselves in a position where the financial burdens and risks of a whole industry are shifted onto them. On the other hand, the tendency of private mine and related financial capital to withdraw from copper mining in underdeveloped countries, especially CIPEC countries, will lead to a substitution of this capital by public national and international capital, which, to an increasing degree, will also come from the copper-consuming countries of the European Community and Japan. This will substitute the dependency of underdeveloped copper-exporting countries on private capital by public capital of Western states and international agencies. Dependence on public capital will accompany and facilitate the new forms of technological dependency fostered by private capital in the copper industry.

THE ROLE OF THE STATE IN THE PROCESS OF INTERNATIONALIZATION OF THE COPPER INDUSTRY

The Basic Concept

The state gives support to the copper industry by activities in two main areas: in underdeveloped countries, and in industrialized countries. The analysis of the role of state in the process of internationalization will here be mainly centered upon the second; the former has been discussed already to some extent above. In this study, the state is viewed as an outcome of the process of internationalization of capital and labor force. This is easy to understand in the case of the newly independent states of underdeveloped countries that have been created as a response to the needs of international capital for new forms of domination in view of the contradictory development of the productive forces during colonial times, with the related emergence of new classes in these countries. But even the older industrialized states may be analyzed with this concept, focusing on the state (comprising both national and international parastatal organizations) in its relation to the copper industry. [30]

By supporting the multinational corporations originating in the national state, the nation-state tends to maintain or better its position within the imperialist chain.* I view the national state as an outcome and as an agent of the process of internationalization of capital and labor force with constantly changing (enlarged) economic functions, according to the needs of worldwide reproduction of capital on enlarged scales. It represents the power of the dominating classes, ever aware of the contradictions within the dominating classes and related changes in the structure of the state. Its overall economic func-

---

*Thus, "state policy is entirely inscribed in this dual character: autonomy-internationalization, with this paradox: push the internationalization as far as possible to preserve the autonomy of some sectors, hence the permanence of the state in the international setting." Nevertheless, this concept has to be criticized since this view may reflect opinions such as, what is the power of a nation-state in view of the multinational corporations, will it lose power in the process of internationalization, and so on, formulations that are rejected by N. Poulantzas, when arguing that they [the authors] "are absolutely false, as it is true that institutions or mechanisms do not 'possess' any 'power' of their own but only express and crystallize class powers."

tion is constantly to better the conditions for the enlarged reproduction of capital, that is, to counterbalance the tendency toward the fall of the profit rate. There are various ways to carry out these functions. Robin Murray distinguishes six concrete functions, which may be assumed by the state, three of which are of special interest for the present study: economic liberalization; input provision; and management of external relations of the capitalist system. The instruments for performance of these functions may be broadly classified as (1) legal and administrative measures that aim at reducing or putting aside hindrances to the flow of capital to regions where the best conditions for reproduction are to be found, or create areas of increased profitability—for example, through tax exemptions; and (2) financial measures that aim at raising profitability by investing public funds in those activities that constitute preconditions for production (for example, infrastructure). In mining, apart from infrastructure, exploration precedes commercially viable copper production.[31]

The Role of State as a Public Financier in the Internationalization of the Copper Industry

In copper there is a tendency for the state to finance more and more of the exploration risk, which in former times was assumed by private capital itself. The United States, Japan, France, the Federal Republic of Germany, and other industrialized countries, have created special institutions that are engaged in the field of mineral and energy exploration, entirely financed by public funds.

The most complete and expensive system in this respect has been developed by the United States. It is the biggest producer and consumer of mine products. By the end of the last century, the U.S. government commenced to create organizations for scientific research in minerals and geology: the U.S. Bureau of Mines and the U.S. Geological Survey. Besides these two, there are the Office of Minerals and Solid Fuels and, for strategic purposes, the Office of Emergency Planning. The two first-mentioned institutions employed 14,000 people by the end of the 1960s and disposed of a budget of more than $182 million.[32]

Japan has developed a comprehensive catalogue of measures to secure its mineral supply. The most prominent feature is the high proportion of public capital in financing exploration of foreign mineral deposits.[33] Indirect financial support to Japanese corporations is given in the form of credits by the Export-Import Bank and by private banks collaborating in the field of raw material business.

Credits that have led to a profitable investment have to be repaid at market terms, whereas the loss in case of unsuccessful in-

vestment is socialized—that is, credits are not to be repaid. For
these "risk investments," some $500 million to $1 billion is foreseen
by the Japanese government annually. Furthermore, the parastatal
organization, Overseas Mineral Resources Development Agency of
Japan, intervenes through direct investment in exploration and mining
by Japanese companies in foreign countries, as well as in the institutions for development aid (especially the Overseas Economic Cooperation Fund). Direct and indirect financial measures are accompanied
by a set of measures, especially insurance, to reduce the risk for
private capital in foreign mining enterprises.

The role of the Japanese public institutions in raw material
policies have often been discussed as a model for future policies for
Europe, which also lacks mineral resources. The European institutions established to service the base metals industry vary considerably. The most successful is certainly the French BRGM (Bureau de
Recherches Geologiques et Minieres), an organization with a double
function working as a state geological service and as a corporation
in mining business operating on commercial lines. In 1969, its budget was 100 million francs, of which some 40 million came from the
private industry. It employs more than 1,000 people. Half of its budget is spent in foreign countries. Its activities as an entrepreneur
are of central importance. It often engages in joint ventures with international private capital and/or the governments of underdeveloped
countries.

The German counterpart of BRGM is the Bundesanstalt fur Bo-
denforschung in Hanover, a public corporation that had comparatively
limited resources at the end of the 1960s (370 employees; budget of
DM 13 million). It has, however, been expanding considerably since.

In view of what happened to oil, there is a growing "concern"
of the German government and the industries with vested interests
that the state should assure raw material provision. This became obvious in a statement by the German minister of Economic Affairs,
who said that it has become necessary to find new forms of international collaboration, as well as finance, to ensure the provision of
raw materials. Thus, it would be necessary to make long-range investments, even if this implied renouncing expanded current consumption. He argued the whole foreign policy should be readapted to this
aim. [34] This proposal, in fact, necessitates increased exploitation
of German workers. Exploitation may be realized in different ways:
higher taxes, deficit spending (inflation), or increasing productivity
without corresponding reduction of the working day and/or wage raise.

In view of competition of multinational corporations operating
on the world market, leading to ever increased concentration and centralization of capital, public financing of preconditions for profitable
production (investments in infrastructure and exploration risk) or of

political risks in underdeveloped countries is compulsory if private
national and international capital reproducing itself within a national
state wants to survive. This is one form in which the national state
itself appears as a product of the internationalization process.

The World Bank may have a role to play here. It is apparent
that the financing philosophy of the World Bank is broadening. Whereas
hitherto its loans have been linked primarily to national infrastructure
projects, it seems now ready to offer finance for specific ventures
such as mine development.[35] Other international authorities such as
the UN Development Program (UNDP) as well as the individual nation-
states will, in the future, finance mine ventures in underdeveloped
countries. Some have done so already, especially bearing the considerable exploration risk, where perhaps seven in a thousand prospects
proves to be commercially exploitable. The old type of dependency of
the satellite state is thus replaced by another form of dependency.

The implications of this structural change from private to state
capital dependence are here analyzed for the case of the World Bank.
The necessity for private capital for (enlarged) reproduction on the
world market has created an international superstructure, of which
the IBRD constitutes a main actor in the performance of relations
between the peripheral state and the multinational corporation.[36] By
its statutes, the World Bank is only entitled to give loans when credit
is not available otherwise on terms considered reasonable by the IBRD.
This implies that the bank does not compete with private capital. By
the obligation (1) to promote by way of loans and guarantees foreign
private investment, (2) to finance only "productive investments," and
(3) to finance only the foreign currency part of investment, it is made
sure that the bank's main function is to support the enlarged reproduction of private capital. The bank mobilizes private capital, yet risks
are socialized internationally, since they are guaranteed by the member
countries. Following the guidelines of operational policies of the
World Bank, the states favored by its financial interventions are those
in which inner stability and foreign credits are considered as corner-
stones for development in the long run.

In view of this, it can be argued that the World Bank as an extra-
territorial international organization assumes far-reaching order and
control functions over the states of underdeveloped countries. It re-
stores their dependent position once more, their functions being reduced to a repressive role in the service of international capital and
to create a favorable climate or favorable preconditions for private
investment. This may be illustrated by the Zambian 51 percent shareholding agreement subject to arbitration and enforcement by the IBRD.[37]

The copper and other base metals mining and processing industries face a period of increased competition due to increased demand
for raw materials, tightening conditions for mining in industrialized

countries and political risks in underdeveloped countries. As far as copper is concerned, conflicts of private corporations with host countries will certainly favor European and Japanese copper industries. The risks of finance will however have to be subscribed by public capital, thus increasing the necessity for the state to raise funds from the noncapitalist classes, thus tendentiously increasing the exploitation of the working class.

For the underdeveloped countries, or to be more precise, parts of their bourgeoisie and the working class, this leads to the following conclusion: "The fruit of dependence is nothing other than dependence, and its liquidation requires the examination of the relations of production which it involves."[38] Given the unity of the process of development and underdevelopment, the indicated developments in the copper and other base metals industries may be expected to contribute to a deterioration of living conditions for large parts of the population in both the developed and the underdeveloped countries.

NOTES

1. "Economic Analysis of the Copper Industry," prepared for the Property Management and Disposal Service, General Service Administration, Washington, D.C., by Charles River Associates, Cambridge, Massachusetts, March 1970, pp. 15-22.

2. From 185,000 tons in 1969 to 320,000 tons in 1973. See "Survey of Free World Increases in Copper Mine, Smelter and Refinery Capacities," Wrought Copper Council, April 1972.

3. See T. Moran, "Transnational Strategies of Protection and Defence by Multinational Corporations: Spreading the Risk and Raising the Cost for Nationalisations of Natural Resources" (Madison: University of Wisconsin, 1973).

4. C. Palloix, Les Firmes Multinationales et le Proces d'Internationalisation (Paris, 1973), p. 35.

5. See G. Adam, "Relationship Between Multinational Corporations and Developing Countries" (unpublished paper, 1973).

6. See Metal Bulletin (London), October 30, 1973, p. 16. Metallgesellschaft holds 40 percent of NA shares. NA added its own technology in order to eliminate the shortcomings of SRC.

7. See Metal Bulletin, January 30, 1973; June 9, 1973; February 27, 1973; December 29, 1973.

8. See Metal Bulletin, February 21, 1974.

9. Kabelmetal, the biggest German producer of semis and a subsidiary of Gutehoffnungshutte, collaborating with Metallgesellschaft, stated that the transfer of production to Poland is under consideration.

10. Mining Annual Review (London), 1973, p. 13.

11. As Palloix points out, "A central hypothesis of our research is the internationalization of one branch or industry, that is, the relationship between production and marketing, the process of production and the process of circulation in an industry no longer develop in a national sphere but in an international one. The process of circulation plays a dominant role in this." He adds: "We speak of the 'dominance' of the circulation process and of the 'determining role' of the production process. The dominance of circulation is only the framework of the 'movement' of productive forces and of the production relations which remain 'determinants' in the final instance." Palloix, op. cit., pp. 15, 19; see also pp. 27, 58.

12. See R. Beck, Japans Rohstoffpolitik (Hamburg, Dusseldorf: Deutsch-japanisches Wirtschaftsburo, 1973), p. 69.

13. Statement by General Bossio, former president of Minero Peru, in Metal Week (New York), November 5, 1973.

14. See Metal Week, February 18, 1974.

15. See Metal Bulletin, December 4, 1973 (hotline).

16. Ibid., July 13, 1973; and October 6, 1973.

17. This does not suggest the OPEC is a powerful cartel. See "Economic Analysis of the Copper Industry," op. cit., pp. 15-22.

18. See J. Bourderie, "Un Metal Rouge, a Reflets d'Or. Les Pays Producteurs Peuvent-ils Realiser pour le Cuivre le Meme Operation que l'O.P.E.C. pour le Petrole?" L'Economiste du Tiers Monde, no. 2 (February 1974).

19. See Moran, op. cit., pp. 131-32.

20. As Poulantzas says, "The most characteristic phenomenon of the internationalization of the cycle of social capital engaged in the sector is to make finance capital dominant in the reproduction of capital, the agent of the internationalization. From this fact it would be convenient to 'tighten' the relations which are established between the internationalization of the sector and also of the capital, and the finance capital (especially the process that it constructs from the articulation of the process of production and the process of circulation)." The financial capital dominates "the process of production in the 'fusion' of parts of capital, and the way they function in this 'fusion,' implying the distinction between industrial monopoly capital and banking monopoly capital."

21. See Palloix, op. cit., p. 44.

22. See Poulantzas, op. cit.

23. See Mining Annual Review, 1970, p. 9.

24. Ibid.

25. Ibid., 1973.

26. See Metal Bulletin, May 11, 1973.

27. Ibid.

28. Mining Annual Review, 1973, p. 9.

29. Andre Gunder Frank, Capitalism and Underdevelopment in Latin America (New York: Monthly Review Press, 1967).

30. The economic role of the state intervening in the process of internationalization of branches is described thus by Christian Palloix: "The relations between state and internationalization of a sector are complex: the state seeks to further the internationalization of the sector to consolidate a dominant multinational firm, guaranteeing—by the national focus of the international cycle of social capital—its autonomy in the international setting."

31. R. Murray, "The Internationalisation of Capital and the Nation State," New Left Review, May–June 1974, pp. 84–109.

32. Sames suggests that putting this huge apparatus in the field of raw materials at the service of private capital has significant political implications. See C. W. Sames, Die Zukunft der Metalle (Frankfurt, 1971), pp. 187 ff.

33. See Beck, op. cit., p. 32.

34. See Suddeutsche Zeitung, January 31, 1974.

35. See Mining Annual Review, 1973, p. 13.

36. See R. Tetzlaff, "Die Entwicklung der Weltbank: Schaffung neuer Produktion-sverhaltnisse oder Pekolonisierung der Dritten Welt," Leviathan, 1973/74.

37. See M. Bostock and Charles Harvey, eds., Economic Independence and Zambian Copper (Washington, D.C. and London, 1972).

38. R. M. Marini, "La dialectique de la dependance," Critique de l'Economie Politique, October–December 1973, p. 27.

5

RAW MATERIALS STRATEGIES OF MULTINATIONAL COPPER COMPANIES BASED IN THE UNITED STATES
Al Gedicks

Amidst all the talk of impending shortages in key mineral resource industries, there has been surprisingly little analysis of the
situation in which the oligopolistic control by the largest U.S. copper
producers, Anaconda, Kennecott, and Phelps Dodge, is being chal-
lenged by the forces of economic nationalism in Third World producing countries and the increasing competition of European and Japanese
capital.[1] This chapter will delineate the consequences of these changing relations of production for the U.S.-based multinational copper
companies and the new strategies being developed to maintain their
control over sources of supply on a global scale. Finally, it will suggest some of the likely consequences of these evolving raw material
strategies for the CIPEC countries, which are seeking, in a variety
of ways, to increase the contribution of their copper industries to
their national economic development.

U.S. FOREIGN POLICY AND THE CONTROL
OVER RAW MATERIAL SOURCES

Even before the continental boundaries of the United States had
been established, far-sighted U.S. businessmen had recognized the
importance of an adequate supply of critical raw materials for the future growth and expansion of the U.S. economy.[2] Rich domestic copper deposits provided the strong domestic position from which U.S.
copper companies expanded into Mexico and Chile in the early 1900s.
For the copper companies to maintain their strong market position,
there had to be a constant search for new sources of supply that might
otherwise pose a threat to market, price, and production control.[3]

Up until World War II, this strategy, intermittently enforced by the establishment of trusts and cartels, ensured the continued leading position of the major U.S. copper producers.[4] The question of oligopolistic control became an important political issue when copper shortages developed during the Korean War. Defense stockpiling was undertaken on a massive scale.[5] Government subsidies encouraged smaller, independent copper producers to exploit inferior domestic ores in hopes of making technological breakthroughs. Other government commissions were soon set up to give early warning of financial or political threats to foreign sources of defense materials. Probably the best-known institute doing continuing research on these questions is Resources for the Future, established with Ford Foundation money.

THE NATIONAL COMMISSION ON
MATERIALS POLICY

In 1970 the U.S. Congress established the National Commission on Materials Policy to develop a national policy[6] that would deal with "national and international materials requirements, priorities, and objectives, both current and future, including economic projections."[7] James Boyd, then chief executive of Copper Range Company, retired that post to become director of this commission. Like its predecessor, the Paley Commission, the National Commission on Materials Policy grew out of a profound concern with the availability of critical raw materials from foreign sources for the ever expanding needs of the U.S. economy.[8] The commission has concluded that "the list of historically politically stable nations where companies can expect to obtain suitable concessions for both exploration and mining as in the past 75 years is not long: principally Canada, Australia, South Africa and Rhodesia."[9]

The recommendations of the National Commission on Materials Policy reflect the concern of the U.S. Congress with reducing U.S. dependence on foreign sources of supply as well as securing future investments abroad against possible expropriation or nationalization moves.

THE REORGANIZATION OF THE DOMESTIC
COPPER INDUSTRY

For the United States, the solution of the material supply problem is the "orderly development of domestic mineral resources to reduce dependence upon potentially unreliable foreign materials."[10] But to accomplish this, the U.S. government will have to convince the

major producers that greater profits can be reaped from investment
in domestic, as opposed to foreign, mining operations.

A number of factors in the domestic mining industry have here-
tofore encouraged greater investment in overseas production: (1)
maintaining control of world supply in the face of domestic as well
as foreign competition; (2) the enormous capital expenditure needed
to replace worn-out plant and equipment in the United States; (3) the
rising costs of pollution abatement control; and (4) U.S. government
enforcement of antitrust laws and the imposition of domestic price
controls.

CONTROL OF SUPPLY AND NEW COMPETITORS

Although the international copper industry has long been characterized by a high degree of concentration and control,[11] there have
been important trends undermining the strength of the producers'
oligopoly at least since World War II.[12] The threat to the oligopolistic control of the large U.S. copper-producers can be seen in both the
domestic and international copper industry. From 1947 to 1954, the
three largest copper-producers—Anaconda, Kennecott, and Phelps
Dodge—were able to maintain their share of domestic mine production
above 80 percent. But by 1960 their share had dropped to 69 percent,
in 1963 that figure was 62 percent, and in 1972 it was 60 percent.[13]
At the same time, there has been a rise in the number of major U.S.
producers from three to eight. The share of the "newcomers" in domestic mine production has risen from 26 percent in 1963 to 30 percent in 1972.

One of the principal factors that has contributed to this dilution
of concentration within the domestic industry has been the encouragement of smaller producers to increase their output with the help of
government loans. With a government advance of $83 million in No-
vember 1967, Duval Corporation was able to finance the Duval Sierrita mine in Arizona, one of the world's largest. In return, Duval
Sierrita agreed to repay the government by copper deliveries at 38
cents a pound. The case of Duval is typical of the pattern of government-financed increases in domestic production that arose out of the
shortage of domestic copper during the Korean War and the Vietnam
War.

The growth in the number of major oligopoly producers is even
more striking at the international level. After World War II, the big
seven, including the Roan-American group, the Anglo-American
group, the Union Miniere group, and International Nickel of Canada,
as well as Anaconda, Kennecott, and Phelps Dodge produced about 65
percent to 70 percent of "free world" copper from 1946 to 1954. By
1969 that figure went as low as 54 percent.[14]

The need for the major U.S. producers to maintain their world market position by investing in overseas production has resulted in the relative slippage of the U.S. industry as a factor in world and domestic supply. Nothing less than major interventions by the capitalist state apparatus will be able to reverse this situation.

CAPITAL FINANCING FOR THE DOMESTIC COPPER INDUSTRY

The National Commission on Materials Policy estimates that plant investment costs for expansion and replacement in the domestic copper industry will be on the order of $16.5 billion or $600 million per year.[15] As a measure of the problems facing the mining industry in raising this financing, the Interior Department's Second Annual Report on Mining and Minerals Policy observes that stock prices for major segments of the industry have not kept pace with the majority of U.S. public companies.[16]

A major consequence of declining profit rates in the domestic mining industry has been the general reduction in domestic mineral exploration.[17] According to the Interior Department, there are five major reasons for this decline in mineral exploration: (1) short-term uncertainties in demand for some metals; (2) uncertainties over future restrictions on land use; (3) prospecting moratoriums in some areas; (4) new pollution-control measures at domestic mines and smelters that have closed some and increased costs at others; and (5) ad valorem tax laws in some states that discourage or limit exploration.

POLLUTION ABATEMENT CONTROL

A major concern of the environmental movement in the United States is the provision of adequate air pollution controls for copper smelters. Uncontrolled, a copper smelter will release large amounts of sulfur dioxide into the atmosphere. The Environmental Protection Agency (EPA) has established guidelines based on ambient air quality that could require the removal of 90 percent of the sulfur in the feed at many of the nonferrous-metal smelters.[18] The Bureau of Mines estimates that the cost of pollution control equipment is approximately 20 percent of the capital costs for a new copper smelter.[19]

Robert N. Pratty, president of Kennecott Sales Corporation, was quite emphatic about the industry's opposition to current air quality standards. If the EPA continues its "current, unrealistic approach to air quality standards, there is no way the domestic copper industry could maintain any semblance of its given capacity."[20]

In a less threatening vein, Ian MacGregor, chairman and chief executive officer of Amax, has called for a national "war on pollution," financed through "tax incentives, accelerated depreciation, and the rapid recovery of capital."[21] The strategy that emerges from MacGregor's call is an attempt to combine the costs for pollution abatement and plant expansion and push for socializing these combined costs. MacGregor called for a massive commitment of funds and energy to revamp the country's industrial plant, which "by today's standards is less productively efficient, and by environmental standards is hopelessly out of date."

ANTITRUST LEGISLATION AND THE
COPPER INDUSTRY

The most persistent attacks on the vertically integrated operations of the major copper producers have come from the independent copper fabricators. Fabricating companies are the principal consumers of refined copper. They work the metal into semifinished forms such as sheet, rod, tube, wire, and extruded and rolled shapes that are the raw materials for a vast industry manufacturing articles or alloys for final consumption. About 35 companies are the most important users of raw copper and are, for the most part, wire mills and brass mills. About one-third of the companies, representing more than 50 percent of the total business, are affiliated with the major copper-producers.[22]

In January 1970 a cabinet subcommittee, headed by Hendrik S. Houthakker, reported that "under the two-priced system [referring to the producers' price and the world price] it is simply too easy for a producer to bias his allocation of low-priced copper toward firms that do not compete with its fabricating subsidiary and away from those that do."[23]

In June of 1970, Triangle Industries, a New Jersey-based fabricator, brought a suit in Philadelphia federal court charging five major copper companies with price-fixing and conspiracy to monopolize the copper-fabricating industry. In October 1970, Reading Industries, another small fabricator, brought a similar suit against the leading U.S. primary copper producers. In May 1972, the Justice Department subpoenaed the leading U.S. copper-producing and -processing firms for a probe that was primarily concerned with nationwide price fixing, dual pricing, and refusing to sell to certain competitors. Meanwhile, amid discussions of a takeover of American Smelting and Refining Company (Asarco) by Pennzoil United, a Philadelphia federal district court held invalid the acquisition by American Metal Climax of the 57.7 percent of Roan Selection Trust it did not already own.[24]

Given this background, one of the principal recommendations made by the National Commission on Materials Policy is to review the "potentially inhibiting effects of antitrust procedures on joint industry-wide research; e.g. anti-pollution efforts; and modifications of present procedures where appropriate."[25] The commission specifically recommended that the government provide rules under the terms of the Sherman Antitrust Act, which would permit "special industry groups to form joint venture corporations for the production of critical industrial materials under economies of scale that cannot be attained by individual companies."[26]

SOCIALIZING THE COSTS OF PRODUCTION

The major thrust of the solution to the problems of the copper industry advocated by the National Commission on Materials Policy is to socialize the costs involved in revamping the industry. "Urgently needed materials developments . . . may be so costly or so predominantly social in nature that industry will look to the Government for support."[27] The case for government-subsidized technological innovations is summarized by the commission as follows:

> Since innovations are inherently risky they are at
> a disadvantage in this competition . . . the Govern-
> ment can induce innovation in an industry when it is
> willing to underwrite the risks by supporting the re-
> search and development of some or all of the pilot
> stages, or when it encourages rapid introduction of
> new technology, e.g., by accelerated amortization.
> This concept must also extend to considerations for
> abandoning obsolete plants by providing incentives
> to do so.[28]

In order to narrow the gap between the large amounts of capital necessary for industry expansion and those funds available for such investment on the market, the commission recommends that "The U.S. Government reestablish and adequately fund a financial institution . . . which can arrange for low-cost investment capital for industry."[29]

The U.S. government has already taken steps in this direction with the reestablishment of the Office of Minerals Exploration, which offers up to 50 percent government participation in the authorized cost of exploration for copper deposits.[30]

THE "ORDERLY DEVELOPMENT OF
DOMESTIC RESOURCES"

The results of government efforts to reduce U.S. dependence
upon foreign suppliers of critical raw materials can be seen in the
following figures showing planned increase in expenditures (in millions of dollars) by U.S. mining companies in domestic facilities as
opposed to overseas investment.

Year	U.S.	Overseas	Total
1972	1,200	1,320	2,520
1973 (estimate)	1,680	1,650	3,330
1974 (estimate)	1,950	1,590	3,540
1975 (estimate)	2,220	1,160	3,380

Source: Engineering and Mining Journal, January
1974.

Included in these totals are expenditures for pollution controls
of $144 million in 1972 and $165 million in 1973, with more than 80
percent of such expenditures being made in the United States. The
figures also include domestic coal-mining projects.

Of particular interest is the participation of Japanese capital in
the Bluebird copper mine of Ranchers Exploration and Development
Corporation of Albuquerque, New Mexico. With demand on the rise
and the lack of capital financing from the private sector, smaller producers are increasingly looking for opportunities for joint ventures
in new mining investments. The participation of Japan's Mitsubishi
Corporation marks a new phase in Japan's stepped-up foreign investment drive to secure adequate supplies of raw materials. The proposed
agreement provides that, following completion of the first phase, Mitsubishi will contribute about $20 million to the venture and will own a
30 percent interest to be paid to Mitsubishi in copper from the mine.[31]
At a recent State Department-sponsored seminar, Kennecott's vice
president for exploration objected that "Against this kind of coalition
United States mining companies have no tools at all, and if this continues for another few decades the newly discovered major copper deposits will be owned by the Japanese."[32]

There is also an increasing trend of oil (Cities Service) and
natural gas (El Paso Natural Gas) companies to diversify to copper
mining. Hugh Liedtke, chairman of the Houston-based Pennzoil Company and a member of the National Commission on Materials Policy,
explained Pennzoil's diversification into copper, sulfur, and molybdenum as follows: "The country's short, the world's short of natural

resources. If Pennzoil has the raw materials—and we do—we'll ultimately profit from them."[33] In 1973, mining accounted for 18 percent of Pennzoil's sales and 20 percent of its earnings.

JOINT INDUSTRY AND GOVERNMENT RESEARCH CONSORTIA

The government and the copper companies are also increasingly turning toward the development of more sophisticated technologies. Some of these new technological development projects are joint industry and government groups. Some are joint projects of two or more international producers.

The Atomic Energy Commission and Kennecott have recently signed a contract to undertake jointly a program of investigation, study, and evaluation of in-place copper-leaching technology that could be applied in an ore body fractured by a nuclear detonation. The process to be studied would involve the possibility of fracturing a deeply buried low-grade copper-bearing deposit with a nuclear explosive, and then percolating an oxidizing leach fluid through the fractured rock. The leach fluid would then be pumped to the surface, where the copper would be extracted. The study will help determine the feasibility of recovering copper from low-grade ore bodies, many of which are deep underground and are not economically recoverable using conventional mining techniques.[34]

With 10 years of research experience in the field of ocean mining technology Kennecott has more knowledge of ocean metal recovery than any other mining company. In March 1974, Kennecott announced a five-year $50 million research and development program to determine the feasibility of mining manganese nodules from the deep sea and extracting the metal content. Included in this international consortia are Rio Tinto Zinc Corporation, Ltd., Consolidated Gold Fields Ltd., Mitsubishi Corporation, and Noranda Mines, Ltd. Kennecott has a 50 percent interest in the program, Rio Tinto Zinc has 20 percent, and the other parties have 10 percent each.[35]

Anaconda, in diversifying its output, has recently agreed with the Aluminum Company of America (Alcoa) to pool their technology to explore ways of producing alumina without bauxite. Anaconda is already operating a successful pilot plant that produces alumina from clay by means of a hydrochloric acid process. It has also purchased substantial reserves of alumina-bearing clay in Georgia for processing. Alcoa, meanwhile, has experimented with alumina production from a variety of substances. These include anorthosite, extensive reserves of which the company purchased in Wyoming about two years ago.[36]

DIVERSIFICATION BY PRODUCT

When the Chilean government of Salvador Allende nationalized the properties of Anaconda in 1971, it deprived Anaconda of its most profitable operation. Anaconda's Chilean operations had accounted for two-thirds of Anaconda's copper production and three-fourths of its earnings. The case of Anaconda was a classic case of failure to plan for eventual displacement of supply source and product line.[37] Although Kennecott was also hard-hit as a result of the Chilean nationalization (21 percent of Kennecott's total profits came from Chile in 1970),[38] the principle of diversification by product and geographical areas was not something new for Kennecott. In 1945, Kennecott's president E. T. Stannard introduced the principle of diversification into Kennecott's budget by allocating $500,000 a year for exploration. Stannard's three principal ventures were oil, gold, and titanium. Frank Milliken, current president of Kennecott, following in the Stannard tradition, began planning for the eventual displacement of Kennecott's Chilean supply a full decade before it happened. Milliken expanded Kennecott's copper production capacity in the United States by more than 20 percent and began moving into other metals like molybdenum, lead, and zinc. Finally, in 1968, at a cost of more than $600 million, Kennecott acquired Peabody Coal Company, the second largest coal-producer in the United States.

Although Peabody has been the most successful of Kennecott's diversification moves, it now appears that Kennecott's three-year battle over the Federal Trade Commission's order to divest Peabody is at an end; the decision was upheld last April by the Supreme Court, but the divestiture is not expected this year. Kennecott now has two basic alternatives regarding Peabody: It can sell it to someone else or arrange some kind of spinoff, possibly involving a swap of new shares of Peabody for old shares of Kennecott. Whatever route is chosen, Kennecott's shareholders are likely to come out very well as a result of Kennecott's diversification. If Kennecott were to sell Peabody now, it would bring a handsome $1.5 billion for Peabody's 8.5 billion tons of reserves.[39]

While Kennecott's efforts to diversify into South African gold mining and oil have fallen through, it continues to invest in titanium. Titanium's chief commercial application is in the armaments field. Titanium is a metal with a weight-strength ratio superior to steel and aluminum, lighter than steel, stronger than aluminum, and superior to both in resistance to corrosion. Kennecott's titanium subsidiary, Quebec Iron and Titanium, has recently announced an $11.4 million plant expansion and pollution abatement program at its Sorel and Havre St. Pierre operations. Kennecott has also purchased a 49 percent interest in a titanium deposit at Richards Bay, Canada. Other titanium deposits are being considered for purchase in South Africa.

Frank Milliken's vision of Kennecott as a "broad-based natural resource company" is shared to a great extent by the leading U.S. copper-producing companies. Anaconda is continuing its diversification into aluminum with the purchase of the Russell Aluminum Corporation, and American Metal Climax has plans to diversify into nickel with a purchase of property in New Caledonia.[40] American Smelting and Refining (Asarco) has recently purchased American Zinc Company's zinc oxide plants in Illinois and Ohio, its mines, plants, and other properties in Tennessee, and its American Limestone Division.[41] With Asarco's purchase of four Illinois surface coal mines from Peabody in November 1970, virtually all of the major U.S. copper-producers now have major investments in energy resources.*

DIVERSIFICATION OF SUPPLY SOURCES

In 1970 the United States was a net importer of primary copper.[42] While pursuing the "orderly development of domestic resources" the National Commission on Materials Policy realizes that "Foreign supplies will . . . continue to satisfy some portion of U.S. needs for some minerals. . . . Inescapably, the United States is concerned with the sufficiency of foreign supplies."[43]

Industry expert Raymond Mikesell spelled out this "concern" with foreign supplies at a State Department-sponsored seminar on the Impact of Economic Nationalism on Key Mineral Resource Industries: "Petroleum and mineral firms, regardless of their degree of vertical integration, desire to maintain their position in world markets, and without control over sources of supply their position is greatly impaired."[44] There appears to be a great deal of flexibility possible in coming to mutually satisfactory agreements between mineral firms and host countries. Robert A. Kilmarx, research principal at the Center for Strategic and International Studies in Washington, D.C., suggests that,

---

*Newmont Mining Corporation (Magma's parent corporation) has its own subsidiary, Newmont Oil Company, engaged in the acquisition, exploration, and development of petroleum and natural gas properties and production of crude oil, distillate, and natural gas. Inspiration Consolidated Copper Corporation has a 50 percent interest in the Arizona Oil Company, owning 160 acres of oil-producing land in the Bakersfield district of California. Amax merged with Ayreshire Collieries Corporation in 1969, and now Amax is a major producer of bituminous coal.

increasingly, U.S. mineral investments abroad will
probably take the form of supplier credits or joint
ventures. U.S. firms may operate facilities under
service or sales contracts. In time, many U.S. companies may function more like Japanese trading companies which have learned how to obtain and wield
economic power without . . . equity ownership.[45]

Nevertheless, the commission warns,

The public interest can scarcely be served without
arrangements giving reasonable certainty that the
capital, technical and managerial requirements for
expanded production will be met. Private capital
is unlikely to be invested in foreign materials industry unless the prospective rate of return is commensurate with the inherent risks.[46]

The commission recommends greater governmental intervention in
offering (1) direct government subsidies, (2) low-interest loans, (3)
tax incentives, and (4) insurance. Each of these recommendations is
already embodied in various programs such as the insurance provided
by the Overseas Private Investment Corporation (OPIC), Export-Import Bank loans to finance the export of equipment, foreign tax credits, depletion allowances, and tax referrals of subsidiary income.
The commission notes that the participation of U.S. corporations in
joint ventures overseas is handicapped by antitrust laws, which do not
restrain their European and Japanese competitors.

A number of projects involving U.S. cooperation in joint ventures warrant our attention as examples of new approaches to the problems of securing adequate supplies of copper from foreign sources.
The most important new development in the international arena is U.S.
government encouragement of joint ventures abroad as well as participation with other foreign interests in new ventures, partly in order to
take some of the burden off the United States if investment disputes
arise. A second new development in the American minerals industry
is Anaconda's service contract with the Government of Iran, another
device for assuring adequate supplies of copper without the enormous
risk associated with ownership. The development of Mexico's La Cari-
dad mine is a case where Asarco reduced its equity shareholding from
49 percent to 34 percent while three private U.S. banks provided the
major share of the capital for the project. In the case of Peru's Pash-
pap and Botswana's Pikwe deposits, major financing for the projects,
which include U.S. companies, are coming from Japan and West Germany respectively.

The number of joint ventures being undertaken by U.S. firms in conjunction with either Japanese or Canadian firms should not be taken as an indication of worldwide cooperation in the development of mineral supplies. For every case of cooperation among these countries many more examples of intense competition may be found. Canadian firms, for example, are moving fast to develop major new deposits of copper in Argentina after successfully outbidding German, Japanese, British, and U.S. firms.[47] The U.S. National Commission on Materials Policy expressed its fear that greater competition for the remaining supplies of copper in the Third World "may evolve into a mutually destructive race for resources when combined with rapidly growing demand for materials."[48]

TECHNOLOGICAL, MARKET AND FINANCIAL DEPENDENCE

Given the long-run pressures on host countries to maximize returns on their natural resource exploitation, the relevant question for the multinational corporation interested in assuring adequate supplies of copper becomes one of specifying the nature of the ties that bind the host country into an alliance with foreign investors. To varying degrees, each copper-producing country is dependent on foreign investors for technology, markets, and international credit.

In an industry where the technology is highly dynamic and where a firm's research and development capability is located outside the host's country, the demands of the host country for control are likely to be low. It would be self-defeating for a country to take action that would cut the flow of badly needed technological know-how. On the other hand, buying and selling technology has become big international business and has given rise to the concept of the "technological balance of payments": the foreign exchange spent on imported technology and that earned by exporting technology. The United States does not have a monopoly on mining technology, and copper-producing countries are free to buy that technology from Japanese, Canadian, and European firms.

Dravo Corporation's marketing of a new process for smelting and refining copper is just one example of the availability of new mining technology on the world market. Dravo has been granted the worldwide license by the International Nickel Company of Canada Ltd. (Inco). Developed and patented by Inco for smelting and refining nickel concentrates, the "top blown rotary converter" (TRBC) process offers several advantages over existing copper smelting and refining techniques, including lower capital and operating costs.[49]

As the case of Chile's copper industry under Allende illustrates, the multinational copper companies are capable of bringing enormous pressures through international credit lines on governments that do nationalize their copper industries.[50] The limited success that the Chileans did have with international financing came from Canadian, Japanese, and Soviet sources. As the competition for available supplies of copper increases, copper-producing countries can reduce their dependence upon any one source of capital and credit for financing the industry and its expansion. Many of the new projects being undertaken in the Third World reflect this realization on the part of producing countries.

IMPLICATIONS OF RAW MATERIAL STRATEGIES OF MULTINATIONAL COPPER COMPANIES FOR JOINT ACTION AMONG CIPEC COUNTRIES

The single most important impediment to the successful operation of a copper producers cartel is the success of the multinational copper companies in securing control over present and potential sources of supply. In a recent article in Foreign Affairs on "Developing Countries and Non-Fuel Minerals" Takeuchi and Varon make the point that

> A priori, scarcity of a resource is not essential for the establishment of a successful cartel. What is required is control over present and potential supply. But the scarcity factor is important in the sense that it strengthens the hand of producing countries in imposing their terms and shaping the ultimate course of supplies, or costs. For it is crucial to the successful operation of a cartel that supply outside the membership be inelastic, i.e., that other suppliers are higher-cost producers with relatively small reserves.[51]

Not only have the multinational copper companies diversified their foreign sourcing of copper with the help of private and public lending institutions, but they have also begun to look toward the more developed economies of Canada, Australia, South Africa, and the Soviet Union as possible sources of supply in the event of any disruption of their supply from the Third World. As a result, nationalist actions that cut the flow of trade with "free world" developed countries would produce growing surpluses in producer countries, not shifts of their trade toward the socialist bloc.

In addition to the question of supply, there is also the possibility of stockpiling, recycling, and substitution open to consumers of copper in the face of any concerted efforts on the part of copper-producing countries to raise prices. Ian MacGregor, chairman of American Metal Climax, has gone so far as to suggest[52] that the industrialized nations beat the producing nations to the punch by helping keep prices stable for copper and other primary commodities by stockpiling these commodities during downturns in the economic cycles and treating these commodities as part of their <u>monetary</u> reserves since they retain their value in inflationary periods. Stockpiling would keep commodity prices from falling when business turns slack in industrial countries, hence attracting more capital to the industry as well.

If all the talk of growing shortages of critical raw materials has raised the hopes of producing countries in terms of extracting greater benefits from the exploitation of their copper reserves it has also spurred a major effort on the part of the consuming countries to reorganize and reconsolidate their positions within the international copper industry. Whether the present members of CIPEC—Peru, Chile, Zambia, and Zaire—and those likely to be admitted in the future—Iran, Mexico, Botswana, and Papua New Guinea—can offset the long-range planning being undertaken by the U.S. multinational copper companies remains to be seen. Conceivably, some of the OPEC countries could come to the aid of a cartel in nonfuel minerals, especially if Iran uses its leverage within OPEC to revitalize CIPEC and bring some of OPEC's enormous financial resources to the service of the CIPEC countries.

SUMMARY AND CONCLUSIONS

The challenge posed by the forces of economic nationalism and intercapitalist rivalries to the hegemony of U.S. copper companies in the world market has resulted in a major effort on the part of the companies and the U.S. government (through the socialization of the costs of production) to reconsolidate their position within the international industry. The elements of this reconsolidation scheme include (1) a reorganization of the U.S. domestic copper industry, (2) a major diversification program both by geographical area and product line, and (3) a greater flexibility regarding ownership and control arrangements with present as well as future supply sources in the Third World.

At the same time, it is important to note that under conditions of intense intercapitalist rivalries for Third World supply sources, "firms are showing more and more willingness to fulfill the functions that the countries assign them according to the development model they have adopted."[53] The possibilities for copper-producing countries to utilize their copper resources for industrialization depends as much upon their ability to specify their "development model" as it does upon the constraints of working within an international economy dominated by the multinational corporations.

NOTES

1. The reader interested in a review of this debate is referred to the following articles: Theodore H. Moran, "New Deal or Raw Deal in Raw Materials," Foreign Policy, no. 5 (Winter 1971-72); C. Fred Bergsten, "The Threat from the Third World," Foreign Policy, no. 11 (Summer 1973); Suhayr Mikdashi, "Collusion Could Work," Foreign Policy, no. 14 (Spring 1974); Stephen D. Krasner, "Oil Is the Exception," Foreign Policy, no. 14 (Spring 1974); C. Fred Bergsten, "The Threat Is Real," Foreign Policy, no. 14 (Spring 1974); Benison Varon and Kenji Takeuchi, "Developing Countries and Non-Fuel Minerals," Foreign Affairs, April 1974; and William Diebold, Jr., "U.S. Trade Policy," Foreign Affairs, April 1974.

2. Angus Murdoch, Boom Copper: The Story of the First U.S. Mining Boom (New York: Macmillan, 1934).

3. Michael D. Tanzer, "Economic Roots of American Foreign Interventionism," mimeo., February 1972, Minneapolis.

4. For an historical account of the industry's efforts at cartel control see the "Chronological Review of 100 Years of Copper Price Controls, 1845 to 1946," in Part I, Section 2 of the Federal Trade Commission's Report on the Copper Industry (Washington, D.C.: Government Printing Office, 1947), p. 83.

5. The Paley commission was created by President Truman in 1952 because of the shortages of materials experienced during and after World War II and the Korean War. William S. Paley was chairman of CBS and life trustee of Columbia University.

6. Final Report of the National Commission on Materials Policy (Washington, D.C.: Government Printing Office, June 1973). Hereafter cited as Materials Report.

7. Public Law 91-512, 91st Congress, H.R. 11833, October 26, 1970.

8. Materials Report, op. cit., pp. 2-3.

9. Ibid., Sec. 9, p. 14.

10. Second Annual Report on Mining and Minerals Policy (Washington, D.C.: Department of the Interior, 1973), p. 59.

11. Federal Trade Commission, op. cit., p. 3.

12. Theodore Moran points to five major factors that account for the gradual breakup of the producers' oligopoly: (1) U.S. govern-

ment intervention during wartime to subsidize small companies; (2) the diversification of companies into copper mining by either forming or buying a mining subsidiary; (3) the preference of companies that have formerly converted their interest almost entirely to smelting or refining to move backward and develop their own source of supply; (4) the willingness of fabricators and consumers of copper to finance the development of new copper sources and to be paid back in output; and (5) the discovery of large new ore bodies. See Theodore H. Moran, "The Multinational Corporation and the Politics of Development: The Case of Copper in Chile 1945-1970" (Harvard University, unpublished Ph.D. thesis, 1970).

13. These computations have been made from the figures given in the Yearbook of the American Bureau of Metal Statistics for the appropriate years.

14. Moran, op. cit., p. 15.

15. Materials Report, op. cit., Sec. 4B, p. 18, Table 4B.3.

16. Second Annual Report, op. cit., p. 67.

17. Ibid., p. 11.

18. An Economic Appraisal of the Supply of Copper from Primary Domestic Sources (Washington, D.C.: U.S. Department of the Interior, Bureau of Mines Information Circular/1973), p. 27.

19. Ibid., p. 28.

20. "Industry Spokesman Says EPA Approach Threatens U.S. Copper Capacity," Engineering and Mining Journal, February 1973, p. 22.

21. "AMAX Executive Proposes a National 'War on Pollution,'" Engineering and Mining Journal, April 1972, p. 144.

22. Mineral Facts and Problems (Washington, D.C.: U.S. Department of the Interior, Bureau of Mines, 1970).

23. Report of the Subcommittee on Copper to the Cabinet Committee on Economic Policy (Washington, D.C.: Council of Economic Advisers, May 18, 1970), p. 16.

24. Metal Statistics: Purchasing Guide to the Industry 1970 (New York: American Bureau of Metal Statistics).

25. Materials Report, op. cit., Section 10, p. 8.

26. Ibid., Section 4B, p. 19.

27. Ibid., Section 10, p. 11.

28. Ibid., Section 10, p. 15.

29. Ibid., Section 4B, p. 18.

30. Minerals Yearbook, 1971, vol. 1 (Washington, D.C.: U.S. Department of the Interior, 1973).

31. Business Week, March 10, 1973, p. 53.

32. Proceedings of a conference on the Impact of Economic Nationalism on Key Mineral Resource Industries, March 20, 1972 (Washington, D.C.: Department of State, Bureau of Intelligence and Research), p. 82.

33. Forbes Magazine, May 1, 1974, pp. 20-21.

34. Engineering and Mining Journal (E/MJ), April 1974, p. 26.

35. E/MJ, March 1974, p. 21.

36. E/MJ, June 1974, pp. 208-9.

37. This point is further developed by Norman Girvan, "Multinational Corporations and Dependent Underdevelopment in Mineral-Export Economies," Social and Economic Studies 19, 4 (December 1970).

38. The implications of the successful nationalization of Anaconda and Kennecott's properties in Chile under Allende are treated in my article "The Nationalization of Copper in Chile: Antecedents and Consequences," Review of Radical Political Economics 5, 3 (Fall 1973).

39. New York Times, October 6, 1974, Section 3. See also Forbes, April 1, 1974, p. 40, and Rush Loving, Jr., "How Kennecott Got Hooked with Catch-22," Fortune, September 1971.

40. Business Week, May 6, 1972.

41. E/MJ, January 1972, p. 133.

42. An Economic Appraisal of the Supply of Copper from Primary Domestic Sources, op. cit., p. 3.

43. Materials Report, op. cit., Section 9, p. 7.

44. "Impact of Economic Nationalism on Key Mineral Resource Industries," External Research Study, U.S. Department of State, Bureau of Intelligence and Research, March 20, 1972, p. 33.

45. E/MJ, March 1972, p. 9.

46. Materials Report, op. cit., Section 9, p. 14.

47. E/MJ, February 1973, p. 21.

48. Materials Report, op. cit., Section 9, p. 15.

49. E/MJ, January 1973, p. 115.

50. James Petras, "Chile: NO," Foreign Policy, no. 7 (Summer 1972).

51. Benison Varon and Kenji Takeuchi, "Developing Countries and Non-Fuel Minerals," Foreign Affairs, April 1974, p. 508.

52. Forbes, February 1, 1974, pp. 44-45.

53. United Nations, Survey of Latin America, 1970, p. 304.

COPPER POLICIES
IN CHILE:
LARGE AND SMALL
MINING SECTORS

6

THE IMPORTANCE OF COPPER IN THE CHILEAN ECONOMY: TWO DECADES OF HISTORICAL BACKGROUND
Ricardo Ffrench-Davis

The aim of this chapter is to offer an overall view of the relative importance of the large-scale copper-mining (LCM) sector in the economy of Chile, and its evolution throughout the period 1952-71.[1] The analysis will apply to the group of enterprises making up the sector, which include the El Teniente, Chuquicamata, Potrerillos, El Salvador, and Exotica copper mines; El Teniente was exploited by Kennecott, and the other four by Anaconda.[2] Up to 1967, the El Teniente mine was operated by the Braden Copper Company. Up to 1969, Chuquicamata was operated by the Chile Exploration Company (Chilex) and El Salvador by Andes Copper Mining Company, both subsidiaries of Anaconda.

This chapter is divided into six sections: The first provides data on the direct impact of the LCM on the national product of Chile on the fiscal budget and on the balance of payments; the second describes the evolution of the structure of copper production—including its structure according to types of copper—and of exports and markets to which they have been destined; some information is included on the international pricing of different types of copper and its influence on exports of copper manufactures. The third section examines the distribution of the income generated by the LCM sector with particular attention to cost structure and its distribution between expenditures in Chile and abroad. The next section briefly analyzes the economic meaning of the system of "partial returns" of the income generated by sales abroad, from which the exports of the LCM sector have benefited.[3] Special attention is given to the successive tax modifications and their effect on the volume of foreign currency inflows into the country, and the distribution of the value of sales among returned and nonreturned values. So that we may gain an overall view, the capital movements (receipts and disbursements) of the LCM sector are given:

own capital and indebtedness of the enterprises, depreciation reserves, and investment in each year. The fifth section analyzes the
evolution of copper prices, their long-run trend, and the fluctuations
with respect to this trend. This material provides the basis for estimates on the impact of price fluctuation on Chile's balance of payments and the fiscal budget. In the concluding section, some comments on the main lessons and experiences provided by the past are
made, and there is a brief summary of what appear to be the most
relevant tasks for the future.

THE RELATIVE IMPORTANCE OF
THE LCM SECTOR

The LCM sector has a dominant place in the Chilean economy
in a variety of ways. It has been the center of foreign investment and
control and constitutes the most important individual item in the national product, in exports, and in services of foreign capital. It has
also, for some years, constituted the main source of fiscal revenue.

Generally speaking, the relative importance of the LCM shows
no definite trend between 1952 and 1971, but it exhibits a notorious
instability, mainly stemming from the fluctuation in copper prices.
This has meant that the exploitation of Chile's rich copper resources
has produced benefits and also caused difficulties for Chile. In fact,
it has generated foreign currency and fiscal receipts, but at the same
time it has been the most notable source of the dependence and instability of the Chilean economy. This is not something unavoidable, a
consequence of the characteristics of this natural resource, but rather
to a large extent it is the result of the shortcomings of Chile's copper
policy. It would be fairer to say that, paradoxical though it may be,
throughout practically the whole period covered, Chile had no copper
policy.[4]

Table 6.1 suggests the importance of the LCM sector. Column
(2) indicates that exports have fluctuated at around 8 percent of the
gross domestic product.* Sudden deviations from this figure may be
seen in 1953 (difficulties in selling production), 1957-58 (drastic price
deterioration), 1955-56, and 1969 (sharp price increase). The close

---

*The figures only include copper exports. The LCM enterprises
also produced by-products such as molybdenum, silver, and gold.
The value of these exports amounted to U.S. $18 million in 1970,
while local sales of molybdenum were worth U.S. $5 million. These
by-products represented around 3 percent of the value of total sales by
the LCM.

TABLE 6.1

Relative Importance of Large-Scale Copper Mining (LCM) as Regards Domestic Product,
Exports, and Fiscal Income
(percent)

Year	As Percentage of GDP			As Percentage of Total Exports		LCM Taxation
	Total Chilean Exports (1)	Exports, LCM (2)	Profits and Depreciation, LCM (3)	Exports, LCM (4)	Profits and Depreciation, LCM (5)	as Percentage of Total Tax Revenue (6)
1952	15.1	8.3	1.2	54.4	8.5	41.4
1953	11.2	5.6	0.6	49.5	5.8	30.6
1954	11.7	6.7	0.8	57.2	7.4	30.7
1955	14.3	9.8	1.9	68.2	13.3	48.2
1956	14.0	9.3	2.5	66.5	17.9	34.5
1957	10.7	6.4	1.4	58.9	13.8	17.3
1958	9.4	5.2	1.2	55.0	12.6	12.8
1959	12.0	7.4	2.1	61.4	17.6	18.9
1960	11.7	7.3	1.8	62.1	15.8	16.4
1961	10.8	6.3	1.2	57.9	11.1	12.5
1962	10.9	6.3	1.4	57.2	13.0	13.0
1963	10.6	6.0	1.4	56.8	13.4	14.9
1964	11.7	6.3	1.5	54.0	12.8	19.3
1965	12.4	6.6	1.4	52.8	11.3	16.5
1966	13.9	8.3	1.8	59.5	13.4	21.8
1967	13.8	8.4	2.2	60.5	16.2	18.7
1968	14.0	8.4	2.2	59.9	15.7	16.8
1969	16.6	10.3	1.8	61.9	10.8	26.7
1970	14.4	8.7	1.0	60.2	7.1	18.7
Average	12.6	7.5	1.5	58.6	12.5	22.6

Notes: All the figures used to estimate the coefficients have been expressed in dollars at 1969 prices. The numerator of column (2) contains FOB copper exports by the LCM enterprises plus exports of semimanufactured copper. Column (3) contains the foreign shares of profits and depreciations of the enterprises. The denominator in the first three is the GDP. Column (4) is equal to (2)/(1) and column (5) is equal to (3)/(1). Column (6) is the income accrued to Chile—in terms of income tax, overprice, profit shares, and exchange rate tax—divided by total fiscal revenue collected. The figures underlined are the minimums, maximums, and averages of each column.

Main Sources: R. Ffrench-Davis, Politicas Economicas en Chile: 1952-1970 (Nueva Universidad, Ceplan, 1973); Balanza de Pagos and Boletin Mensual of the Central Bank; and Corporacion del Cobre (Codelco) reports.

relation between the variations in Chile's total export and the LCM exports (compare columns 1 and 2) shows that the instability of the total exports is basically due to the volatile nature of prices in the international copper markets. Representing an average of nearly 60 percent of Chile's exports (column 4), deliveries by the LCM sector faced less stable prices than other exports of the country. On the average, one-fifth of the value of copper exports remained abroad in the form of profits and capital depreciation of the LCM. This has been the visible price that Chile has had to pay for the contribution of foreign capital to LCM. (The sum of these two items in the gross book value of capital earnings. The book depreciation contains an element of net earnings, due to the system of "accelerated depreciation," which has been applied to this activity. This means that charges to costs have been made in a period shorter than that required for the actual deterioration of equipment and machinery.)

Changes in the amount of capital services are closely linked to copper prices. On four occasions, however, state action has modified this relationship. During the early years of the period under analysis, remittances were drastically restricted by the application of taxes on the part of the prices of copper exceeding 24.5 cents per pound and on production expenditures performed in domestic currency (mainly purchases of domestic inputs and salary and wage payments). In 1955, both these duties were eliminated, with the result that the net profits of the enterprises increased. In 1967, a generous tax reduction, which drastically diminished the rate paid by Kennecott to nearly half, again increased remittances of profits abroad. Lastly, the establishment in 1969 of a progressive tax on high prices and the acquisition by Chile of a 51 percent share in the profits of the Anaconda group sharply increased the proportion of earnings attracted by the state.

In the tax field, the duties levied on LCM have constituted a sizable item, but while they contributed an appreciable quota in the fiscal budget, they have been a source of instability in the year-to-year revenue. The erratic nature of the international market has been accompanied by the almost permanently critical state of public finance. Thus, in the majority of cases, any appreciable improvement in copper prices, even when obviously temporary, has immediately been spent by the government, postponing necessary efforts as regards taxation reform and fiscal savings required by public finance.

On numerous occasions during the 1950s, taxes on the LCM have represented the major source of fiscal receipts.* Furthermore,

---

*Very frequently, the contribution of copper to public finance is underestimated in fiscal accounts. This is mainly due to two factors:

since the taxes were collected in foreign currency, they covered a
large proportion of the country's imports. The vicissitudes of copper
exports therefore also had repercussions on the level of custom duties. Especially during the period 1952-55, the instability of exports,
as a result of the foreign trade policy pursued, was transmitted directly to imports. Consequently, both the taxation from the LCM and
the revenue provided by import duties, an important source of fiscal
receipts during these years, suffered fluctuations linked with those
of copper exports.

During the next few years, the impact of copper on public finance
was reduced, first in 1957-58, owing to a sharp drop in prices on the
international markets. In the next few years, prices recovered and remained stable until 1963. However, while LCM production remained at
a standstill, the secular growth of fiscal income reduced the relative
importance of the funds coming from copper. In contrast, during the
latter years of the period, copper taxes increased notably as a result
of the rise in the price of this metal; taxes on the domestic economy,
however, also grew rapidly after 1965. This explains why the incidence of the LCM in the tax burden did not rise drastically during these
years, despite the high copper prices. (The tax domestic burden consists of all types of fiscal receipt divided by the gross national product.
This ratio, excluding the increase in copper prices, rose by 48 percent
between 1964 and 1970.)[5] Following the creation of a tax on the overprice in 1969, however, the importance of public receipts from the
large-scale copper mining sector grew considerably. The new duty allowed Chile to retain a large fraction of the extraordinary level reached
by copper prices.

Price instability had only slight repercussions on imports and
customs receipts, as a result of the new foreign trade policy during
the last five-year period. Between 1965 and 1970, the tendency of
prices to fluctuate mainly influenced the level of foreign exchange reserves.[6] Thus, the supply of imported goods was not affected in a
significant form by abrupt changes in the value of exports.

PRODUCTION: ITS STRUCTURE
AND ITS MARKETS

The level and structure of copper production also had a significant influence in fiscal and balance of payments receipts during the
two decades. (See Table 6.2.)

---

one, that some earnings have not been entered in fiscal accounts;
second, that the book exchange rate has sometimes been lower than
the actual rate. This last partly conceals a subsidy on imports of
certain goods.

TABLE 6.2

Structure of Copper Production
(thousands of metric tons of fine copper)

| Year | Large-Scale Copper Mining | | | | Small and Medium-Scale Mining |
	Electro-lytic (1)	Fine-refined (2)	Blister (3)	Total (4)	(5)
1952	151.5	156.6	65.6	373.7	32.7
1953	89.5	121.2	114.4	325.1	34.3
1954	109.9	67.6	145.1	322.6	32.6
1955	127.6	113.0	150.5	391.1	41.8
1956	140.1	100.4	203.1	443.6	46.0
1957	154.9	66.3	213.7	434.9	44.6
1958	128.4	59.4	230.8	418.6	46.8
1959	177.0	82.6	237.5	497.1	47.6
1960	147.0	78.6	253.6	479.2	52.7
1961	153.5	62.3	265.4	481.2	64.8
1962	180.1	66.4	263.7	510.2	75.5
1963	178.9	61.8	266.7	507.4	92.5
1964	177.8	78.9	270.9	527.6	94.5
1965	190.3	77.4	210.8	478.5	105.3
1966	243.4	74.1	207.4	524.9	99.8
1967	224.9	70.2	241.3	536.4	123.8
1968	232.3	56.7	230.3	519.3	139.0
1969	275.8	68.4	196.0	540.2	147.7
1970	280.4	65.4	188.9	534.7	157.1
1971	292.9	48.1	176.6	517.6	190.6

Notes: The LCM blister copper refined in Ventanas since 1966 is included in column (3), which for 1970-71 also contains small volumes of concentrates. Column (5) corresponds to the exports of the sector; it includes the output of Andina mine as from 1970.

Source: Compiled by the author.

Production increased by 41 percent between 1952 and 1971, a cumulative annual growth of barely 2 percent.[7] In international terms, the expansion was extremely modest. Consequently, the share of Chilean large-scale mining in total world production of copper dropped from 13 to 9 percent between 1952 and 1971,[8] while the growth of output in what is known as the small and medium-scale copper-mining sector sextupled during the period. The cumulative annual growth of 10 percent enabled this sector to climb to a quarter of total output of copper in Chile by the end of the period studied.[9]

LCM output comprises three varieties of copper: electrolytic, fire-refined, and blister. Although these three types are more than 99 percent pure, there is a notorious price differential among them, on account of their different uses and economic value.

The structure of Chilean production deteriorated during the 1950s. During this period, the output of fire-refined copper dropped drastically. Production of electrolytic copper was maintained in absolute terms, with sharp variation from year to year. Blister production absorbed the fall of fire-refined copper and captured all the expansion of total output. During the following decade, however, expansion was concentrated in electrolytic copper, as a result of the enlargement of refining capacity in Chuquicamata. In addition, part of the output of blister began to be refined in Chile in the Ventanas refinery, owned by the state mining enterprise, Empresa Nacional de Mineria (Enami), which went into production in 1966.*

A notable change appeared in the destination of copper exports after 1954. A significant proportion was diverted from the United States, the predominant destination until then, to Western Europe. During the 1960s various other markets, such as Japan, were developed. By the end of the period (see Table 6.3) a quarter of the exports of the LCM sector was sold outside the United States and Western Europe.

Copper sales within Chile serve to illustrate improvisation in public action. The copper-manufacturing enterprises are supplied with raw materials, mainly electrolytic, by the LCM companies at a discount on the ceiling price obtained by the LCM abroad. The rate of discount, fixed by the Ministry of the Economy and subsequently by the State Copper Corporation (Corporacion del Cobre, Codelco) has been as much as 10 percent. Since the value added to the raw material by the processing industry is quite small in products such as copper wire rod, this discount represents a substantial effective subsidy to enterprises.

---

*In 1969, the LCM enterprises paid Enami $3.9 million for refining nearly 60,000 tons in its Ventanas refinery.

TABLE 6.3

Markets for Production of Large-Scale Copper-Mining Sector
(percentages in respect to total sales)

Year	Sales Within Chile		Sales Abroad to:				Millions of Dollars at 1969 Prices
	For the Domestic Market (1)	For Export (2)	United States (3)	Western Europe (4)	Latin America (5)	Other (6)	(7)
1957	0.9	0.6	16.2	81.2	0.4	0.7	296.3
1958	1.2	5.4	5.7	86.7	0.1	0.8	261.3
1959	1.9	6.2	21.3	70.3	0.1	0.2	357.8
1960	1.5	0.9	10.5	86.3	0.8	0.0	373.7
1961	1.6	0.8	8.4	86.3	1.0	1.9	330.2
1962	1.8	0.7	15.1	76.9	3.7	1.8	356.3
1963	2.4	0.6	16.5	71.7	5.9	3.5	357.9
1964	2.5	10.0	15.7	62.9	5.1	3.8	392.9
1965	1.9	11.6	18.5	60.9	3.1	4.0	420.7
1966	1.6	5.8	15.7	72.1	1.1	3.7	564.1
1967	1.3	1.4	18.9	71.9	2.0	4.5	592.1
1968	1.7	2.2	18.8	70.4	1.8	5.1	612.3
1969	1.4	1.2	17.6	72.3	2.6	4.9	778.2
1970	1.9	1.5	15.7	74.1	2.7	4.0	673.9
1971	2.2	1.5	7.3	63.1	6.3	19.6	505.6

Note: Columns (1) and (2) show sales by the LCM to the processing enterprises. Columns (3) to (6) record exports of copper to its final destination. For the years previous to 1957, the data available refer to the intermediate destinations. The main difference is constituted by shipments for refining to the United States, subsequently exported to Europe. Column (7) records total CIF copper sales by LCM.

Source: Compiled by the author.

Manufacturing output, in different forms, such as wire rod, profiles, tubes, and alloys, has been directed to both domestic consumption and exports. Sales of raw material for processing for domestic consumption show a relatively stable upward trend. Sales of raw material for manufactured export, however, reveal extreme instability, due to speculative movements tied to differences between the selling price of the LCM (price listing of large U.S. producers up to 1966) and the price on the London Metal Exchange (LME).*

When the latter price has exceeded the former by more than a few cents per pound, the domestic processing enterprises have brought pressure to bear to increase their purchases from the LCM companies, in order to sell their exports, after a brief processing, on the basis of the higher LME prices.[10] In fact, since exports of copper with a rather higher degree of processing remained stable at levels of around $1 million per year, exports of "semiprocessed" copper were the source of all the fluctuations.** This has mainly consisted of "wire rod," which, as compared to electrolytic copper, includes as added value only 2 or 3 U.S. cents per pound. This speculative phenomenon was particularly strong in 1951, 1955, and 1964-66. Thus, for example in 1961-63, years of stable copper prices, exports of semiprocessed products amounted approximately to U.S. $2 million per year. Over the next three years, with the growing split between LME prices and the still frozen prices of LCM, exports of semiprocessed products increased sharply to an average of $42 million per year. (Deliveries by the LCM to the processing enterprises for export by them increased from 4,000 to 48,000 tons per year. After speculation opportunities were suppressed in 1966, deliveries remained at between 8,000 and 11,000 tons per year. This level reflects an increase in real manufacturing achieved by Chile.)

---

*The first governmental institution designed to regulate the LCM was created in 1955: the Copper Department (Departamento del Cobre). In 1966 its powers were enlarged, and it was renamed Corporacion del Cobre, Codelco. In general, we will use this latter name throughout this chapter.

**These fluctuations are offset by those of direct export by large-scale copper-mining enterprises. In fact, any increase in the local sales to manufacturing enterprises is necessarily deducted from the potential volume of sales abroad by the LCM. For this reason, and because of the low value added contained in semiprocessed copper, the latter is included in Table 6.1 together with sales abroad by the LCM.

THE DISBURSEMENT OF LCM INCOMES

The value of sales by the LCM enterprises is counterbalanced by their various disbursements. These include expenditure on imports of inputs, depreciation of equipment and machinery, purchase of inputs in Chile, payment of wages and salaries in domestic and foreign currency, diverse expenditures abroad, and miscellaneous expenses.* The sum of all these items gives an estimate of "production costs" per pound.** Residual earnings or taxable income (sales minus the costs mentioned) is distributed between the Treasury and owners of the enterprises, according to the tax system in force.

The costs appearing on the balance sheet of the enterprises depend not only on the quality of the ore and the efficiency of its exploitation but also on the depreciation rules laid down by the Internal Revenue Office and by the decrees authorizing investment by the enterprises and the level of the exchange rate applied to foreign currency returns.† (See Table 6.4.) In 1952-55, for example, the indirect tax applied to LCM expenditure made in Chile, via the tax on the foreign currency proceeds, which the enterprises had to sell to the Central Bank, was considerable. During this period, the average rate of taxation on the purchase of inputs and the payment of wages in domestic currency rose to around 100 percent.‡ The size of the tax on the domestic component in production costs dropped sharply af-

---

*The value of sales by the LCM also include sales of molybdenum, silver, and gold. The production costs of copper usually include those of its by-products.

**In Chile's balance of payments, exports are registered net of "expenditures abroad," which includes freights, insurance, sales commissions, and refining costs abroad. To calculate the CIP cost of electrolytic copper, expenditures abroad should also be included.

†Also influencing the costs and the relative intensity of labor and capital as registered in the accounts of the enterprises is the authorization to consider as cost items that are not so. In fact, the book costs only include part of the capital cost: depreciation and the interest accruing from debts to the companies with third persons. Moreover, up to 1955, 7 percent of the assessment of real estate was included in the costs. Hence, a variable proportion of capital costs has been considered, while all labor costs are included.

‡ The tax was estimated by comparing the exchange rate imposed on the LCM with that applied to the rest of exports. It constitutes an underestimate, since the latter exchange rate was also undervalued, in varying degrees, during practically the whole period (see Chapter 8).

TABLE 6.4

Distribution of Receipts of Large-Scale Copper Mining
(millions of dollars, 1969 prices)

	Disbursements in Dollars				Disbursements in Escudos			
Year	Im-ported (1)	Other Ex-penditures Abroad (2)	De-precia-tion (3)	Expendi-tures in Chile (4)	Net of Exchange Tax (5)	Ex-change Tax (6)	Foreign Profits Interest (7)	Fiscal Re-ceipts (8)
1952	31.3	11.2	6.2	9.6	42.6	51.1	40.8	127.0
1953	24.6	13.7	9.8	7.4	48.6	47.2	18.5	81.3
1954	17.6	17.3	11.3	8.0	46.9	40.3	25.9	89.3
1955	25.6	17.8	12.4	8.0	51.1	27.8	67.4	190.5
1956	30.3	19.3	15.6	9.1	68.9	13.4	92.0	151.0
1957	38.8	26.2	18.4	11.2	79.8	0.1	44.7	83.8
1958	34.3	27.1	19.3	13.9	70.8	0.1	32.3	61.0
1959	41.1	35.1	27.6	15.2	98.1	0.1	66.1	99.1
1960	35.0	35.5	29.5	15.2	115.0	0.2	57.5	98.3
1961	24.3	32.3	16.7	15.0	113.7	0.2	44.5	83.1
1962	24.1	35.7	23.7	14.4	124.4	0.0	50.5	93.2
1963	27.3	36.0	31.0	14.0	109.8	7.3	45.1	95.2
1964	28.6	37.9	30.0	11.3	117.9	21.1	53.5	111.6
1965	33.6	29.9	35.0	12.6	139.8	11.2	47.6	132.2
1966	27.9	34.5	35.5	16.0	155.5	4.9	85.8	227.4
1967	29.8	34.5	20.3	18.5	170.7	3.6	128.4	213.2
1968	35.8	36.0	21.5	24.8	199.0	4.1	133.1	202.7
1969	36.4	36.9	20.7	28.8	233.2	4.8	120.2	353.0
1970	34.8	31.8	19.7	n.a.	250.1	5.0	86.0	278.6
1971	41.8	n.a.	17.5	n.a.	n.a.	n.a.	n.a.	n.a.

Note: Column (1) records imports of inputs for operation. Column (2) includes marketing, freight, and refining costs abroad. Column (4) contains salaries paid in dollars to executives and supervisors, some freights, and as from 1966 payments to Enami for refining services. The sum of columns (5) and (6) gives the purchase of inputs in Chile, payments of remunerations and custom duties; it equals the sale of foreign currency to the banking system plus sales in domestic currency of copper and molybdenum. Column (7) contains interest paid on foreign loans plus profits obtained by Anaconda and Kennecott. Column (8) gives the sum of income and overprice taxes, and profit shares accruing to Chile.

Source: Compiled by the author.

ter 1955. These modifications distorted the data on production costs and their structure. In order to avoid the distortion, an estimate was made of the tax in column (6). A similar although smaller distortion was caused by the practice of accelerated depreciation.

Together with the suppression of the exchange tax in 1955, the then newly created Copper Department began to apply a policy directed to the further integration of the companies into the national economy. Both factors produced a significant change in the structure of costs to the advantage of the domestic component, with the increase in both purchases of inputs and payments of wages and salaries in Chile—compare columns (5) and (1) of Table 6.4.

SYSTEMS OF FOREIGN CURRENCY RETURNS
AND THEIR EVOLUTION

Throughout the period analyzed, all receipts generated by sales abroad had to be returned to the country. Foreign enterprises sold the total value of their exports to a bank. They also had to purchase through the banks the currency needed to cover their imported inputs, after approval of their import application by the Central Bank. Once their accounting periods were completed, they could purchase currency in order to remit profits and depreciation to the parent office abroad.

The LCM enterprises, however, benefited from a concession known as the "system of partial returns."* They were required to bring only enough export proceeds into the economy to cover disbursements in Chilean currency. The other sources of returns from the LCM were the taxes paid to the Treasury. The legal system applicable to both sources of inflows of foreign currency to Chile, stemming from copper production, underwent substantial changes during the decades under consideration.

The historical development of the measures that have affected LCM returns is interesting. In 1932, following the world crisis, the first exchange controls were established. Only then was differentiated treatment for large-scale mining attempted. Shortly afterward, the government fixed the minimum quarterly returns the companies were to make in order to cover their operating costs in Chile.[11] This measure, however, had no effect, since the companies, whether they liked it or not, had to return foreign currency to cover their costs in Chile. The practical results of the measure appeared when a spe-

---

*Large-scale iron mining (Bethlehem Steel Company) and nitrate were also included in the large-scale mining sector.

cial exchange rate was established for these returns, lower than that in force for the rest of the country's exports. Thus the ratio of 19.37 pesos per \$1 appeared for the returns made by the LCM companies to cover the production costs in Chile and remained in force until 1955.*

In 1952, all the returns for production costs were acquired by the Central Bank at 19.37 pesos, while the average import rate of exchange in force in the country was 70 pesos per dollar. As from June of the same year, Condecor, then the public institution in charge of the Chilean foreign trade policy, authorized the companies to sell part of their returns to the banks. In this market, the exchange rate was free to float; in the course of the year, it reached levels six times the special rate in force for the returns by the LCM enterprises. Consequently, by this means, the heavy exchange duty applied to purchases within Chile was attenuated, but the proportion of returns that were to be sold at each of these two rates was not determined beforehand nor according to clearly defined rules. On the other hand, the volume of returns actually exchanged by the companies in the banks reached a considerable level and gave these enterprises influence in this market.[12] This was reinforced by the fact that the funds brought in to cover investment expenditures were also sold outside the Central Bank, in markets subject to supply and demand. Lastly, on account of the distressing situation of the balance of payments prevailing during these years, the rate in force in the "free" areas was abnormally high and exceeded the rate that would have been in force had there been a single exchange market.[13] Consequently, the result was a confused and contradictory situation, in which part of the costs of production were covered by an excessively low rate of exchange, while the remainder plus capital inflows benefited from an excessively high exchange rate.

The mixed system of foreign exchange returns by the LCM enterprises lasted until 1955. In this year, under the so-called new treatment act,[14] the special exchange rate was eliminated. At the same time, the sale of returns for production costs was centralized in the Central Bank. On the basis of the exchange reform implemented by the Klein-Saks mission in 1956, transactions had to be made at the single exchange rate for foreign operations. (Sales of investment cash inflows were partially maintained in the free exchange market— that is, outside the Central Bank and on the basis of a fluctuating rate

---

*For several years, this was also the "official rate." Only in 1942 did it come to be called "special rate." However, as from 1933, there existed an "export exchange rate," which was higher than the official one. (See <u>Memoria</u>, Central Bank of Chile, 1952.)

different from that applied to foreign trade.) This system was maintained without substantial innovations until the end of the period. (Following the balance of payments crisis of 1961, a duty was reestablished on the purchases of the LCM in Chile. This consisted of submitting large-scale mining to the "spot" bank rate, which was lower than the "future" bank rate, applied to other exports. The highest implicit tax rate, amounting to 18 percent, was recorded in 1964.)

The tax system underwent further variations, both in the definition of taxable income and the rates applicable, consequently affecting the volume of returns the enterprises had to make.

The income tax system, which lasted until 1964, was established in 1924. The LCM sector was left subject to the general mining and metallurgy rate and to the tax on profit remittances (additional tax, in force from 1925). Only in 1939 was a special duty imposed on copper with a view to providing finance for the then recently created Development Corporation (Corfo).*

In 1942 a new "extraordinary" tax, of a single rate of 50 percent on income obtained by the LCM enterprises over and above specific price levels, was levied.[15] In certain respects, this was a forerunner of the overprice taxes. In 1947 this duty was temporarily increased to 60 percent; the overcharge, after having been suspended for some time, was reinstated in 1952 and remained in force until 1955.

Subsequently, in 1951, the so-called Washington Agreement represented Chile's first attempt to intervene in fixing the price of its main export product. The result of this step was that the production exported to the United States was to be sold at a price three cents per pound higher than that of the major U.S. producers, a price equivalent to that of the London market.** The price differential was to the exclusive benefit of the Treasury.[16] In 1952 the Central Bank began to acquire the entire output of the LCM enterprises at a fixed price of 24.5 cents per pound. Consequently, when the bank was

---

*Total surcharge was 15 percent on both taxes under Act 6,334 of April 1939. (See "Balanza de Pagos," Central Bank of Chile, 1952.) If a single tax were to be applied to the LCM sector at a rate equivalent to the "normal" levels of taxation applicable to the rest of the country's activities, the "economic rent" deriving from the extraordinary quality and geographical position of Chile's large mineral deposits would unjustifiably accrue to the foreign investors.

**The price in London, which underwent frequent fluctuations in 1952 and reached peaks of over 50 cents, achieved averages of 27.5 and 32.4 cents per pound in each of the two calendar years.

able to place the output at a higher price in international markets, Chile retained this entire difference for itself. At this stage, actual taxation consisted of the 60 percent duty, the exchange rate differential, and the overprice in favor of the Treasury.

In 1955, when the "new treatment" act came into force, the tax system of the sector was completely modified. (See Table 6.5.) The different forms of taxation were recast into a single income tax, consisting of a basic rate of 50 percent and a variable surcharge of up to 25 percent. This last fluctuated according to the volume of production; it reached the maximum level when actual production equaled "basic production" and dropped to zero when that production was double the basic level. The system aimed, by that incentive, to promote the increase of production capacity.

In order to understand the meaning of the tax scheme, it becomes necessary to define "basic production" and analyze some key implications of the system created at the time. Basic production, individually fixed for each of the enterprises, amounted to 333,000 metric tons per year for the three companies making up the sector.* The basic figure represented, for Chuquicamata in particular, a sizable underestimation of the installed capacity of production—that is, without extra investment, the output of Chilex increased by more than 50 percent and actually did so after only a brief lapse of time. (During the next few years, the production of Chilex fluctuated between 1.5 and 2 times the basic level.) Even in 1955 itself, the output of the sector as a whole was more than one-fifth higher than the legally fixed basic level.

The second important aspect of the new tax system consisted in the form of the surtax that implied that any increase in production over and above the basic figure was subject to very low marginal tax rates. When production was increasing up to 1.5 times the basic level, the additional taxable income was subject to a rate of only 37.5 percent. As production increased further until it doubled the basic level, the effective rate of taxation on the marginal taxable income dropped to a mere 12.5 percent. (The surtax was applied only to taxable income generated by basic production. When production increased, therefore, taxation rose by 50 percent of the additional taxable income

---

*The basic level was equal to 95 percent of the average annual output registered in the period 1949-53. The initial phase of this period coincided with crisis years in the international copper market; toward the end of this five-year period, world demand was increasing, but Chile found difficulty in selling its production, and this had negative repercussions on the use of the installed capacity of the sector.

TABLE 6.5

Tax Rates Fixed by "New Treatment" Act

Production (1)	Taxable Income (2)	Average Rate of Taxation (percent) (3)	Tax Receipts (2) × (3) (4)	Marginal Tax Rate (4) : (2) (percent) (5)
100	20	75.0	15.00	75.0
150	30	62.5	18.75	37.5
200	40	50.0	20.00	12.5
250	50	50.0	25.00	50.0

Note: Column (1) is an index of annual production: The figure 100 corresponds to the basic level of production. Column (2) is hypothetical and assumes that taxable income increases in proportion to the level of production. Column (5) is the ratio of marginal tax receipts to marginal taxable income for each interval of production.

Source: Compiled by the author.

minus the fall in the proceeds of the surtax. In determining the amount of taxation, the Internal Revenue Service assumed that the taxable income corresponding to each production interval was proportional to output.)

The definition of taxable income was also changed. The same act authorized a cost charge of one cent per pound of expanded refining capacity.* The accounting estimate of wear to equipment and machinery was modified, and a system of accelerated depreciation was brought into use that benefited investment programs directed to expanding the installed capacity.** That mechanism was of particu-

---

*During those years, the value added in the refining process was of the order of 1.2 cents per pound of copper. (Copper Department, El Cobre en Chile [Santiago, 1959].) Consequently, assuming a tax rate of 60 percent, the government was granting a 125 percent effective subsidy (0.6/0.4/1.2) to refining in Chile.

**Chilex already had an accelerated depreciation system for its sulfides factory on which construction had begun in 1948. The benefits accruing to the enterprises were proportionately larger the greater

lar benefit to the new mine of El Salvador, which from 1959 replaced the exhausted Potrerillos mine. El Salvador was considered a "new enterprise," although it was obviously carrying on from Potrerillos, being subject only to the basic tax of 50 percent. (Additionally, the greater part of investment was financed with loans, the interest of which was charged to costs, with a consequent reduction in taxable income.)

The "new treatment" act maintained the system of partial returns but established that annual taxation should be paid regularly in four quotas at the end of each quarter. This was a positive feature of the act, since it increased the frequency and regularity of fiscal receipts.

In sum, the "new treatment" applied to the LCM sector contained the positive features of greater regularity of fiscal receipts, elimination of the exchange tax that kept down local purchases, creation of the State Copper Department, and an endeavor to stabilize the tax scheme. The previous instability of the tax system did not benefit either Chile or the foreign companies. However, for Chile to reap benefits from stability, it was necessary for the taxes to be sufficiently high to capture the rent deriving solely from the natural richness of the copper deposits, but this issue was totally absent from the government approach. Furthermore, the new system came into force when a spectacular boom began on the copper market, in circumstances in which no tax mechanism on the overprice had been considered.* Consequently, the increase in profits obtained by the enterprises between 1955 and 1956 enabled them to finance the whole of the net investment made under the commitments they had contracted in compensation for the advantages accruing from the "new treatment" act.** It might be said that the shortcomings of the "new treatment" act were under-

---

the incidence of fixed assets with respect to production costs. The system thus stimulated the mechanization of production and depressed to the same extent the absorption of manpower.

*Another important negative feature was the exclusion of any participation of Chile in the marketing of the output of the LCM sector. Even though there is a lack of consensus on the meaning of the measures contained in this respect in Act 11.828, in practice the government's intention to restore this faculty to the foreign enterprises predominated.

**Gross investment, financed with credit or with own funds of Anaconda and Kennecott, amounted to a total of U.S. $151 million between 1956 and 1959, while depreciation was U.S. $81 million. In all the remaining years of the period 1954–66, depreciation allowances clearly exceeded the gross investment made in each accounting year.

standable since it had resulted from the first negotiation of broad
scope and complexity between Chile and the enterprises, and exceeded
the experience, knowledge, and capacity for negotiation of the Chilean
government. An important aspect was the exclusion of any participation of Chile in the marketing of the output of the LCM sector. However, the negative outcome should have served as a warning so that
the same gross mistakes would be avoided in future negotiations.

The tax system established in 1955 remained unchanged for
several years, until in 1961 two additional duties were established:
an increase of 8 percentage points in the tax rate and a surcharge of
5 percent on the taxes thus calculated.* A new modification to the
system took place with the aim of "Chileanizing" the LCM sector,
when Codelco acquired 51 percent of the Braden Copper Company,
which came to be called the Sociedad Minera el Teniente.** In addition, the tax rate was drastically reduced: A 44 percent duty replaced the basic 50 percent rate, the surtax of the "new treatment"
act, and the two surcharges established in 1961.† The tax situation

----

*Acts 14,603 and 14,588. Two additional steps forward were
taken by the parliament that affected the item "Expenditures abroad"
(which includes the cost of refining abroad, U.S. import tax, sales
commissions, and freight). See Law 15,575 of 1964; on the one hand,
it ended the practice of the copper enterprises of recording the import tax levied by the United States as a production cost; and, on the
other, it taxed exports of blister copper as long as there was nonused
installed refining capacity in Chile. The first measure meant that
the U.S. tax began to be fully paid by the U.S. investors; the second
represented a relative protection to Chilean refineries and contributed to the fact that, by the end of the period, Enami was refining
60,000 tons per year of copper produced by the LCM.

**The government used the powers granted by Act 16,425 of Jan-
uary 1966, later merged with 11,828 in Act 16,624 of May 1967. By
these same powers, mixed enterprises were set up between Codelco
and Anaconda (Compania Minera Exotica and the Compania Minera
Andina) and between Codelco and Cerro Pasco. The former was, financially, an appendage of Chuquicamata.

†The rate of 44 percent corresponds to a rate of 20 percent on
net profits accrued and to 30 percent on remittances of profits (net of
income tax). The tax concessions granted were so considerable that
prior to the increase of production derived from the expansion program, and with a price of only 29 cents per pound, Kennecott's 49
percent share of profits would exceed the earnings received when it
possessed the entire capital. It is of interest to note that the income
tax rate on profits was appreciably lower than that affecting corporations in the United States.

of the two subsidiaries of Anaconda underwent no significant change:
In the case of Chilex (Chuquicamata), there was a slight reduction*
and in the case of Andes Copper some increase.

When this negotiation took place, there was again, as in 1955,
an increase in copper prices. The combination of tax reductions
and the appreciable excess of the real price of copper over that assumed by Chilean negotiators meant that the annual earnings reaped
in Chile by Anaconda and Kennecott more than doubled its 1965 level
(see Table 6.6). This indicated that the same errors were being repeated, despite the passage of 10 years and the distinctly more favorable conditions for Chile's negotiating position.

The excessive profits obtained by Anaconda and Kennecott gave
a new impulse to public opinion in favor of the nationalization of these
enterprises. In part, this took the form of several endeavors to obtain for Chile a larger share of the profits of the LCM.[17] Proposals
were put forward to apply a sales tax to exports of copper and to tax
the part of the price of copper in excess of 40 U.S. cents per pound.**
However, only when "concerted nationalization" took place in June
1969 was a duty established on the overprice.† This consisted of a
progressive scale of maximal taxation of between 54 and 70 percent
for prices ranging from 40 to 50 cents per pound. (The scale shifted
upward yearly as production costs increased.) In addition, Chile acquired a share of 51 percent of the profits of Chilex and El Salvador.

The different tax changes occurring during the two decades had
important implications for the volume and stability of tax receipts.
The 1969 modifications had a significant positive impact, but at the
same time they increased the sensitivity of fiscal and foreign currency
inflows vis-a-vis fluctuations in output, and especially in copper
prices. This was to become inescapably stronger after the nationalization of the sector. This fact required a modification in fiscal norms
of behavior. As progress was made toward the total availability for
Chile of the surplus generated by the sector, it was necessary for
the Treasury to regulate its level of expenditure in accord with the
trend of copper prices, accumulating reserves in periods of high in-

---

*One of the curious aspects of the tax system designed for Chilex
was that, probably unnoticed, negative "marginal" taxes appeared in
some intervals of production.

**As from 1968, a forced loan was imposed on the taxpayers at a
rate of 13 percent of tax accruals. The sum paid by LCM companies
amounted to around $20 million per year.

†It consisted of a preferential share of the enterprises in favor
of Codelco. Income tax underwent formal changes but showed no variations in its fiscal yield.

TABLE 6.6

LCM Capital Investment and Rates of Return of Large Copper-
Mining Enterprises
(millions of dollars at 1969 prices)

Year	Capi-tal Stock (1)	Depre-cia-tion (2)	Net Prof-its (3)	Gross Invest-ment (4)	Foreign Loans Stock (5)	Rates of Return (percent) (6)
1952	200.0	6.2	40.8	38.2	93.5	20.4
1953	209.4	9.8	15.4	23.1	97.4	7.4
1954	202.1	11.3	22.5	10.5	104.0	11.1
1955	221.2	12.4	64.7	5.9	78.8	29.3
1956	268.6	15.6	89.7	25.8	74.4	33.4
1957	266.5	18.4	42.1	37.2	92.8	15.8
1958	290.3	19.3	30.7	47.8	93.0	10.6
1959	323.7	27.6	63.3	40.6	109.9	19.6
1960	341.6	29.5	53.4	10.0	99.4	15.6
1961	369.7	16.7	40.9	12.7	75.0	11.1
1962	390.3	23.7	47.6	8.9	56.9	12.2
1963	397.8	31.0	43.0	13.9	46.4	10.8
1964	395.7	30.0	51.9	9.6	35.6	13.1
1965	376.4	35.0	46.3	11.0	30.7	12.3
1966	399.4	35.5	84.3	14.1	26.0	21.1
1967	425.8	17.9	126.9	61.9	31.4	29.8
1968	451.6	18.1	129.2	139.2	148.4	28.6
1969	349.8	17.4	109.2	141.3	336.9	31.2
1970	255.5	9.7	65.2	78.7	433.0	25.5

Note: Column (1) contains the book capital of Anaconda and
Kennecott at the end of each year in the five enterprises (El Salvador
and Potrerillos, Chuquicamata, Exotica, and El Teniente). Columns
(2) and (3) measure the depreciation and net annual profits of the foreign partner. Column (4) gives the total investment made each year
(except for the Anaconda housing plan); part of this was usually financed with credit. The stock of external credit at the end of each
year in the liabilities of LCM companies is recorded in column (5).
Column (6) is the ratio of net profits to capital [column (3): column
(1)]. For various reasons, the real rate of return is underestimated
—for example, because of the treatment for accounting purposes of
part of the profits as depreciation, and because of the U.S. $80 mil-
lion paid to Kennecott for 51 percent of Braden's capital. In column
(1) in 1967 only 51 percent of Braden's own capital was imputed as a
negative investment; that is, approximately equal to one-half of what
was actually paid by Chile.

Source: Compiled by the author.

ternational prices to cover needs during periods of worsening terms
of trade.

Furthermore, owing to the nonrenewable characteristic of natural resources, the surpluses generated by copper ought to be ear-
marked for financing the creation of new production capacity in the
different sectors of Chile's economic activity, instead of covering
current public expenditures. (In this area a key role must be played
by an estimate of the "economic rent" of the deposits.)

This section began by describing the systems of "partial" and
"total" returns as they operated in the 1950s. During the two decades
under consideration the privilege of partial returns for LCM enterprises continued.* It is therefore relevant to examine more closely
the meaning of "total returns" and "partial returns." Probably two
categories of "total returns" should be distinguished.

There exists a concept that implies the retention in the country
of the total value of exports. This clearly requires that the copper
deposits should be developed, in their entirety, using national resources, so that all factor payments remain in the country. Polemics
in Chile, however, took another stand. It was claimed that once the
exports had been made, the companies should return the whole sales
value and only at a later date make remittances as they needed them
to cover import payments, other disbursements abroad, and the remittances of profits.

This second concept of "total returns" was applied to most exports, even to other enterprises controlled by foreign capital. As
compared with partial returns, it has two advantages for the host country: a control and training aspect, and a financial aspect. The former consists in the closer control levied on expenditures abroad, especially in the case of relations between the subsidiary and the parent
company, which allows savings in actual payments abroad and provides valuable information on how the enterprises are run. The financial aspects are connected with the right to use cash balances, which
the system of total returns grants the host country, instead of granting them to the foreign enterprises. The gains from such a system
arise from the productivity of interest provided by these balances
while they remain in the hands of the country. An estimate made in
1965 showed that the establishment of a system of total returns would
have meant a net annual income of about U.S. $10 million for Chile.[18]

---

*As from 1955, all returns from the LCM sector were channeled
through the Central Bank. There were, however, some attempts,
fortunately unsuccessful, to authorize the enterprises to sell or deposit their ample returns of foreign currency in the commercial
banks.

This is not a negligible figure, but greater concern should have been devoted to numerous other aspects of greater importance that were almost unanimously ignored. As an obvious example, the large profit obtained by the enterprises after 1955 and 1965 was the result of the absence or incoherence of a copper policy and improvisation in the negotiations, in a country whose balance of payments and fiscal budget depended so crucially on their systematic and coherent design and application.

COPPER PRICES, BALANCE OF PAYMENTS, AND FISCAL BUDGET

Copper accounts for a large proportion of Chilean export receipts. At the same time, the pricing of copper in the international market is very unstable. During the period 1952-71, the real price received by LCM exports suffered extreme fluctuations. If the percentage variations of the average annual price are measured year by year, it may be seen that the fluctuations averaged 12 percent annually. Each year's average, however, underestimates the intensity of the fluctuations, since these were also large within each year. The price instability becomes a macroeconomic problem since the balance of payments is extremely sensitive to copper prices. To evaluate the economic policies and their effects on the balance of payments, the impact on it of changes in the conditions prevailing in the international markets must be identified and isolated.[19] The economic policy should concern itself with the long-term "equilibrium" of the balance of payments. In this context, it is important to distinguish the permanent (long-term or normal trend) from the transitory components of copper prices.

The copper market is stratified into several semimonopolistic compartments, in which extremely divergent prices have coexisted.[20] Changes have taken place in Chile's trade policy that have produced large-scale differences between the price actually received by the LCM and the world price. In 1967, for example, the average price obtained by Chile increased, while prices in the international markets dropped. The explanation of this apparent paradox lies in the removal of Chilean exports from a market dominated by the price of the large U.S. producers to the LME, where the price was almost double.

Chile's first attempt to intervene in the marketing and price of copper took place on the signing of what was known as the Washington Agreement of 1951.[21] Under this agreement Chile obtained a higher price for exports to the United States and timidly began to export part of the output of the sector to the European market. Later, in 1952-53, another important change took place, when Chile took over the control of the marketing of all the production of the LCM and, for a short period, was able to fix on a price 45 percent higher than that in force in the United States.[22]

Between 1954 and 1964, the price of exports by the LCM tended to be that fixed by the major U.S. producers or by the government, manipulated within the limits that conditions existing in the various markets permitted.

Following the notable stability of copper prices between 1959 and 1963 at around 29 cents per pound, a rise in prices began, which, with some ups and downs, was to continue until 1970. The initial price increases did not spread to all markets. Thus, for example, in April 1965 the prices on the LME were 75 percent higher than that of the "major producers." This latter price was the one that from 1955 had, generally speaking, ruled the sales of the LCM enterprises. After much hesitation, the Chilean government decided to take its own path in fixing prices and established a level slightly above that of the U.S. market. All the major world producers moved together with Chile, except those from the United States, where the government obliged them to freeze the price at 36 cents. Meanwhile, the prices on the LME continued their upward trend, reaching a record level of nearly U.S. $1 per pound. By the end of 1966, the major producers outside the United States, without exception, had liberated themselves from their close dependence on a price common to that applied in the United States. In fact, in November, they were all operating on the basis of the LME prices.* Since then, the LME has unquestionably become the most important price in the international copper markets. LME quotations have been more unstable than those in the United States. As a consequence, the U.S. market price is above that of the LME in periods of strong demand and below it in periods of troughs. (See Table 2.2.)

A comparison of Chile's export price and the price trend in the world market gives an indication of the incidence of short-term fluctuations on the fiscal budget and on the balance of payments. Thus, for example, in 1955 the Treasury obtained extraordinary revenue by an amount equal to U.S. $58 million; in 1958, there was a loss of U.S. $55 million.[23] Effects of similar intensity may be observed in other years. This phenomenon enhances the importance of action aimed at reducing the instability of the international markets and also at designing improved foreign trade and fiscal policies so that the fluctuations in copper prices, which may continue to exist in the future, do not spread toward the domestic economy.

---

*In 1966, they operated on the basis of the "forward" price. In 1968, owing to an excessive difference between this and the "spot" price, Chile adopted the latter.

CONCLUSIONS

Despite the importance of the LCM sector in the Chilean economy, no efforts of a permanent and systematic nature have been laid out by the governmental, political, and academic media. This generalized inertia is reflected in many aspects and can be summed up in five points: lack of statistical data, inadequate organization for the action of the public sector, partial implementation of the policies adopted, absence of definitions in key aspects of the strategy to be followed in the development of copper, and improvisation in Chile's negotiations with the foreign counterpart.

First, the lack of coherent statistical data causes problems, as do the problems the public meets in seeking access to the material available: It is not reasonable that more may be learned about Chilean copper in a foreign library than in a Chilean library or that for "confidential" reasons bureaucratic blocks are placed in the way of national researchers regarding data that, for example, at least two multinational competitor enterprises possess in detail. There is an obvious need for systematizing the data available, which involve immense contradictions not easily observed because of the scattered form in which the data are presented. A complete system of accounts should be prepared common to all the enterprises in the sector, with data and definitions of variables appropriate for making correct decisions on how much and what to invest in, selection of technologies, design of economic policies, and control of results. It is also necessary to promote the awareness and interest of public opinion and of the research media toward the main natural resource of the country.

Second, the type of organization adopted for the nationalized sector is crucial. Generally speaking, three levels may be distinguished. There are decisions that are so important for national activity—of a macroeconomic nature—that they should be taken in a centralized form. These include the marketing of copper, overall aspects of programs to expand production, and technological development. At a second level may be found the planning of specific economic policies for the sector—for example, exchange-rate policy for returned values, profit taxes, the charge for the use of capital contributed to the enterprises by the country, and rent charges on the basis of the variable quality of the various copper deposits.* The third level is connected with determining the degree of autonomy to be assigned to the enterprises in order to carry out their operations efficiently. The first two levels provide this sector with the framework within which the maximization of surpluses (or the minimization of

---

*For details, see Chapter 9.

costs) enables a full potential contribution to the development of
Chile, providing the appropriate quota of financial resources to the
Treasury and of purchases from the national industry. It is a matter
of developing each function within the sector at the level of the organization having the greatest probability of efficiently adopting the
specific type of decision involved. Approaches that lead to excessive
centralization or excessive decentralization are equally dogmatic and
lacking in realism.

Third, it is not enough to define a policy, enact a decree, or
make a law. Too many times, laws or policy definitions have fallen
through because of governmental inability to implement them. One
good example is the inadequate and precarious form in which Chile
engaged in marketing the output of the LCM sector between 1951 and
1955.

Fourth is the question of how much copper production should
be expanded, of which varieties, and to what degree of processing.
Because of the quality of the deposits and the magnitude of the reserves, the installed capacity should be considerably increased. Although they create little employment, the major deposits, because of
their high productivity, generate resources, which permit the expansion of productive occupations in other sectors of the economy, where
the most efficient technological options may be labor-intensive. The
technological progress that has been introduced into the transport of
concentrates[24] is an element that offers new alternatives and renders
advisable a heterogeneous expansion in the variety of copper products.
The development of the production of mining capital goods, within the
framework of the Andean Pact and that of the member countries of
CIPEC, is an important means of access to the adaptation and development of the mining technology most appropriate to the endowment
of human, financial, and mining resources in the developing countries.

Fifth, the area of the relations with the multinational enterprises
is not closed for Chile. Their exclusion from the LCM sector is no
bar to their presence in other sectors or processes of production, as
continued to be the case after 1971.

In order that relations with foreign enterprises may benefit
Chile, within a framework of genuine reciprocal respect, it is necessary to implement the code on the common treatment of foreign capital agreed upon among the six Andean countries. It is a matter of
major importance to develop negotiating capacity—which demands an
accurate knowledge of the realities of the situation and of what the
foreign counterpart really can contribute—and to acquire experience
in negotiations.

It is true that the development of real negotiating capacity will
result in fewer inflows (or even no inflows) of foreign investment in
some sectors. Since the myth that foreign investment is per se good

for developing countries is manifestly untrue, improved negotiating capacity will guarantee that foreign enterprise inflows will, without interfering in domestic affairs, really contribute to an enlarged technological and management knowledge and to the conquest of new markets abroad.[25]

Finally, it should be pointed out that the policy of the public sector toward the LCM sector has, with considerable setbacks on some occasions, made considerable progress over the last few decades. In this context mention should be made of the import policy established by Codelco in 1955; the first steps toward Chile's own marketing of copper in 1951 and then systematically since 1968; and the tax on the overprice applied to the LCM companies after 1969, and since 1970 to the rest of the copper-mining sector.

In 1971, with the unanimous support of the country, a long-expected step forward was taken with the full nationalization of the LCM enterprises. This step increased the country's (and its government's) responsibility in designing and applying a copper policy that could contribute decisively to satisfying the needs of the majority of Chilean people.

NOTES

1. The author wishes to express his indebtedness for the comments received in the seminar on copper held by Ceplan during 1973. This article was prepared for a collection of essays on copper policy in Chile, edited in Spanish. See Ernesto Tironi and Ricardo Ffrench-Davis, ed., El Cobre en el Desarrollo Nacional (Santiago Ediciones Nueva Universidad, Ceplan, 1974). In some cases, the data in this chapter only go up to 1970, due to the extremely provisional nature, or absence, of more recent figures.

2. The main features of the activities of these multinational enterprises and their relations with Chile are described in N. Girvan, "Las Corporaciones Multinacionales del Cobre en Chile," in Tironi and Ffrench-Davis, op. cit.

3. An analysis is to be found in "The Integration of Large Copper Mining in Chilean Economy· The Role of Economic Policies," in Tironi and Ffrench-David, op. cit.

4. Compare R. Ffrench-Davis, Politicas Economicas en Chile: 1952-1970 (Santiago: Ediciones Nueva Universidad, Ceplan, 1973).

5. See ibid., p. 185.

6. Ibid., Ch. 4, Section 7.

7. The comparison is between actual production in 1952 and 1970. Both figures underestimate the installed capacity of production. See Mario Vera, La Politica Economica del Cobre, p. 98; and Eduardo Noveo, La Batalla por el Cobre, p. 101.

8. The figures refer to the LCM in relation to world output of primary copper. These percentages contrast with the estimates that by 1970 Chile had nearly a quarter of the estimated world reserves. In recent decades, it was only during World War II that Chilean output approached a percentage of this magnitude. See R. Bentjerodt, "El Mercado Internacional del Cobre," in Tironi and Ffrench-Davis, op. cit.

9. More data may be found in Chapter 7, below.

10. Compare R. Tomic, "Primeros Pasos hacia la Recuperacion del Cobre: los Convenies de Washington de 1951," in Tironi and Ffrench-Davis, op. cit.

11. Authority granted by Act 5,185 of June 1933. See Copper Department, El Cobre en Chile (Santiago, 1959).

12. See Memoria (Santiago: Central Bank of Chile, 1954), pp. 53-55; and Balanza de Pagos (Santiago: Central Bank of Chile, 1954), p. 76.

13. See Ffrench-Davis, Politicas Economicas en Chile, op. cit., pp. 69 and 72.

14. Act 11,828 of May 1955.

15. Act 7,160 of January 1942, modified by Act 8,758 of May 1947. See Balanza de Pagos (Santiago: Central Bank of Chile, 1947).

16. See Act 10,003 of October 1951; and Tomic, op. cit.

17. A description of the various proposals formulated appears in Luis Maira, "El Camino a la Nacionalizacion del Cobre," Revista de Derecho Economico (Santiago, University de Chile), no. 28 of 1969 and no. 29 of 1970.

18. R. Ffrench-Davis, "Estimacion de un Sistema de Retorno Total: Gran Mineria del Cobre y Andina," mimeo., 1965. During the discussions in 1965 of the copper agreements, a small step was taken toward total returns, with the establishment of the monthly payment of the profits tax. This represented capturing a tenth of the financial benefits from total returns. Despite its small volume, it was of importance in stabilizing receipts and the balance of payments with the more uniform distribution of payments throughout the year.

19. This section is based on R. Ffrench-Davis, Politicas Economicas en Chile, 1952-1970, App. IV, Sections 4, 5, and 7; and Ch. 4, pp. 93 and 103.

20. See Bentjerodt, op. cit.

21. See Tomic, op. cit.

22. The difficulties Chile faced in selling its production in 1953-54 are described in Balanza de Pagos and Memoria, Central Bank of Chile, 1954. Another view is found in Theodore Moran, "The Multinational Corporation and the Politics of Development: The Case of Copper in Chile (1955-70)" (Ph.D. thesis, unpublished, Harvard University, 1970).

23. Ffrench-Davis, <u>Politicas Economicas en Chile</u>, op. cit., Table 51. All the figures reflect dollars of 1969.

24. See Bentjerodt, op. cit.

25. See the author's lecture in a Conference of the International Economic Association, "Foreign Investment in Latin America: Recent Trends and Prospects," in V. Urquidi and R. Thorp, eds., <u>Latin America in the International Economy</u> (London: Macmillan, 1972). Also reproduced as Ceplan, Document no. 14.

7

THE INTEGRATION OF THE LARGE COPPER MINES IN THE CHILEAN ECONOMY: THE ROLE OF ECONOMIC POLICIES

Ricardo Ffrench-Davis

This chapter will analyze the way in which certain state policies have influenced the degree and type of integration into the Chilean economy of the large-scale copper-mining (LCM) enterprises, a sector that until 1971 was to a large degree controlled by foreign interests.

Integration with the rest of the country can be measured by different kinds of variables. From the economic point of view, integration can be described in simple terms as an expression of the country's other productive activities and of the benefits accruing therefrom. In concrete terms, this may mean the tax revenue that the sector brings to the Treasury; the use of manpower and the purchase of domestic inputs; the contribution to technological research, development, and dissemination; or the degree of raw material processing within the country. Of these several aspects, only the first two named received some systematic attention by the state during the period 1952-71—namely, the tax treatment of the LCM and the ratio of domestic to foreign components in operating costs. This chapter will concentrate on an analysis of the behavior of the enterprises with respect to their purchases in Chile and on the influence of economic policies on the substitution of imported inputs. The tax aspect will be dealt with only insofar as it is relevant to the foregoing.

An improvised approach to taxes may have a negative impact on other relevant variables. This is apparent from the exchange-rate policy that has been applied in Chile to large mining enterprises, particularly during the first half of the 1950s. The specific form in which exchange-rate policy was used as a means of tapping resources from the LCM actually slowed down the integration of the enterprises into the national economy. By contrast, a second aspect, the control by the Copper State Corporation (Corporacion del Cobre, Codelco) of the imports of the sector, represents one of the most valuable contributions that the institution has made since its creation.* Codelco's import regulations opened up new and encouraging prospects, successfully promoting a process of import substitution within the sector.

This chapter will concentrate on the exchange policy as it applied to the export returns of the LCM sector and on the import control exercised by Codelco in sections analyzing the following issues: (1) the expenditure trends of the LCM enterprises in Chile and abroad; (2) import policy; (3) exchange-rate policy and its effect on the level and composition of production costs of copper; and (4) impact of exchange-rate policy on foreign currency returns, under alternative foreign-exchange systems. Finally, the last section draws conclusions from the study of this aspect of Chile's copper policy and shows how the best policy would be a combination of exchange-rate and tariff regulations and the direct control of certain imports.

EXPENDITURE TRENDS IN CHILE AND ABROAD

Two decades ago, the LCM sector in Chile corresponded to a considerable degree to the stereotyped definition of an enclave, whose most striking feature was the high incidence of imported components in its costs of production. The latter included—in varying degrees over the years—refining and marketing costs abroad, depreciation, and profits on foreign capital and sizable imported inputs. The sum of these items adds to what is commonly known as nonreturns or unreturned export proceeds. (In general, official statistics include foreign inputs as an export return.) Alternatively, returns are constituted by taxation and expenditures in Chile.

Up to that date, the Chilean state had made no systematic attempt to remedy the situation described above, save partially with respect to the tax system. The tax provisions made from 1939 to 1955 contained a number of controversial loopholes and were subject to frequent changes; despite the lack of a clear and systematic national policy toward the copper sector, taxation did enable the Treasury to retain 52 percent of the value of production for the three-year period 1952-54. (This percentage is the ratio between the tax revenue accrued by Chile—income tax, overprice, and exchange tax—and the total value of the production of the LCM; it does not include customs

---

*Departamento del Cobre, created in 1955. In accordance with Act 16,425 of January 1966, the body was renamed Corporacion del Cobre (Copper Corporation), the term employed throughout this chapter.

duties. All figures are expressed in terms of 1969 purchasing power and, unless otherwise stipulated, refer to the comparison between the annual averages of the periods 1952-54 and 1970-71. Only in 1951 did tax proceeds begin to exceed production costs in Chile.)

In later years, after 1954, taxation declined, while expenditures in Chile increased steadily until they exceeded the former (Table 7.1). From 1966 onward, however, taxation once again came to represent the larger share of returns, because of the increase in the price of copper, the acquisition of a 51 percent share in profits and the subsequent introduction of the overprice tax in 1969. (Between 1966 and 1968, however, the increase of the foreign enterprises' profits was far more spectacular than that of fiscal revenue, owing to the reduction in taxation taking place in 1967 for Braden Copper Company, the subsidiary in Chile of Kennecott.)[2]

In the two decades since 1952, while output rose by only 55 percent, domestic production costs increased sixfold.* The behavior of each of the two most important components of expenditures in local currency—wages and salaries, and domestic inputs—was, however, very different.

Table 7.2 summarizes the trend of the variables measuring employment and labor costs. The number of workers rose by 50 percent;** this rise occurred simultaneously with rapid legal reclassification of specialized blue-collar workers into the category of white-collar workers. (This implied a change in the social status of those workers and a more favorable social security system. Most reclassifications took place according to specific laws approved by the parliament.)[3] The total amount of wages and salaries paid tripled. (The wages bill figures are for the period 1956-70, as information on pre-

---

*At the beginning of the period under consideration, the costs represented a bare 44 percent of total production costs; by 1970, this figure had reached almost two-thirds of the total. The ratio is equal to column (6) divided by the sum of columns (1) to (3) of Table 7.1 plus dollar expenditures in Chile. Both terms exclude the exchange-rate tax, which has been merged with the profits tax in column (7). Generally speaking, accounting or book production costs do not include the alternative cost of the capital itself; hence, because of the external origin of the capital, the result is an overestimate of the relative share of domestic costs.

**The larger increase in employment took place during the last years of the period, when output stagnated. The figures include some workers employed in the investment programs launched in 1967 and some "redundant" labor.

TABLE 7.1

Large-Scale Copper Mining: Expenditures in Chile and Abroad
(annual averages, millions of dollars at 1969 prices)

Years	Imported Inputs (1)	Other Expenditure Abroad (2)	Depreciation and Interest (3)	Profits (4)	Nonreturns 1 to 4 (5)	Expenditure in Chilean Currency (6)	Income and Exchange Taxes (7)	Returns (6) + (7) (8)
1952–54	24.5	14.1	11.3	26.2	76.1	46.0	145.4	191.4
1955–57	31.6	21.1	18.0	65.5	136.2	66.6	155.5	222.1
1958–60	36.8	32.6	28.3	49.1	146.8	94.6	86.2	180.8
1961–63	25.2	34.7	26.7	43.8	130.4	116.0	93.0	209.0
1964–66	30.0	34.1	35.0	60.8	159.9	137.7	169.5	307.2
1967–69	34.0	35.8	26.3	121.8	217.9	201.0	260.5	461.5
1970–71	38.3	31.8	40.5	65.2	177.8	250.1	283.6	533.7

Notes: Column (1) contains the CIF value of imports of inputs for operating purposes; column (2) includes sales, costs, freight, and refining abroad; column (3) comprises interest and depreciation remittances; column (4) records the net profits earned by Anaconda and Kennecott. The sum of columns (1) to (4) corresponds to un-returned values. Column (6) includes all expenditures paid in domestic currency, minus an estimate of the exchange tax; the latter is contained in column (7), along with income and overprice taxes, and Chile's share in profits. The table does not include dollar expenditures in Chile. During the period 1967–69 this expenditure amounted to $24 million per year; its main components are wages and salaries paid in foreign currency to executive and supervisory staff. Except in the case of column (1), the last line refers only to 1970.

Source: Compiled by the author.

TABLE 7.2

Large-Scale Copper Mining: Employment, Wages, and Salaries
(annual averages)

	Number of Workers			Workers' Earnings in 1969 Prices			Cost per Pound of Copper (cents)
Year	White and Blue Collar	Supervisors and Executives	Total	White and Blue Collar (dollars per worker)	Supervisors and Executives	Total	
	(1)	(2)	(3)	(4)	(5)	(6)	(7)
1952/54	15,275	802	16,077	n.d.	n.d.	n.d.	n.d.
1955/57	15,883	929	16,812	2,584	11,278	3,077	5.9
1958–60	16,378	1,139	17,517	3,160	11,744	3,735	6.8
1961–63	17,130	1,068	18,198	3,898	12,245	4,393	8.0
1964–66	19,939	1,161	19,100	4,330	12,081	4,787	9.1
1967–69	20,357	1,351	21,708	4,800	12,568	5,282	10.3
1970–71	22,683	1,473	24,156	5,352	14,161	5,958	12.2

Notes: Columns (1) to (3) are based on data supplied by the Department of Labor Relations of Codelco and comprise the permanent staff, including persons employed in investment activities. Column (4) has been estimated by the cost in escudos of personnel remunerated in that currency, deflated by an index of Chile's inflation and then converted into dollars on the basis of the 1969 import exchange rate; this figure was then divided by the average number of workers employed on a permanent basis. The figures tend to measure the real gross income of the workers and not the real cost to the enterprise. Column (7) represents the cost to the enterprises per pound of copper produced, after deduction of the exchange rate tax. Production was measured in terms of the equivalent of pounds of electrolytic copper. Columns (4) to (7) cover solely the period 1956–70.

Source: Compiled by the author.

vious years is very unreliable.) The consequent increase—of around
95 percent—in the per capita income of workers employed in the LCM
sector was substantially greater than that obtained by the average
Chilean worker. During the same period, the national product per
member of the labor force rose by no more than 30 percent. (If
the figure is corrected for changes in terms of trade, the increase
drops to about 25 percent.) As a result, there was an increase in
the volume of foreign currency returned by the copper companies in
order to cover manpower costs.* In 1970 expenditure under this
heading represented about 40 percent of production costs. In absolute terms, it amounted to around 150 million dollars.

In the case of physical inputs, in contrast, there was a progressive subsitution of domestic intermediate goods for imported goods
(see Table 7.3). This phenomenon has two explanations. It was
partly the natural consequence of the development of domestic industry, particularly in the metal-mechanics sector, and, to an important degree, the outcome of policies applied by Codelco and by the
Central Bank. As a consequence, in 1970 domestic inputs accounted
for 72 percent of total physical inputs as compared to 40 percent in
1955-57.

Since its creation in 1955, the Copper Corporation pursued a
policy designed to promote import substitution. At the same time,
changes in the exchange-rate policy of the Central Bank in the last
years of the period altered the approach whereby, consciously or un-
consciously, it taxed expenditure in the country. These two mechanisms were the main causes of the extraordinary increase in the LCM
sector expenditures in Chile.

DIRECT IMPORT CONTROLS ON THE LARGE
COPPER MINING ENTERPRISES

During the 20 years covered by this study, the sector's imports
were subject to a system known as "Internaciones con cambios prop-
ios" (imports financed with their own foreign exchange). This system
permitted the copper companies to use the foreign currency they
earned from their sales abroad to pay for imported inputs. As a result, their payments abroad were unaffected by the frequent shortages
of foreign currency and bureaucratic restrictions on imports that the
rest of the country suffered.

---

*The increased inflow of foreign currency that the Central Bank
obtained from this source, was, however, partially offset by the reduction in tax revenue resulting from the corresponding rise in production costs.

TABLE 7.3

Domestic and Imported Inputs and Rate of Exchange
(1969 prices)

Years	Inputs in Cents per Pound			Domestic Inputs as Percent of Total	Exchange Rate for LCM (escudos/U.S. dollars)	Exchange- Rate Tax (percent)
	Domestic	Imported	Total			
	(1)	(2)	(3)	(4)	(5)	(6)
1952-54	n.a.	3.6	n.a.	n.a.	3.39	114.6
1955-57	2.6	3.7	6.3	39.9	6.84	25.4
1958-60	3.1	3.9	7.0	43.9	8.77	0.1
1961-63	3.9	2.5	6.4	60.5	8.70	2.3
1964-66	4.5	2.9	7.4	60.6	8.05	9.7
1967-69	6.4	3.1	9.5	67.5	8.88	2.0
1970-71	7.7	3.0	10.7	71.6	8.93	2.0

Notes: Columns (1) and (2) correspond to a series of expenditures in escudos and CIF dollars, respectively, deflated by estimates of their respective price indexes; subsequently, real expenditure on domestic inputs has been converted into dollars, when being multiplied by the 1969 average import rate of exchange. Column (5) includes the annual average exchange-rate applied to returns for operating costs of the LCM. The tax rate estimate shown in column (6) is measured as the tax in escudos per dollar divided by the rate of exchange applied to the LCM. The tax is equal to the difference between the rate of exchange for Chilean nontraditional exports and the LCM exchange rate.

Sources: "Substitucion Entre Insumos Nacionales e Importados: Une Estima Econometrica" (mimeo.; Ceplan, 1974); and author's compilations.

Until 1955, the enterprises were entirely free to make their purchases abroad; among others, prohibitions, quotas, and exchange controls established under the general import system did not apply to large-scale mining. All the sector's imports, however, were subject to customs duties. These duties were levied both on imports described as operating costs and those intended for investment purposes. Until that year, therefore, each enterprise determined the volume, composition, and origin of its imports as it saw fit. The public sector was involved only in applying the relevant customs duties and in preparing the import figures, recorded by the Customs Department, to be used in drawing up the annual balance of payments accounts.

The New Deal Law (ley de Nuevo Trato) created Codelco, a public institution set up to supervise and control the production and marketing conditions of copper. After 1955, Codelco was made responsible for supervising the purchases of the mining enterprises abroad and had to approve the imports of inputs of the LCM sector. [4] The framing of the policy was a gradual process. Its main lines of action were as follows:

1. For the most part it concentrated on imports subject to customs duties, which under the "New Deal Law" consisted of goods classified under operating costs and replacement investment.* Imports of equipment and machinery under productive capacity expansion programs, which became exempt from all customs duties, were not covered by the import substitution policy until 1966.

2. Goods controlled by Codelco were classified in two groups: The first group of goods was subject to the approval only after a comparison with the prices and quality of domestic production. (Some 30 percent of the value of imported inputs was subject to this control process during the two-year period 1957–58.)[5] The second group was freely imported, subject to approval by Codelco after the arrival of the goods to the customs area for price and data control. Import duties applied similarly to both categories of goods. The lists of goods in both groups were modified as time went by as indicated by the experience accumulated by Codelco.

----

*The term "replacement investment" is used to indicate investment not effected in accordance with the provisions of "investment decrees." The enterprises themselves classified each item in one category or another, and this was subsequently checked by the accounting department of Codelco. The importance of this classification derives from the fact that investment expenditure, whether for replacement or expansion purposes, increases the capital and assets of the enterprise, whereas operating expenditures are recorded as costs and therefore reduce the taxable income.

3. No import requests were to be authorized when a substitute article was available on the local market, which could offer reasonable quality, price, and terms of delivery.[6] "Reasonable price" was understood to mean that the price of the domestic input should not exceed that of the foreign product after payment of the corresponding customs duties. (This measure does become necessary if the enterprises installed in Chile buy inputs from subsidiaries abroad that are attached to the same parent company; if, given equal conditions, the companies show a preference for the imported goods; and if there is a tendency to continue operating with the same supplier without considering inputs available in the domestic market. This tendency is encouraged by the <u>centralization</u> of the subsidiaries' purchases in New York and the fact that statistics, requests, and orders were drawn up in the English language and U.S. system of measurements.)[7]

4. An attempt was made to supply potential domestic producers with information on imports effected, especially with respect to their volume, technical specifications, and prices. A listing of volumes, prices, and specifications was drawn up periodically. Contact was made with the domestic enterprises in the branch concerned in order to promote an interest in the substitution of these imports.

5. The large mining enterprises received instructions to advise the domestic producers in advance of their future demand, so they had time to prepare to compete with established foreign producers. Major lines of goods were substituted following the preparation of purchasing programs from 1961 onward.

6. The sanctions that Codelco applied in the event of the large mining enterprises' failure to conform to its instructions involved the refusal to approve the entrance of those particular imports and, from time to time, appropriate fines.

Codelco's policy of control and information was pursued in conjunction with the prevailing system of tariffs;* hence, protection granted to domestic producers was a relative one, in contrast to the frequent absolute protection received by other import substitutions taking place in the Chilean economy. Judging from the fragmentary information that is available, there was no significant change in customs duties after 1955. The average rate fluctuated around 50 percent, a rate lower than that applied to the same type of products when

---

*The taxes were payable on imports for both operational requirements and replacement of equipment. Investment for purposes of increasing the productive capacity, being covered by government decrees, were exempt from all duties.

imported by domestic enterprises.* The reason for the difference
is that the LCM enterprises were only subject to the general tariffs,
being exempted from several surtaxes and prior deposits, which
were frequently imposed on the imports of the rest of the country.
(All these additional charges applied to the general import system,
together with prohibitions and other nontariff restrictions on trade.
Other import systems, such as free ports, also benefited from this
and other tax privileges. After these import duties were merged in
1967, the LCM received a stated discount, which provided an estimate of the incidence of import surtaxes.)[8]

The share of Chile's total imports from the United States was
about 50 percent at the beginning of the period and dropped to less
than a third by the end. In 1957, of the imports of the LCM for operational requirements and investment, excluding petroleum imports,
92 percent came from the United States.[9] Codelco's policy did not
affect the geographical origin of the imported inputs, the vast majority of which, 80 percent to 90 percent, came from the United States.
The percentage was higher in the investment programs than in operating expenditure. Furthermore, part of the process of import substitution was carried out by foreign enterprises that installed themselves in Chile in order to continue supplying the market of the large
mining enterprises. Similarly, the national producer sometimes
operated under license of the previous U.S. supplier. These two forms
of relationship with foreign countries—external investment and licensing—have perhaps contributed more efficiently (or less harmfully,
if one prefers) to import substitution in the case of the large-scale
copper-mining sector than in the other sectors of the economy.
Codelco took considerable care to ensure an acceptable level of quality and price in the domestic substitutes, whereas foreign investors
producing import substitutes in other sectors in Chile were, generally
speaking, able to operate under conditions of absolute protection,
without being subjected to any direct or indirect effective control of
costs and prices.[10]

---

*An estimate carried out by the Import Section of Codelco for
1967 indicates that the average rate was somewhat higher than 50 percent. The author's own estimate for the five-year periods 1956-60
and 1966-70 is that the average tax was 48.1 percent and 49.4 percent, respectively. It must be noted that the rate varies according
to the item imported; consequently the average rate has fluctuated
from year to year according to the composition of the actual imports.

EXCHANGE-RATE POLICY AND PRODUCTION
COSTS OF COPPER

The imports of the large mining enterprises have also been influenced by the exchange-rate policy applied to the sector. Exchange-rate regulations for the LCM sector have differed from those for the rest of the country's exports, as regards the system of partial returns of its export proceeds and the rate of exchange applied to those returns. In essence, LCM was only required to return and exchange that portion of foreign currency earnings required to pay for domestic inputs.

The rate of exchange for these partial returns influenced the composition and amount of the production costs. A reduction in the exchange rate, for example, increases the cost in dollars of domestic inputs and thus raises the production costs of the LCM sector and reduces its taxable income. In order to ameliorate the influence of this variable on costs, the enterprises tend to diminish the relative share of the domestic component, by replacing it by imported inputs whose price in dollars has not been altered by the economic policies. Each of these headings contains countless items.* Consequently, the process of substitution occurs among those items that are more easily interchangeable.

Under a system of partial returns, in other words, the application of a rate of exchange to the foreign currency sales of the large mining enterprises lower than what prevails for other foreign trade deals represents a tax on production whose magnitude varies in direct proportion to the incidence of costs incurred in local currency.

In practice, the exchange-rate policy applied to the LCM has frequently differed from that applied to the rest of the economy, as a means of tapping resources for the Treasury. During the years 1952-54 (see column 6 of Table 7.3) the exchange-rate tax on purchases in Chile amounted to a surcharge of over 100 percent. Toward the end

----

*The domestic component includes manpower. First, it is probable that the increase in the rate of exchange in the course of the past 20 years has facilitated the rise in the real wages and salaries earned by LCM workers. Second, in the short run, a very low rate of exchange provides a greater incentive to increase the number of employers (supervisors and executive personnel) paid in foreign currency; in this way, the enterprise avoids selling the foreign currency at the official price. Finally, in the long run the rate of exchange affects the composition of investment: a low rate promotes investment involving a more intensive use of capital goods, most of which are imported.

of the period under consideration, the tax disappeared as the rate of
exchange of the LCM sector and for other Chilean exports became virtually the same.

The evolution of the exchange-rate policy applied to the LCM
sector in the framework of the partial return system, together with
the qualitative import control imposed by Codelco, explains the remarkable change in the cost composition, as regards domestic and
imported components, from the 1960s to 1970.[11]

After nationalization in 1971, a system of total returns was introduced. Both the economic and bookkeeping repercussions differ
substantially from those experienced under the partial returns system.

As regards the economic effects, under a system of total returns, a low rate of exchange for foreign currency earnings represents a tax on exports,* which reduces the profits of the enterprise
without altering the relative prices of the various cost components.
Available empirical data indicate that the negative impact on the level
of production of copper is very slight.[12] By contrast, the negative
effect on the composition of costs, governed by the exchange-rate tax
under the partial returns system, is considerable.

Under a system of total returns, the exchange rate has a direct
impact on imports of inputs on remittances of depreciation allowances
and profits and on sales of copper in the domestic market to be processed by manufacturing enterprises. If the same undervalued rate
of exchange is applied to all these operations, the outcome is similar
to that obtained under the partial returns system: a tax on the use of
labor and domestic inputs. Unfortunately, this has been by and large
the policy applied after 1971.

The combination of exchange rates to be selected must depend
on the objectives being pursued. When the aim is to retain in the
form of fiscal revenue the high natural productivity of the Chilean deposits, the tax should not affect the composition of production and
sales. If, consequently, because of some difficulty in implementing
other mechanisms, the tax is applied on the exchange rate, it should
be payable on all export proceeds while "normal" rates of exchange
ought to be applied to all other purchases and sales (that is, imports
and interest payments, capital inflows) of foreign currency effected
by these enterprises. This recipe is valid whether they are foreign
or state-controlled enterprises.

Major differences also arise from the bookkeeping standpoint,
according to the return system that is utilized. The tax prevailing un-

---

*Unless indicated otherwise, the following considerations are
based on the assumption that the remaining foreign exchange operations on the enterprises take place at a normal rate.

der a system of partial returns is levied on production costs, which is not the case under a system of total returns. In order to avoid the obvious distortion of costs and their composition, figures must be depurated of the incidence of the rate of exchange. In this way, the sudden artificial fluctuations that the production costs of these enterprises experienced as a result of modifications in the size of the exchange-rate tax can be avoided.

FOREIGN-EXCHANGE POLICY AND
EXPORT RETURNS

The effect of the rate of exchange on the volume of foreign currency that the enterprises returned to Chile each year depended not only on the system of partial returns but also on the possibility of substitution between domestic and imported components, the impact of each on costs, and the tax rate on profits. As time went by, various repercussions took place that produced an effect on the volume of export returns pointing in the opposite direction from that produced in the short run by the partial returns system.

In the first place, the tax on expenditure in Chile raised production costs, consequently diminishing taxable profits. The establishment of an exchange tax thus automatically brought with it a reduction in the yield of profit taxes. Considering that the average rate of tax between 1952 and 1971 amounted to over 60 percent, * the net yield of the exchange tax represented a mere 40 percent of the gross amount of the exchange-rate differentials. ** On the few occasions when a statistical control was made of this type of taxation, the figure recorded was the gross amount.

Second, as mentioned, the enterprises tried to reduce the effect of these taxes on costs by replacing their purchases in Chile by imported inputs. This phenomenon was particularly noticeable in the substitution between physical inputs. Estimates available for the period indicate that, on average, a reduction of 10 percent in the real

---

*The figure refers to profit taxes and Chile's share in profits. The overprice is excluded insofar as it is independent of the level of profits.

**To illustrate, if the initial costs are 100 and the value of sales is 140, the profit tax is 24. An exchange-rate tax of 30 percent, on a domestic component that amounts to 60 percent of total production costs, raises them to 118; as a result, the taxable income is reduced to 22; and the corresponding yield of the income tax is 13.2. Consequently, the total fiscal revenue amounts to only 31.2.

rate of exchange (or relative price of imported inputs) led to a decline
of approximately 2.6 percent in purchases in Chile and an increase
in those effected abroad of about 3.3 percent, and vice-versa. (These
figures do not take into account the effect of the level of production
on the demand for inputs—that is to say, they only measure the substitution effect.) In other words, in response to such a policy, the
ratio between domestic and imported inputs decreased (increased) by
6 percent.[13] In practice, the increase in returns originating from
the introduction of an exchange-rate tax was approximately balanced
by the subsequent reduction in returns corresponding to purchases in
the country and profit taxes. These compensatory effects, particularly the first, only take place after some time, so it is essential to
consider the role of different time horizons in the exchange-rate policy.

The foregoing considerations show categorically that the exchange tax applied, particularly under the conditions prevailing up to
1971, was not an effective way of tapping for Chile the excess profits
obtained from the production of copper.* Various alternatives would
have proved more successful: an export tax, additional taxes on profits, an overprice tax, or a fixed tax related to the quality of the copper
deposits. However, the unsuitability of the exchange-rate tax mechanism does not mean that, where it exists, it should be done away
with suddenly. Such a step would mean an abrupt reduction in production costs (and fiscal revenue) without there being any possibility of
an immediate shift from purchases abroad to acquisitions in the country. This takes time; therefore, the exchange-rate policy must be
clearly defined and stable and relative adjustments brought into effect
gradually. Between 1965 and 1967, for example, it was decided to do
away with the exchange tax in order to remove the discrimination
against purchases in Chile that it fostered. It was, however, done
away with gradually so as to prevent the enterprises from obtaining
undesirable capital gains.**

---

*It may be interesting to note that, in most cases, the revenue
generated by the exchange-rate tax on purchases in Chile was used to
subsidize [promote] imported consumer goods.

**This policy, obviously, could not offset the excessive profits
that the companies made thanks to the unduly low rate of profit taxes
and the long delay in introducing an overprice tax. Despite the excessive profits received by the foreign owners of LCM in those years,
the representatives of the interests of those enterprises exerted constant pressure, though with little success, for the rate of exchange of
the sector to be raised abruptly.

CONCLUSIONS

Economic policies can effectively modify the behavior of the units of production. This is equally true of foreign and of national enterprises. In fact, the exchange-rate policy of the Central Bank and the qualitative control over imports exercised by Codelco substantially affected the composition of LCM expenditure as between domestic and imported components.

The items that the enterprises utilized in their productive process amount to many tens of thousands in number, owing to the wide variety and different specifications required. Consequently, the direct control of imports is neither feasible nor sufficient by itself. Similarly, when the enterprises are owned and controlled by foreign capital, such instruments as exchange-rate and tariff policies are necessary, but insufficient, to regulate imports.

In the 16 years between 1955 and 1971, considerable progress was made in the substitution of imported inputs; moreover, the process was considerably more efficient in this sector than in the rest of the Chilean economy. The important thing today would seem to be not to undo that effort, but, rather, to increase Codelco's supervision and to extend it to other sectors where international enterprises are present in the Chilean economy.

With regard to the production of capital goods, on the other hand, the field is still virtually unexplored. Not only was Codelco's qualitative control policy not applied to them, but, in addition, imports of equipment benefited from total exemption from customs duties. This major resource in copper production has thus remained outside of either direct or indirect mechanisms of regulation. The failure of the state to take the necessary steps has been aggravated by the absence of any systematic and coherent industrial development policy. For many years, the policy applied by the Imports Section of Codelco was an oasis in the desert created by the absence of any clearly defined industrial policy.

The selective production of capital goods for the mining sector can be justified on two counts in countries like Chile. It affords a large market insofar as it is a major purchaser of capital goods for the expansion of its productive capacity.* The production of capital

---

*For certain items, domestic demand alone is in keeping with the scale of production of plants installed in developed countries. Naturally, this is not a sufficient reason to start producing such goods in the country, but it is a significant factor. It must not be forgotten that the imports of capital goods have for the most part been exempt from all customs duties; when the rest of a country's foreign trade is

goods is one of the main vehicles for the development of appropriate
technological know-how and its adaptation to the economic characteristics of the country. Both aspects, however, pose serious problems.
If the process is to be efficient, it has to be selective, and this means
appropriate specialization; similarly, although local demand may be
considerable it may be irregular owing to the discontinuity of programs of investment in large deposits. These difficulties could probably be minimized if Chile were to embark upon such a venture in
collaboration with other countries with similar national aspirations
and characteristics. The Intergovernmental Council of Copper Exporting Countries (CIPEC) and, more particularly, the countries of
the Andean Pact offer fertile ground for such collaboration.*

The nationalization of the LCM sector is only one step toward
the full development of the sector so that it can best serve the national
interest. The biggest challenge is yet to come.

Even with state enterprises, the LCM sector requires a flexible
and imaginative combination (such as is not usually found in textbooks)
of indirect instruments and direct action, so as to avoid the pitfalls
of the simplistic view of both bureaucratic intervention and laissez-
faire. The combination of instruments that was a feature of the import policy of the LCM sector during the 1960s is a good, though modest, example of this necessary and efficient combination of various
mechanisms of state action.

NOTES

1. Lengthy historical series, albeit without the necessary adjustment to take account of the exchange tax, are reported in C. Reynolds, "Development Problems of an Export Economy: The Case of
Chile and Copper," in Marcos Mamalakis and Clark W. Reynolds,
Essays on the Chilean Economy (Homewood, Ill.: R. D. Irwin, 1965),
statistical appendix, columns 13 and 25. An adjustment for the exchange rate tax for the years 1950-54 can be found in Theodore Moran,
"The Multinational Corporation and the Politics of Development: The

---

subject to severe restrictions, that exemption acts as a negative effective protection for the domestic production of those goods.

*See, for example, Decision no. 57 on the sectoral industrial
development program of the metal-mechanics sector, which was
adopted in October 1972, by the Commission of the Cartagena Agreement. The commission is the political authority of the integration
process of six South American countries (Bolivia, Colombia, Chile,
Ecuador, Peru, and Venezuela).

Case of Copper in Chile 1945-70," Ph.D. thesis, Harvard University, 1971.

2. A complete analysis can be found in Luis Maria, "El Camino a la Nacionalizacion del Cobre," Revista de Derecho Economico (University of Chile), nos. 28 and 29 (1969 and 1970).

3. See J. Barria, "Organizacion y Politicas Laborales en la Gran Mineria del Cobre," in Ernesto Tironi and Ricardo Ffrench-Davis, eds., El Cobre en el Desarrollo Nacional (Santiago: Ediciones Nueva Universidad, Ceplan, 1974).

4. See "Funciones del Departamento del Cobre," in Panorama Economico, no. 169 (June 1957): and Pedro Rios, "La Experiencia Chilena en la Integracion de Sectores Importadores al Mercado Interno," in Panorama Economico, nos. 280 and 209 (December 1959 and February 1960).

5. See Rios, op. cit., p. 416.

6. See Seccion Importaciones, Codelco, "Estadisticas de Adquisicion" (mimeo., 1967).

7. See Rios, op. cit., pp. 31-32.

8. A full discussion of foreign trade policies in the period 1952-70 can be found in R. Ffrench-Davis, Politicas Economicas en Chile: 1952-70 (Santiago: Ediciones Nueva Universidad, Ceplan, 1973), Ch. 4 and Apps. 3 to 5.

9. Rios, op. cit., p. 32, Table 5.

10. See R. Ffrench-Davis, "Foreign Investment in Latin America: Recent Trends and Prospects," in V. Urquidi and R. Thorpe, eds., Latin America in the International Economy (New York: Macmillan, 1973).

11. That most of the production is sold abroad in foreign currency is another influential factor. R. Ffrench-Davis, "Dependencia, Sub-Desarollo y Politica Cambiaria," in Estudios Monetarios II (Santiago: Banco Central de Chile, 1970); and Trimestre Economico, no. 146 (Mexico, 1970).

12. In other words, the elasticity of supply is fairly low. References can be found in R. Bentjerodt, "Produccion y Mercados Mundiales del Cobre," in Tirone and Ffrench-Davis, El Cobre en el Disarrollo Nacional (Santiago: Ediciones Nueva Universidad, Ceplan, 1974). Original estimates appear in Fisher, Cootner, and Baily, "An Economic Model of the World Copper Industry," in the Bell Journal of Economics (Autumn 1972).

13. R. Ffrench-Davis, "Subsitution of National for Imported Inputs in Copper Production: An Econometric Study" (mimeo.; Santiago; Ceplan, 1974).

THE SMALL-AND
MEDIUM-SCALE
COPPER-MINING
SECTOR IN CHILE
Ernesto Tironi

Toward the end of the 1960s, small- and medium-sized copper-mining enterprises were generating approximately one-sixth of Chile's total exports and one-fifth of its production of copper. They also employed at least half the workers engaged in this economic sector.

Despite the sector's evident importance, few studies provide an overall analysis of its basic nature and characteristics. The lack of information available in this field has resulted both in a number of misconceptions on the part of the general public and, worse still, a serious lack of responsibility on the part of the state in the introduction of appropriate policies for the sector. It is a common belief, for example, that the sector is made up entirely of small-scale Chilean enterprises, whereas, in fact, almost half of its production has been controlled by a handful of foreign companies earning very high profits.

The development and present characteristics of this sector is an example of the relevance of the "duality theories" of economic development in their modern versions, which stress the "structural heterogeneity" of less developed countries (LDCs) rather than just duality.[1] Several authors implicitly treat this entire sector as the traditional or backward section within the Chilean copper-mining economy, while in fact some very important firms in it really belong to the modern sector. The lack of consideration to the institutional and organizational characteristics that account for the differences between several types of producers often leads to very inefficient and socially undesirable policies for the development of the sector.

Juridically, the small-scale mining industry is defined as that productive activity that is conducted in mines or in smelting plants whose owners are natural persons or mining corporations, with registered capital of not less than the equivalent of U.S. $50,000.[2] Medium-

scale mining refers to the activity conducted by companies whose
capital exceeds the amount stated, but whose annual production is
not more than 75,000 metric tons of copper bars. The thousands of
own-account workers operating in this sector necessarily belong to
the small-scale mining industry, but there are companies that may
belong to either sector. Many of the companies in the small-scale
mining sector do not differ significantly from firms in the medium
sector in terms of method and volume of production.

This chapter will concentrate on the five principal private enterprises that without doubt belong to this medium-scale category
and that have been operating during the entire decade of the 1960s.
They have been responsible for more than half of the total production
of the small and medium mining enterprises. The remaining medium
enterprises either have a considerably smaller volume of production—
in some cases even lower than that obtained by certain independent
producers in the small-scale mining sector—or are subsidiaries of
the five principal enterprises in this category.

Apart from legal distinctions with regard to capital and levels
of production, the main differences between small, medium, and
large copper-mining enterprises have to do with the type of productive process carried on, the way in which the enterprises market the
metal, the system of taxation that applies to them, and the state
agencies that supervise or participate in their activities.

The productive processes of the various subsectors of the copper-mining industry vary. The large enterprises not only extract
minerals but also concentrate, smelt, and refine it in their own
plants. Medium enterprises usually go as far as the concentration
stage (though two of them have their own refining facilities), whereas
the small enterprises are predominantly engaged in extraction only.

Most of the small-scale producers sell their production to the
state company (Empresa Nacional de Mineria, Enami), which exports
that copper in different forms. The medium- and large-scale producers, on the other hand, usually export their production directly,
even though in many cases their production is refined by Enami.
(Both sectors have, however, always been under the general system
of exchange control: compulsory repatriation of the total sum of foreign currency earned by their exports.)

The small-scale mining sector pays taxes on its gross sales,
while the medium enterprises pay on their profits. The small firms
are highly dependent on Enami for their supply of inputs, credits, and
capital. By contrast, Enami's intervention in, and control over, the
activities of the medium-scale mining sector is very limited—nor has
this function been exercised by the State Copper Corporation (Corporacion del Cobre, Codelco), which has always controlled mainly the
large-scale sector (and now administers the nationalized enterprises).

This study, essentially covering the 1950s and 1960s and the first years of the 1970s, deals with (1) the recent growth in importance of the small and medium copper-mining enterprises and some of the general characteristics of the sector, including the main features of the copper-marketing system; (2) the nature of the activities of the small mining enterprises, and the state's participation in them, with particular emphasis on Enami's role; (3) the pattern of ownership and operation of the principal medium copper-mining enterprises, including production, taxation, costs, and profits; and (4) the critical problems and prospects of small and medium mining in Chile.

PRODUCTION AND MARKETING IN THE SMALL- AND MEDIUM-SCALE MINING SECTOR

The share of the small- and medium-scale mining sector in the total exports of the country increased from 7 percent in 1960 to 16 percent in 1970,* as can be seen in Table 8.1. Its relative share of the country's total copper exports rose from 10 percent in 1960 to 21 percent in 1970.

Production

Table 8.2 shows the volume of production of the small- and medium-scale copper-mining sector, according to type of copper processed in the sector's own plants. During the 1950s, the small and medium enterprises generated an average of one-tenth of the national supply of copper, which rose to about 19 percent in the second half of the 1960s. The large mining enterprises' share declined steadily from 90 to 75 percent over the same period.

The fluctuations in the production of the small and medium mining enterprises were more accentuated than for the large mining enterprises, probably owing largely to the sector's greater sensitivity to variations in price. The sudden changes in production figures are also partly attributable to the entry into operation of new mines and to the expansion of smelting capacity of Enami. Enami's Ventanas refinery, for example, opened in 1966 and put a volume of electrolytic

---

*It should be noted that the percentage of the small- and medium-scale sector's exports in 1970 represents an absolute value of $178 million, more than the country's total industrial and agricultural exports ($148 million) and more than that of the whole of the exports of the noncopper sector ($111 million).

TABLE 8.1

Composition of Chilean Exports

(percent)

Exports	1960	1965	1966	1967	1968	1969	1970
Mining exports*	87.1	89.8	88.4	88.7	88.2	89.4	86.7
Copper mining	68.9	70.7	73.4	75.5	76.1	79.4	76.7
Large mining enterprises	62.1	52.8	59.5	60.5	59.9	61.9	60.7
Small and medium mining enterprises	6.8	17.9	13.9	15.0	16.2	17.5	16.0
Other mining	18.2	19.1	15.0	13.2	12.1	10.0	10.0
Industrial exports	7.8	6.9	9.1	8.6	9.0	8.3	10.4
Agricultural and sea-produce exports	5.1	3.3	2.5	2.7	2.8	2.3	2.9
Total	100.0	100.0	100.0	100.0	100.0	100.0	100.0

*Including exports of semimanufactured copper that are added to those made by the large mining enterprises. This procedure is followed because in fact such exports include a very modest manufacturing process—that is, the value added to the product by the national manufacturing firms that buy the primary copper to the large enterprises is very small. The value of these exports varies widely from year to year, largely because of its speculative nature.

Source: R. Ffrench-Davis, Politicas Economicas en Chile: 1952-1970 (Santiago: Ediciones Nueva Universidad, Ceplan, 1973), Table 43, p. 273.

TABLE 8.2

Annual Production of Copper in Chile by Type and Sector of Origin
(thousands of metric tons of copper content)

| | Small- and Medium-Scale Copper Sector[a] | | | | | Large-Scale | |
Year	Electrolytic	Blister (1)	Mantos Blancos Refined[b] (2)	Ores, Cements, Concentrates, and Other (3)	Sub-total (4)	Copper Sector (5)	Total (6)
1956	—	15	—	31	46	443	489
1957	—	16	—	29	45	436	481
1958	—	20	—	27	47	417	464
1959	—	20	—	28	48	497	545
1960	—	26	—	27	53	479	532
1961	—	32	10	23	65	482	547
1962	—	31	17	28	76	510	586
1963	—	31	18	44	93	507	600
1964	—	38	21	35	94	528	622
1965	—	57	20	28	105	479	585
1966	14	31	24	31	100	537	637
1967	37	36	21	30	124	536	660
1968	34	42	27	35	138	519	657
1969	35	45	27	41	148	540	688
1970	35	49	28	45	157	535	692
1971	31	51	26	83	191	518	709

[a]Including Andina during the years 1970 and 1971.
[b]Refers to the special type of refined copper produced by Minera Mantos Blancos.

Source: Central Bank, Balance of Payments and Monthly Bulletin, on the basis of information supplied by
Codelco. The figure for the small and medium mining enterprises correspond to exports effected each year.

copper on the market equal to approximately 20 percent of the total for the sector. (Enami was also responsible for the smelting and refining of part of the production of the large mining enterprises, but this fact is not taken into account in the figures given here.)

A third important feature of the production of small and medium mining enterprises is that, unlike the large mining enterprises, a significant fraction of their copper is sold without being refined or smelted. Principally under Enami's initiative, however, the fraction of copper that the sector places on the market in this form declined from 62 percent to 26 percent between the second half of the 1950s and the second half of the 1960s.

The bulk of the electrolytic copper produced by the small and medium mining enterprises derives from Enami's purchases from small-scale producers; the remainder comes from ore produced by independent exporters for treatment under contract, plus a smaller volume produced directly by Enami itself.

Table 8.3 shows that the growing share of small and medium mining enterprises in the country's supply of copper can largely be explained by the expansion of the latter subsector. Between 1960 and 1965, the production of the medium-scale mining sector rose from 34,000 to 86,000 tons, coming close to 120,000 tons by the end of the decade. The sector's share in the total supply of Chilean copper rose from 6 percent to 17 percent between 1960 and 1970 and to 21 percent in 1971.

The number of producers and miners engaged in small and medium mining enterprises varies from year to year according to the economic prospects of the sector, especially the price of copper and the availability of credits and inputs for the small producers. At the end of 1970, the files of the Servicio de Minas del Estado listed approximately 115 medium and small enterprises or independent entrepreneurs and around 180 processing or smelting plants with a daily capacity of over 20 tons each. (A plant with a 20-ton capacity annually produces approximately 90 to 120 tons of refined copper. Several producers or enterprises in this sector own more than one plant.)

These global figures, however, disguise the vast differences that exist within the sector. In 1970, the two largest medium-scale mining enterprises accounted for 47 percent of the production of small and medium mining enterprises (excluding Andina). The five largest accounted for almost 57 percent. Ten smaller private enterprises, exporting directly between 3,000 and 4,000 tons of refined copper in that year, made up a further 8 percent of total production. The remaining 35 percent was produced by much smaller enterprises or independent producers, which sold their minerals to Enami. (See Table 8.5.) In 1970, Enami purchased ores from 4,590 different mines, less than 4 percent of which produced more than 30 tons of re-

TABLE 8.3

Composition of Chilean Copper Production by Sector of Origin
(percent)

| Sector | Mining Enterprises | | | Total |
	Large	Medium	Small	
1960	90.0	6.0	4.0	100.0
1965	82.0	14.1	3.9	100.0
1966	84.3	12.1	3.6	100.0
1967	81.2	13.9	4.9	100.0
1968	79.0	15.3	5.7	100.0
1969	78.5	15.6	5.9	100.0
1970	77.3	16.7	6.0	100.0
1971	73.1	21.2	5.7	100.0

Source: Calculation made by the author on the basis of figures
supplied by Codelco and of the Anuario de la Mineria en Chile, published by the Servicio de Minas del Estado de Chile, Ministerio de
Mineria.

fined copper a year. This group of sellers accounted for about 74
percent of the volume of ore Enami acquired. By contrast, 96 percent of the mines (a total of 4,400), all of which were obviously very
small, provided the remaining 26 percent of ore purchased.[3]

Marketing

All the production of the small and medium mining sector is exported. Therefore, the distribution of exports among firms follows
the same pattern as shown above for production. The bigger firms
and independent producers export their copper directly, sometimes
after having it smelted or refined at Enami's smelters. By the end of
the 1970s about 20 private firms figured in the records of the Central
Bank as active direct copper exporters. Enami ultimately exports
the production of small miners, after it is processed in its own
plants, and this explains why, statistically speaking, the copper appears as being produced by that enterprise. The bigger firms usually
sell under relatively long-term contracts, while the smaller ones sell
on the spot and on a cash basis.

Small and medium mining enterprises have followed a standard exporting procedure in recent years. Enterprises are authorized to sell directly to foreign customers but are subject to strict control by Codelco. There are two types of sales: (1) sale of refined copper to final consumers, and (2) sale of cements or concentrates to foreign foundries and refineries. Codelco must approve each item of the sales contracts, including the purchaser himself, the quantity and the provisional sales price, the method of determining the final price, the sales commission, insurance, transport cost, means of payment, and other technical details. In the case of the sales of ores to foreign foundries and refineries, the cost of smelting or refining and drops in metal content must be included in the contract.

The minimum base price since 1966 has been the quotation of Chilean producers determined by Codelco. For example, the definitive price of a contract may be an average of the seller's cash price during the days the ore is being refined, during the contractual month of embarkation, and so on. Codelco advises the Central Bank of the amount of foreign currency that each exporter should earn. The producers have 90 days in which to bring into the country the foreign currency earned from exports of refined copper and 180 days for that earned from unrefined copper, both of these periods being measured from the moment of embarkation.

Traditionally, the small and medium copper-mining exporters argue that they obtain higher prices for their sales abroad than the large copper companies.* Nonetheless, revised and more up-to-date statistics indicate that this view is not absolutely correct in every case. (See Table 8.4.) The biggest difference between prices obtained by Chile's two sectors was in 1965, when the small and medium producers began to sell at the London Metal Exchange spot price, while the large companies, owned by U.S. corporations, sold most of their production at the price fixed by the major U.S. producers. After that year, the Chilean government decreed that the exports of large mining enterprises should be based on the LME "future" prices, and both sectors began to sell at more similar prices, although the spot price was higher. Only in June 1968 was the latter price also adopted by the large enterprises, which in 1970 managed to obtain a higher average price for their copper than the small and medium producers got.

---

*This might occur because the small and medium enterprises enjoy greater flexibility in selling their production, but would imply that these exporters possess a full knowledge of market trends at all times, which is hardly likely.

Following the nationalization of the large copper companies, an attempt was made to introduce major changes in the system of marketing small and medium mining enterprises' copper. By placing several enterprises under state control and establishing a relatively lower exchange rate for copper exports, Enami began to handle a far greater proportion of the sector's sales. (The exchange rate is here defined as the number of local currency units per foreign currency.) It became more convenient for former private exporters to sell their production to Enami in local currency, at the price it established, than to try to export directly. (During the period 1968-70, Enami controlled 36 percent of the exports of the small and medium mining enterprises. According to figures provided by the Department of Studies of the Sales Office of Codelco, the percentage rose to 40 percent in 1971 and to 47 percent in 1972.)

A second major change attempted was to make Codelco responsible for all the copper sales of the small and medium mining enterprises, including those of Enami, as well as of the large firms. This measure aimed to strengthen Chile's negotiating position, which was essential in such an oligopolistic market as that of copper.* However, this policy was reversed in 1974.

THE SMALL-SCALE MINING SECTOR
AND THE ROLE OF ENAMI

In practice, the small-scale sector is formed by individual producers or enterprises that employ from one to about 50 workers. Their mining methods differ widely and can be classified into four main categories:[4]

1. Mining by self-account workers, who generally work with their family and operate mines that either belong to them or are rented for a fixed annual or monthly sum. In such cases, the worker is both a small entrepreneur and capitalist, since he provides his own factors of production.

_____

*The existence of an oligopolistic market means that it is necessary for each seller to channel his sales as far as possible through a single negotiating agency. In the case of copper, certain isolated sellers from small and medium copper-mining enterprises can at times obtain higher prices, but this could hardly become a permanent, long-term state of affairs without prejudicing the interests of the large mining enterprises and the attainment of other national foreign trade objectives. Moreover, it means an unnecessary increase in the country's overall marketing costs.

TABLE 8.4

Price of Copper: Annual Average
(U.S. cents per pound)

Year	CIF Price of Electrolytic Copper (medium mining enterprises) (1)	CIF Price of Electrolytic Copper (large mining enterprises)* (2)	Major U.S. Producers (3)	London Metal Market (4)
1965	58.6	37.5	35.0	58.6
1966	54.2	47.8	36.2	69.5
1967	49.6	48.9	38.2	51.1
1968	52.6	51.7	41.8	56.1
1969	68.2	65.4	47.5	66.6
1970	58.8	60.3	58.3	64.2
1971	49.2	48.2	51.4	49.3

*Does not include prices of deliveries to the national market.

Sources: Column 1, Codelco; columns 2, 3, and 4, R. Ffrench-Davis, La Importancia del Cobre en la Economia Chilena: Antecedentes Historicos, Estudios de Planificacion No. 34 (Santiago: Catholic University of Chile, Ceplan, 1973). Table 7.

2. The al pirquen or "free-lance" system, which consists of operations by independent miners (known as pirquineros) who work a mine for which they pay the owner or lease-holder royalties based on a percentage of the gross production. The degree of control of the owner or lease-holder over productive operations varies. In some cases the pirquineros are completely independent and entirely free to organize their work, obtain the necessary inputs, market their production, and distribute their profits as they see fit—that is to say, they are themselves small entrepreneurs and capitalists. At the other end of the scale, there is what is known as pirquen apatronado, where the operation is under the strict control of an administrative foreman, who may or may not be the owner or lease-holder of the mine, who organizes production. He distributes the work to the crews, determines the "points" of the deposits from which the ore is to be extracted, supplies the inputs, and transports and sells the product. Sometimes these small entrepreneurs and/or capitalists possess their own plants for concentrating the ore.

3. Another type of organization is that of small enterprises employing wage-earning workers who receive a monthly or daily remuneration that is mostly fixed—that is to say, only part of the remuneration is related to the volume of ore extracted. Under this system, the entrepreneur or capitalist provides his workers with their work tools and inputs and, at times, with a certain contribution to their social security, housing, and other like benefits.

4. Finally, there is a cooperative mining system mainly formed by ex-pirquineros. Despite all efforts and all the resouces that Enami has invested to promote it, especially since 1966, the system has not developed to any great extent.

Of these four productive systems, the free-lance system (al pirquen) has predominated. According to Enami estimates, of the 10,000 persons engaged in the small-scale mining sector during the period 1970-71, from 6,000 to 8,000 workers operated in this manner.[5] Many people, however, only work in copper mining when conditions are favorable; at other times they are engaged in other mining activities, like the extraction of gold and silver, or in agriculture.

The capital goods employed in small mining operations are usually of a fairly rudimentary nature, ranging from a winch (sometimes combined with a combustion engine and/or a compressor) and a truck for transporting the ore, to simple processing plants in the case of the larger units. The shortage of capital in these mines means that a far more intensive use is made of labor per ton of refined copper produced than in the medium or large-scale mining sector.* It is

---

*Enami has estimated that toward the end of the last decade the annual production per worker varied between 2 and 10 metric tons of

thus a major source of employment, particularly in certain regions of the North of Chile. (About 58 percent of the total production of copper of the small-scale mining sector is generated in the province of Atacama alone.)[7]

Small copper-mining enterprises are very dependent on Enami, which is an autonomous state enterprise engaged in different kinds of mining activities. In practice, it has to do almost exclusively with copper-mining and, more specifically, with small-scale copper-mining.

Toward the end of 1971, Enami had 21 agencies responsible for purchasing ore, 8 smelting plants, 2 foundries (Paipote and Ventanas), and one refinery (Ventanas). In 1970, it began to take control of a number of productive enterprises that were brought under state ownership, and thus Enami began to engage in strictly productive activities.* Even before this, it was often considered as another medium copper-mining enterprise because of the volume of its capital. Up to 1970, Enami paid for its purchases of ores from small and medium producers a rate that was based on the Chilean producers' price of electrolytic copper (established by Codelco in accordance with the world market price), transformed into double currency (escudos) at the official rate of exchange, less the cost of processing the ores. From 1971, the rates began to be established on the basis of an index devised by Enami, which was geared not to world copper prices but, essentially, to the variations in domestic costs of production, particularly of certain key mining inputs. Enami administers its own processing plants, smelters, and the Ventanas refinery. It is in charge of selling directly in the international market the copper it buys from national producers and processes in its plants. Enami is also responsible for promoting small and medium mining enterprises by means of the granting of special credits and the rental or sale of inputs and machinery. It provides technical advisory services and assists in the administrative organization of the sector, as well as the establishment of cooperatives, the training of its members, and so on.

---

refined copper in the small mining sector, between 12 and 20 metric tons in medium enterprises, and between 30 and 32 in large enterprises.[6]

*However, as early as 1966 Enami had begun to move into the production field by means of the establishment of subsidiaries and joint ventures with private national and foreign companies. Nine of the former and 12 of the latter were established during the period 1966-70, in addition to three companies established with foreign enterprises to prospect for new mines.[8]

With respect to taxation, this branch of activity is exempt from corporation tax, additional tax, and personal income tax. Instead, it is subject to a single flat rate of 2 percent on the volume of sales of ores, which is retained by the purchaser (usually Enami).[9] Moreover, for the last 20 years, small mining enterprises have benefited from a set of general tax exemptions and other advantages (exemption from import taxes, drawback, and other specific advantages for certain regions). They have also been exempted from the payment of tariff duties and other taxes on the importation of machinery and inputs.

THE MAIN ENTERPRISES OF THE MEDIUM-SCALE COPPER-MINING SECTOR

The study of the medium-scale mining sector will concentrate on the five enterprises that, because of their installed capacity and their volume of production, were the most important during the last decade.* Two other companies were formed during that decade but started production in this one, and, thus, they deserve to be mentioned briefly in advance.

The Compania Minera Andina entered into production in 1970, followed in 1972 by Sagasca. Both are joint ventures of foreign companies with the Chilean state represented by Codelco (in the same way as were the large mining enterprises). Partly for this reason, they are customarily considered as belonging to the large-scale mining sector, although, legally speaking, both belong to the medium sector.

The Compania Minera Andina mines the Rio Blanco deposits in the province of Aconagua. The company was established in 1966, by virtue of the copper "Chileanization" agreement, with 70 percent of the capital contributed by the Cerro de Pasco Corporation, of the United States, and the remainder by Codelco. (Originally the share of the Cerro de Pasco Corporation was to be 75 percent, but this was reduced in 1969.)

The total authorized investment amounted to $157 million, with $15 million of that in the form of capital provided by the Cerro de Pasco Corporation, and more than a third contributed in the form of Eximbank credits. The Japanese consortium Sumitomo financed a further 20 percent, and the remainder came from credits for the Cerro

---

*Other smaller enterprises include Chile Canadian Mines S.A., Cima Mines Ltds., Compania Minera Chanaral y Taltal (Chatal), Carolina de Michilla, and Panulcillo, the two latter being subsidiaries of Sali Hochschild and Tocopilla respectively.

de Pasco Corporation, Codelco, and others. The investments were practically completed in 1971, approximately half being spent in local currency. The taxes applied to this company consisted of 15 percent of profits and an additional 30 percent of remitted dividends. Its production was 6,000 metric tons of refined copper in 1970 and 53,600 metric tons in 1971. In this way, the share of the total production of small and medium mining enterprises generated by the six main enterprises of the sector amounted to 68 percent in that year.

In 1971, the Compania Minera Andina was explicitly included in the law nationalizing the large-scale mining sector. It thus came under the full control of the Chilean state, or, more precisely, of Codelco, along with the rest of the nationalized enterprises.

The Compania Anomima Cuprifera de Sagasca was established in 1968 to mine the Sagasca deposit in the province of Tarapaca, with foreign capital provided by the Continental Copper and Steel Industries Inc. The following year, it was transformed into a joint-venture in which Codelco controlled 25 percent of the capital, Continental approximately 59 percent, and the International Finance Corporation about 15 percent. Total planned investment was $32.5 million, mostly in the form of credits. The taxes applied were essentially the same as those applied to Andina; the enterprise began production in 1972 and expects to reach a volume of 24,000 metric tons of fine copper in form of coments by 1974.

General Background of the Five Major
Enterprises of the Medium-Scale
Mining Sector

The following section will describe the total assets, capital (domestic and foreign), the owners of it, and the plant capacity of the five main medium-sized enterprises.[10] This information is summarized in Table 8.5.

Compania Minera Disputada de las Condes was formed by French investors who established it in 1916. Toward the middle of 1971, of all shares, 86 percent were under the control of Societe Miniere et Metalurgique du Penarroya, the rest belonging to various private national shareholders. During the government of President Salvador Allende, however, most of the shares of the company owned by Penarroya were purchased by the state through Enami, which came to control close to 70 percent of the shares.

Disputada has three separate plants in the central part of Chile, with an aggregate capacity of 10,200 tons of ore a day.* The effective

---

*The capacity of the plants is linked to the quantity of ore that they can receive rather than the volume of refined copper that they

TABLE 8.5

Copper Production of Principal Private Enterprises of Medium-Scale Mining Sector and Enami
(thousands of metric tons)

| Year | Principal Private Enterprises | | | | | | Enami[b] | Total Small and Medium Mining Enterprises[c] |
	Disputada	Mantos Blancos	Sali Hochschild[a]	Pudahuel (ex-Santiago Mining)	Toco-pilla[a]	Sub-total		
1965	27.6	23.7	2.7	7.2	3.3	64.5	38.7	105
1966	26.6	27.4	3.5	3.8	3.6	64.9	20.0	100
1967	29.7	25.0	5.2	4.6	3.5	68.0	46.0	124
1968	33.2	30.3	5.6	4.3	3.0	76.4	50.0	137
1969	33.6	31.3	6.6	3.5	3.5	78.5	53.6	148
1970	38.4	34.1	6.1	4.4	3.8	86.8	53.4	151
1971	34.3	32.5	6.2	3.0	3.5	79.1	51.6	137
Annual average	31.9	29.2	5.1	4.4	3.5	74.0	44.8	129.9
Percentage of the sector's total[d]	24.7	22.6	4.0	3.4	2.7	57.4	34.8	100

[a]Not including production of subsidiaries.

[b]Calculated on the basis of exports effected each year; note that what is shown here as Enami's production corresponds strictly to purchases from small producers.

[c]Excluding the production of Andina in 1970 and 1971.

[d]Refers to production of each enterprise as a percentage of the total production of the small- and medium-scale mining sector during the whole period 1965-71.

Source: Annual reports of each company, except in the case of Disputada, Pudahuel, and Enami, for which data supplied by Codelco on the basis of each enterprise's exports were utilized.

average annual production for the two-year period 1970-71 was 36,000 tons of refined copper. (See Table 8.6.) This enterprise also owns the Chagres foundry, in the province of Aconcagua, whose annual production capacity is approximately 24,000 metric tons of blister copper, as well as a sulfuric acid plant in the same location.

Empresa Minera Mnatos Blancos S.A. is controlled by foreign investors, although Corfo, the Development Corporation, owned slightly less than 10 percent of the shares toward the end of 1971. Fifty percent of the shares belong to Marvis Corporation S.A. and 21 percent to Empresas Sudamericanas Consolidadas S.A., both of which have their headquarters in Panama.* These enterprises are owned by the family of Mauricio Hochschild H., of Bolivian nationality, who founded the enterprises in 1955. Mantos Blancos was the only medium copper-mining enterprise that the government of the Unidad Popular did not even attempt to bring under state control.

The enterprise works the Mantos Blancos deposit, near Chuquicamata in the northern part of the country, which has very special characteristics. It produces a special copper (known simply as "Mantos Blancos refined copper") whose content of fine copper is almost equal to that of electrolytic copper although the principal refining process is chemical in nature. The first processing plant was built in 1957 and started production in December 1960. Its effective average annual production bordered on 32,000 tons of refined copper during the two-year period 1970-71.

Compania Minera y Commercial Sali Hochschild S.A.I.C. is a completely Chilean firm. It was formed in 1937 and established juridically in 1960 as an incorporated company. It was responsible for the mining business of Sali Hochschild, who was the company's founder and first president. (This enterprise engaged in commercial activity as well as strictly mining activities. It had a service station for motor vehicles in the city of Copiapo and acted as import agent and supplier for small mining enterprises in the area. However, the commercial activities only accounted for a tenth of the company's revenue.) At the beginning of 1971, almost all the shares in the enterprise were

---

produce. The latter is also dependent on the copper content of the ore and of its metallurgical properties. Thus, the capacity of an enterprise's plant may be greater than that of another and yet its volume of production can be similar or smaller, although the costs of production of the former will be correspondingly higher. This is the case of Disputada compared to Mantos Blancos.

*The remainder is in the hands of U.S., Canadian, and international financial corporations such as the International Finance Corporation (IFC).

in the hands of members of his family.* Around that time, the company had three ore-processing plants with an aggregate daily capacity of 1,340 tons of ore. Part of the plants were used for the processing of ores purchased from small producers from the surrounding area (in a manner similar to Enami), but the bulk of their production came from their own mines and amounted to an annual total of almost 6,000 tons of refined copper during the period 1970-71.

Sociedad Minera Pudahuel Ltda. was established in 1969 following the purchase of the assets and liabilities of Santiago Mining Company, a subsidiary of the Anaconda Company. Santiago Mining had been founded in 1918 and owned the La Africana and Lo Aguirre deposits in the province of Santiago. It only works the first of these deposits, however, for which it has the second largest processing plant in the sector, with a daily capacity of 7,200 tons, but with an effective annual production that stood at a daily average of less than 4,000 tons of refined copper during the years 1970-71.

The mining assets and liabilities of this enterprise, including the deposit, were bought by a group of Chilean investors for approximately $11 million. (The book value of the enterprise's capital at the moment it was bought was approximately $4.4 million, which is the value shown on the last balance sheet of the Santiago Mining Company.)

Finally, Compania Minera Tocopilla S.A. was founded in 1919 as an ordinary Chilean joint-stock company. Toward the end of 1971, however, its list of shareholders included four foreign financial corporations, which controlled 57 percent of the company's shares, three national banks with 42 percent, and other minor shareholders with the rest. The founder of the enterprise, however, was Mauricio Hochschild H., the same owner of Mantos Blancos, and there is information that suggests that his family still held control of the company until 1971. (In 1972, for example, one of the members of its board of directors was the general manager of Mantos Blancos.) During the government of the United Popular, negotiations were started for the purchase by the state of a majority of the shares of the enterprise, but these never came to fruition. The company owns two plants in the province of Antofagasta, with an effective production bordering 3,500 tons per year in 1971.

---

*Sali Hochschild S.A. or its shareholders own a number of subsidiary enterprises, such as the Compania Minera Delirio de Punitaqui and Compania Minera Carolina de Michilla S.A., which is partly a state-owned company established in 1969. The latter has a plant with a production capacity of 6,000 tons of refined copper per year.

Production and Sales

Table 8.6 shows the production of four of the five major enterprises of the medium-scale mining industry over the past few years. (The balance sheets of the fifth, Santiago Mining, do not contain necessary information, and so the company is not included in the table.)

The sharpest increases in production during the period were by Sali Hochschild. Santiago Mining (now Pudahuel) registered a drop in production. The output of small enterprises marketed by Enami rose from 28 percent of the total in 1965-66 to almost 36 percent in 1970-71.

Almost all the major private enterprises of the medium-scale mining sector broke their production records in 1970, when copper prices and the rate of exchange were particularly favorable. The most striking increase was in the production of Disputada, which by the middle of the year was ending the expansion programs that it began in 1969. In 1971 there was a general drop in production owing to a number of factors, including the fall in the price of copper on the world market, the lowering of the rate of exchange in real terms, and the uncertainty caused by the nationalization policy of the new government.*

Disputada smelted approximately half of its concentrates in its own Chagres foundry, transforming them into blister copper for subsequent refining in Europe and the United States. More than a third was exported in the form of concentrates for smelting and refining in the Norddeutsche Affinerie in Germany; another smaller but increasingly large part was smelted and refined electrolytically by Enami. Ultimately, however, Disputada itself is responsible for selling the electrolytic copper obtained from its various refineries.

Mantos Blancos sells almost its entire production in the form of refined copper, which it obtains from its own refinery. Only between one-sixth and one-eighth of its production is exported in the form of cements.

The smaller enterprises, such as Sali Hochschild and Tocopilla, exported almost their entire production in the form of cements or concentrates. Pudahuel, on the other hand, exported only electrolytic copper smelted and refined under contract by Enami in Ventanas.

---

*The reduction in the production of Disputada—which had already completed its plans of expansion—would seem to have been due also to labor problems in the Disputada mine and technical difficulties in the Chagres foundry.

TABLE 8.6

Financial Statistics of Principal Enterprises of Medium-Scale Mining Sector
(millions of dollars)[a]

	Capital Plus Reserves (1)	Total Assets (2)	Profits Before Taxes (3)	Income Tax[b] (4)	Over-price Tax	Net Worth Tax	Profits after Taxes (5)	Dividends (6)	Additional Tax[c] (7)	Rates of Return Of the Enterprise Itself (5)/(1) (8a)	Rates of Return For the Shareholders[a] (5)−(7)/(1) (8b)
Disputada											
1965-66	11.16	51.97	4.35	1.46	—	—	2.89	1.37	0.51	25.9	21.0
1966-67	18.15	54.17	3.93	1.69	—	—	2.23	1.45	0.47	12.3	9.5
1967-68	21.34	64.02	2.73	0.98	—	—	1.75	1.54	0.49	8.2	5.9
1968-69	21.24	63.41	5.18	0.96	—	—	4.22	3.31	1.07	19.9	14.8
1969-70	25.00	81.70	26.51	6.79	—	—	19.71	16.77	5.03	78.8	47.0
1970-71	25.67	61.53	10.67	2.43	0.14	0.18	7.90	2.38	0.71	30.8	6.5
Promedios 1965-71	—	—	—	—	—	—	—	—	—	31.6	15.1
Mantos Blancos											
1965	12.9	30.2	6.4	—	—	—	6.4	6.0	1.8	49.6	35.7
1966	14.3	32.7	19.2	—	—	—	19.2	18.0	5.5	134.3	95.8
1967	13.1	32.3	15.1	—	—	—	15.1	10.0	3.4	115.3	89.3
1968	14.3	27.3	20.9	—	—	—	20.9	18.0	6.6	146.2	100.0
1969	15.3	31.5	27.6	—	—	—	27.6	20.5	6.8	180.4	135.9
1970	17.5	30.8	32.4	—	5.0	—	27.4	23.6	8.5	157.7	108.0
1971	17.8	32.4	7.6	—	—	0.3	7.3	—	1.1	41.1	34.8
Promedios 1965-71	—	—	—	—	—	—	—	—	—	117.8	85.7
Sali Hochschild											
1965	2.12	5.46	0.63	—	0.09	—	0.54	—	0.12	25.4	—
1966	2.62	8.28	0.56	—	0.12	—	0.46	—	0.12	17.6	—
1967	2.91	8.09	0.58	—	0.13	—	0.44	—	0.08	15.2	—
1968	3.16	7.86	0.55	—	0.18	—	0.37	—	0.07	11.7	—
1969	3.52	9.11	1.99	—	0.33	—	1.66	—	0.33	47.1	—
1970	4.84	9.88	0.87	—	0.27	—	0.60	—	n.d.	12.3	—
1971	4.47	7.97	(0.22)	—	—	0.03	(0.19)	—	—	(4.0)	—
Promedios 1965-71	—	—	—	—	—	—	—	—	—	16.4	—
Tocopilla											
1965	2.24	5.13	2.15	—	0.06	—	2.09	—	1.52	93.3	—
1966	2.45	5.87	2.11	—	0.07	—	2.06	—	1.53	84.2	—
1967	2.60	5.38	1.85	—	0.04	—	1.81	—	1.43	69.6	—
1968	2.58	4.42	2.58	—	0.10	—	2.48	—	2.29	24.2	—
1969	2.34	6.14	1.93	—	0.07	—	1.85	—	1.71	79.2	—
1970	2.55	4.79	0.84	0.02	0.01	0.21[e]	0.60	—	0.62	23.6	—
1971	2.42	3.92	(0.37)	—	—	0.01	(0.39)	—	—	(16.0)	—
Promedios 1965-71	—	—	—	—	—	—	—	—	—	61.1	—

[a]All companies, except Mantos Blancos, carry their accounting books in escudos. Those values were converted into dollars at each year's average exchange rate.

[b]Corresponds to company provisions to pay each year's taxes.

[c]In the case of Mantos Blancos it refers to taxes effectively paid, but in the case of Disputada the additional tax has been calculated applying the 37.5 percent rate to the dividends reunited abroad.

[d]Not counting the additional tax.

[e]Corresponds to a 2 U.S. cent tax per pound of copper exported without being refined in the country. (Act Nos. 15.575 and 17.272.)

Source: Company balance sheets.

Taxation

The medium-scale mining sector is ruled by a very complex legislation, which in practice leads ultimately to relatively low tax rates and very discriminatory (1) among medium firms; (2) with respect to small-scale producers; and (3) with respect to ordinary firms in other areas of the economy. In some cases these firms have paid even lower taxes than the small independent copper producer. Surprisingly, in general, this discrimination favors more the foreign firms and the relatively bigger ones within the sector.

This situation arose as a consequence of the accumulation of several laws granting special exemption and concessions—presumably to foster mining development "attracting" foreign investment and for regional development.

A law passed in 1943 "authorizes the President of the Republic to grant new Chilean enterprises, whose object is to produce or process copper, iron or steel and which utilizes national ores, the following tax benefits: a) Total or partial exemption from all income tax and from taxes on ordinary or extraordinary profits that are applicable to company earnings; b) Exemption from all fiscal payments on real estate; c) Exemption from taxes on the export of goods."*

Article 24 of Act No. 12.937 of 1958 and other supplementary articles stipulate that, for a period of 15 years from the first of January 1959, the income tax payable by established industries or industries to be established in the provinces of Tarapaca, Antofagasta, and Atacama are reduced by 90 percent of the rate or amount specified by general legislation.

Law Decree 258 of 1960 (the Status of Foreign Investors) authorizes the president of the republic to approve or reject inflows of capital covered by the decree and authorizes him to grant profit tax reduction and a number of other concessions for a period of up to 20 years. The latter include (1) exemption from tariffs on the various kinds of capital goods imported; (2) a guarantee not to increase the existing income tax and additional tax rate on dividends reunited abroad and not to introduce new taxes; (3) special periods and conditions for amortization and assets revaluation; and (4) the right to utilize foreign currency deriving from exports to pay remittances on profits. DFL 258 further guarantees free access to the foreign currency market for purposes of withdrawing from the country the capital invested and the profits and interest that have accrued and of effecting new inflows of capital.

---

*These exemptions are granted for a period of 20 years from the date of the decree that authorizes each investment (Act no. 7947 of December 1943).

The only nondiscriminatory tax legislation affecting all medium mining enterprises was the extension of the "overprice" tax to this sector, by virtue of Article No. 17,272 of December 1969. The rate of this tax is 50 percent applied on the overprice, which is defined as the difference between the actual price per pound and a base or reference price of copper exported in any form whatever. The base price is the sum of the cost of production per pound of copper incurred by each enterprise, plus 15 U.S. cents per pound. A second major innovation under the same act was the abolition of a series of exemptions from the payment of a tax of two cents per pound of fine copper exported unrefined. This provision has existed since 1965 but had never been applied.

Thus, the general legal framework already mentioned gave rise, in practice, to a situation whereby the various governments negotiated with each company the preferential tax system that would supposedly permit the eventual establishment of each enterprise and that, in most cases, implied the application of very low tax rates. Hence arose the discrimination between different enterprises that makes it necessary to describe the system and rates of taxation applicable to each company separately. (The point should be made also that the general legislation referred to does not contemplate any formal active role on the part of Codelco, or Enami, in providing information to determine the tax rates applicable to different firms, nor in the supervision of the complete fulfillment of the tax laws.)

The two largest enterprises, Disputada and Mantos Blancos, enjoy tax exemptions and other advantages incorporated in the Status of Foreign Investors. The former has only benefited fully from these exemptions since 1969, when the company obtained that concession in exchange for the investment of additional foreign capital. The new system of taxation consists of a flat rate of income tax, amounting to 20 percent for the years 1970 to 1974 inclusive, and 25 percent from 1975 to 1985.* The enterprise is also subject to the 5 percent Corvi tax. It further received a guarantee that a flat rate of additional tax of 30 percent between 1975 and 1985 on dividends distributed and remitted to foreign shareholders.

----

*The general system of taxation of Chilean enterprises included a rate of income tax, which, during the period 1965-70, varied between 30 and 35 percent. In addition, a separate tax of 5 percent on income has existed since 1963 for the construction of housing (Corvi tax). Furthermore, enterprises operating with foreign capital are subject to what is known as the "additional tax" (37.5 percent since 1967) on the remittance of profits abroad.

Previously, Disputada's various installations were subject to different systems of taxation: The Disputada mine and the San Francisco plant were exempt from corporation tax, in accordance with article 17 of Act no. 7.747; its other operations were subject to the general Chilean tax system. Furthermore, the entire enterprise was required to pay the additional tax on dividends remitted abroad. However, the effective rate of income tax, according to the enterprise's balance sheets, would seem to have averaged 26 percent for the period 1965-69 (excluding the Corvi and additional tax), compared with a mere 19 percent for the three-year period 1969-71. (See Table 7.7.) Consequently, the new system applicable to the enterprise has apparently reduced the rate of income tax payable. The same occurred with the additional tax, which was directly reduced from 37.5 percent to 30 or 32 percent. This represents, therefore, another case of tax concessions being granted in exchange for new investment and eventual expansion of production.

In 1969, Disputada benefited from another set of discriminatory privileges contained in the DFL 258, which included total exemption from taxation on profits payable but not distributed to its shareholders and on interest paid on foreign loans, a guarantee not to apply new methods of calculating taxes or to modify the tax advantages granted, a right to special tax-free revaluations of capital, and other general advantages that have already been referred to.

The Empresa Minera Mantos Blancos is covered by the Status of Foreign Investors and by Act no. 7.747, under which it is exempt from income tax. It pays only the Corvi tax and additional tax on dividends remitted abroad, at the regular rate of 37.5 percent. It also benefits from the various advantages laid down in the DFL 258, to which reference has already been made.

Sali Hochschild benefits from the 90 percent reduction in regular income tax provided for under Act no. 12.937 in respect of its plants in the province of Atacama; the rest of its operations are subject to the same system of taxation as any ordinary Chilean incorporated company. On average, the effective rate of income tax paid during the period 1965-70 was around 23 percent.

The Compania Minera Tocopilla also enjoys the 90 percent tax reduction, except that since it operated only in the province of Antofagasta, that reduction means that the enterprise paid an effective rate of approximately 3 percent for the period 1965-70. In 1970, however, the company was required to pay the tax of 2 U.S. cents per pound of copper exported that was not smelted in the country, which implied the payment of an amount equal to 25 percent of its profits before taxes. (See Table 8.7.)

As a subsidiary of Anaconda, the Santiago Mining Company had an extremely complex tax system under which its various operations were governed by different provisions. On the whole, however, the taxes paid by this enterprise were the highest of any paid by medium mining enterprises, insofar as it was subject to income tax and, in addition, paid the additional tax on the remittance of dividends at the normal rate.

The Compania Minera Pudahuel, by contrast, pays taxes like any Chilean limited company—that is to say, a tax of 17 percent on income plus the Corvi tax. However, this type of company is a special case in that its individual shareholders are subject to income tax on the profits they earn, even if these are not distributed, which is not the case of incorporated companies.

Finally, in 1970 all small and medium copper-mining enterprises became subject to the overprice tax, regardless of the tax advantages they may have obtained previously. Owing to the way in which the tax was calculated, however, the only enterprises that were affected by it were Mantos Blancos and Disputada, which paid out the equivalent of 15 percent and 1.5 percent respectively on profits before taxation. (The reason for this considerable difference is that the base price on which the overprice is calculated relates to the cost of production per pound, which in the case of Mantos Blancos was less than two-thirds that of Disputada in 1970. Moreover, in the case of Disputada, the balance sheet referred to the second half of 1970 and first half of 1971, during which latter period, the sales prices dropped so sharply that not even Mantos Blancos was required to pay the tax.)

During 1971, no enterprise was required to pay the tax, owing to the drop that occurred in the price of copper on the world market. That year, however, a separate tax was introduced based on the net worth of the enterprises, but this had little effect except in the case of Disputada.

Costs, Employment, and Imports

The figures available on costs, employment, and imports in this sector are few and not very trustworthy. (Codelco does not control or even compile data on costs in the large enterprises of the medium-scale mining sector as it does for the LCM.) However, on the basis of information contained in the balance sheets of the main enterprises, estimates have been made of production costs during the period 1965–71 (see Table 8.7). An analysis of this statistical data suggests the following conclusions.*

---

*Table 8.7 does not include Santiago Mining (or Pudahuel) whose balance sheets do not contain the necessary information. Sali Hoch-

TABLE 8.7

Unit Production Costs of Principal Private Enterprises of Medium-Scale Mining Sector, 1965–71
(U.S. cents per pound of copper)*

Unit Revenues and Costs	Principal Private Enterprises				
	Disputada	Mantos Blancos	Sali Hochschild	Tocopilla	Average
Revenues	50.5	53.5	53.8	48.2	52.1
Total costs	37.8	27.8	49.5	31.8	33.8
Operation costs	29.7	18.9	37.5	2.42	25.0
Depreciation	3.8	3.8	2.9	1.8	3.7
Other Costs	4.3	5.1	9.2	5.8	5.1
Profits before taxes	12.5	28.8	5.8	22.8	19.8

*The figures in escudos have been converted into dollars at each year's average official exchange rate.

Source: Companies' annual reports, with some corrections introduced by the author.

1. The average total cost of production per pound, including depreciation but excluding an alternative cost of capital, was 34 U.S. cents for the six years. This figure is approximately 50 percent higher than the average cost of production for large-scale mining enterprises during the same period. The two sets of figures, however, are not strictly comparable because, to a certain extent, the enterprises produced different goods. The medium mining enterprises sell much of their production in the form of cements and concentrates whereas the entire production of the large mining enterprises is sold in the form of blister or refined copper produced by their own plants. The production costs of the latter include the smelting and refining processes, while those of the medium mining enterprises only include that part that is processed in Disputada's and Mantos Blancos' own plants, since the cost of contract work elsewhere (either by Enami or abroad) is deducted from revenue and does not appear as a production cost. Furthermore, extension of the alternative cost of capital itself contributes to a greater undervaluation of the costs of large mining enterprises since, generally speaking, they are more capital-intensive, and the smelting and refining processes are even more so. All in all, the total real production costs of medium mining enterprises would seem to be considerably higher than those of large mining enterprises.

2. Of the production costs, approximately three-quarters (25 U.S. cents per pound) represent operating costs, while slightly more than 15 percent (5 cents) goes on general administrative and financial expenditure, special provisions for employees and workers, and so on. The remainder, approximately 4 cents per pound, corresponds to depreciation.

3. For the reasons given above, especially because of the different types of copper produced, it is also difficult to compare the production costs of the individual medium mining enterprises. Allowing for this factor, Mantos Blancos would appear to be the enterprise with the lowest average total operating costs (28 cents per pound),

---

schild engages in both commercial and mining activities, and the former accounted for approximately one-tenth of its income and costs during the period under examination. The balance sheets are far too inadequate for it to be possible to make categorical deductions. Many of the cost items appearing in them are arbitrary, such as the depreciation and revaluation of assets. Furthermore, definitions often differ from one enterprise to another, not to mention the difficulties involved in attributing the costs for each year to the production of that period and the flexibility the enterprises enjoy in anticipating or postponing the record of some revenues and expenditures.

followed by Tocopilla with 32 cents, and Disputada with 38 cents.
Sali Hochschild has the highest production costs, but the figure is not
altogether reliable since its operations include commercial as well
as mining activities.

4. Disputada and Mantos Blancos show higher depreciation
costs per pound than the rest because they have their own foundries
and Disputada has a refinery, which entail a considerably higher capital investment than strictly extractive operations. However, the fact
that these enterprises have produced their own blister and refined
copper has earned them a higher income per pound of copper sold,
as is clearly apparent from Table 8.7.

5. Finally, average costs during the period under consideration have shown a persistent tendency to rise, with a few isolated exceptions. Tocopilla registered the highest increase from 1967-68 onward, whereas Disputada, at the other end of the scale, tended to
maintain its costs relatively constant and even to reduce them to a
certain extent during the period 1969-70. In the case of all the companies, however, costs rose sharply in 1971, owing mainly to the increase in wages in real terms.

With respect to the employment of manpower, it can be estimated that some 4,300 workers were engaged in the five enterprises
under consideration at the beginning of the 1970s. Only the two largest enterprises—Disputada and Mantos Blancos—published statistics
on the subject in their annual reports. According to these figures,
both enterprises employed about 1,800 workers; the labor force had
been gradually declining since 1965 in the case of Disputada and
sharply increasing in the case of Mantos Blancos, hence the average
product per worker in these two companies moved in opposite directions.* On the whole, these productive units employ more workers
per ton of copper produced than do the large mining enterprises but
less than the average small and medium enterprises.

As regards the proportion of production costs represented by
imported inputs, the few statistics that are available point to their
being fairly high. The imports of Disputada and Mantos Blancos during the period 1966-67 ranged between 20 and 25 percent of their production cost, while in the large-scale mining enterprises they fluctuated between 9 and 21 percent. Statistics supplied by Codelco on
imports of inputs for the whole of the small and medium mining sector

---

*The annual production per worker for the period 1965-68 was
13.5 tons for Disputada and 21.6 tons for Mantos Blancos. During
the period 1969-70, however, these figures had increased to 19 tons
in the case of Disputada and fallen to 21 tons for Mantos Blancos,
with an average of 20 tons.

in 1966 indicate that they amounted in value to $50 per ton of refined copper produced. (This figure is an underestimate, since part of these inputs are bought directly on the national market or are imported under the special provisions governing the "free ports.")[11] The large-scale mining sector uses the same amount of imported inputs per ton of copper despite the fact that the copper produced by the latter receives a greater degree of processing in its own smelting plants and refineries, which requires more imported inputs, and despite the fact that the process of extracting the ore itself involves the use of considerably more mechanized techniques. It is therefore all the more surprising that the small and medium mining enterprises should require such an intensive use of imported inputs. Since Codelco exercises an administrative control over the import operations, as it does for the large-scale copper sector, it is probable that this situation derives mainly from the tariff exemptions enjoyed by the imports of the small and medium enterprises.

Just as in the case of the large mining enterprises, a section of Codelco must expressly authorize each import transaction made by the producers. Codelco ensures that the articles to be imported are not available in Chile at similar prices and that the cost of freight and insurance is normal for this type of operation. However, the serious bureaucratic limitations that these direct controls pose in the case of the large mining enterprises are more in evidence where small and medium enterprises are concerned. Hence the need to introduce such indirect methods as tariffs to limit the excessive imports of this sector.[12]

Profits and Rate of Return

In view of the sector's relatively low taxation and production costs—deriving mainly from the great richness of the Chilean copper deposits—the profits and rate of return of the enterprises under examination are, on average, extraordinarily high. The rate of return is here defined as net profits after income tax (but without deducting the additional tax on dividends in the case of foreign enterprises) divided by the capital and reserves as shown on the enterprises' balance sheets. This reached an average of 66 percent for the whole of the period under consideration (1966-71). In 1970, the figure was 122 percent (despite the introduction for the first time that year of the tax on the overprice of copper, which, for Mantos Blancos, signified a disbursement of $5 million).

Mantos Blancos undoubtedly has the highest profit. As a simple annual average between the years 1965 and 1971, its rate of return after taxation (but without deducting the additional tax) was 118 percent. In 1969 it was 180 percent and 158 percent the next year, dropping to 41 percent in 1971 owing to a sudden decline in the price of copper and the real rate of exchange. After deducting the additional tax, the rate of return for Mantos Blancos foreign stockholders, during the period 1965-71 was approximately 86 percent of the book value of its capital and reserves and 136 percent in 1969. All this means that the remittance of profits alone, which Mantos Blancos sent to its shareholders abroad in the form of dividends (that is to say, not counting additional tax) represented an accumulated value of $57.7 million, from its entry into operation to the end of 1970. This is nearly eight times greater than the effective inflow of external capital that the company had received since its establishment.[13]

The rate of return of Disputada, though not as spectacular, has also been very high. The average for 1965-71 was 32 percent. Between July 1969 and June 1970, it went up to 79 percent but dropped the following year to 31 percent for reasons that have already been mentioned.[14] No precise information is available on remittance of profits abroad since not all the enterprise's investments are covered by the Decree on the Status of Foreign Investors. The accumulated profits remitted according to the provisions of that decree alone amounted to $1.2 million from 1964 to the end of 1968, rising to $2.9 million during 1969 and 6.9 million the following year.[15]

Sali Hochschild's rate of return was somewhat lower, at an annual average of 16 percent for 1965-71. The best year was 1969, with 47 percent, and the worst, 1971, with a loss equal to 4 percent of the capital. The dividends paid out by the enterprise were fairly low, the tendency being to concentrate on reinvestment of profits and thus to increase substantially the enterprise's capital during the period.

The Sociedad Minera Pudahuel has not drawn up any balance sheets since its establishment. The profits and rate of return of the Santiago Mining Company after deduction of the additional tax on the remittance of dividends is only known for the years 1965-67. The simple average rate of return to its shareholders amounted to 6.6 percent, considerably lower than that of the other enterprises studied.

Tocopilla is the second most profitable enterprise in the sector, showing an annual average rate of return for the period 1965-71 of 61 percent. In 1965-67, the figure rose as high as 73 percent. Its total profits in 1967 approached that of the Disputada, which had a volume of production and capital that was 10 times higher. In 1968, the company's rate of return dropped to 24 percent, rising again the following year to 79 percent. Finally, in 1971, the enterprise suffered a loss equal to 16 percent of the capital for that year, despite an additional 2 million escudos in revenue it received from Enami in payment for part of the shares of its subsidiary company, Compania Minera Panulcillo.

To sum up, the average rate of return of the main enterprises of the medium-scale mining sector was extraordinarily high during the second half of the 1960s but varied greatly from one company to another and from year to year. The highest average rate of return of the whole group was in 1970 (thanks to the high price in copper), when profits after income tax for that year exceeded enterprises' capital and reserves. The lowest figures were registered the following year mainly because of the drop in the rate of exchange in real terms, the introduction of new taxes (on overprice and net worth), the labor and administrative difficulties arising from the efforts of the government of that time to bring some of the enterprises under state control, and the decline of the price of copper.

CONCLUSIONS

The most obvious conclusion from the foregoing is the need for state agencies to pay more attention to and exert more control over the operation and development of the small- and medium-scale copper-mining sector. This branch of activity should receive special attention, not only because of the foreign currency it generates and may generate for the country in the future—insofar as it is a sector in which Chile has evident comparative advantages—but also because of the foreign control to which it has traditionally been subjected. It is vital to design a suitable, homogeneous, and stable policy for the sector, taking into account the broad interests of the country as a whole.

A choice must be made about the fundamental role that the small-scale mining sector should play in the long run. Should it be supported with its present characteristics, given the volume of employment it generates? Should a major reorganization of the traditional type of productive operation take place, transforming it into more modern medium-size firms?

In the long term, it probably will prove undesirable to spread investment resources and credits in an attempt to maintain a vast number of small producers and isolated enterprises in business; on the contrary, such resources should be channeled into the establishment of larger enterprises, which would employ the workers who are currently operating on an independent basis and who could thus be guaranteed stable employment and higher remuneration as a result of the foreseeable increase in productivity. However, important social considerations must be taken into account. Previous experience has shown that the formation of mining cooperatives is a possible but insufficient solution. It is necessary to place more emphasis on efficiency and on the autonomous character of each mining enterprise. Labor-managed enterprises organized along modern business lines, paying the state a reasonable rent on the deposits being worked and an interest on the capital utilized in its operations, would seem to be a suitable arrangement ensuring more efficient utilization of the limited national resources.

A second problem on which a fundamental decision must be taken has to do with the role of foreign investment in the sector, especially in medium-scale mines. First, there can obviously be no justification for foreign capitalists to continue keeping for themselves an inordinately high share of the rents obtained from the exploitation of Chilean deposits, which bears no relation to the contribution that they have made to the development of the sector. This does not mean, however, that such capital should be discarded a priori. New types of association with foreign investors may be devised in the form of contracts for advisory services, design and construction of plants, and so on, when the necessary technical know-how is not available in the country.

More immediately, the taxes on the enterprises in the sector should be raised to a level that at least does not imply discrimination against enterprises in other national economic activities and among copper companies. The system in force during the 1960s permitted absurd situations like having medium enterprises like Tocopilla and Sali Hochschild pay less taxes per ton of copper produced or per escudos of sales than the small producers, while foreign companies, like Mantos Blancos, did not pay income taxes at all, except for the additional tax on remittances of dividends abroad.

New and more efficient systems of taxation should be designed, or existing systems improved. First, modification of the tax on the overprice, that is, on the difference between the sales price and a given "base price," if properly defined and applied, could be one of the most effective means of capturing some of the revenue from abnormal increases in the world's copper price. The main improvements of the tax on overprice should include (1) the introduction of a rate, which increases with the rise in price, and (2) the definition of a more or less homogeneous "base price" for all enterprises, taking into account the richness of all deposits and a reasonable long-term return on capital, rather than only production costs indicated in the enterprise accounts, as is the case at present. The latter provision would provide an incentive for relatively more efficient operations.

A second complementary provision could consist of the collection of a rent on deposits, varying according to the quality of each deposit, but established for fairly long periods. This would encourage the enterprises to work the deposits more efficiently and more intensively and, also, reduce the difference in profits deriving from the varying quality of the deposits.

It is particularly important to determine which state bodies are
to be responsible for adopting and implementing policies for the small-
and medium-scale copper mining sector.* It would probably be advisable for Codelco to take over the general planning of the sector,
while Enami could concentrate on the more efficient administration
of its processing plants, its own mining operations, and the purchase
of ores. The latter functions should be separated from the general
promotion of small mining enterprises and the provision of credits
and social services for them. Enami's activities should be stepped
up in respect of prospecting, project design, the installation of plants,
and the provision of technical services to make better use of Chile's
limited resources in this field.

In the marketing of copper, a move has to be made toward centralizing sales through Codelco to take full advantage of Chile's monopolistic position in the world market. A stable price has to be guaranteed to national producers for exports purposes, since small mining
enterprises in particular have been highly sensitive to fluctuations in
price.

The policy relating to imports of inputs and machinery must be
revised. Reasonable import duties or tariffs should be introduced to
encourage greater and more efficient substitution for national inputs,
while ensuring an adequate and stable supply of all inputs.

These measures, combined with a set of policies conducive to
a comprehensive planning system of the entire national economy and
its various productive sectors, could generate a considerable increase
in the benefits accruing from the small- and medium-scale mining
sector. The potential and reserves of this sector are immense. A
large number of deposits are already known, and feasibility studies
exist, demonstrating that medium-scale enterprises may profitably
exploit them.

As to the reserve of deposits currently being worked, the picture is also encouraging, despite the fact that in many cases no serious studies have been carried out. The enterprises may tend to underestimate their reserves to claim that they deserve more favorable
tax treatment because the period available to them to recover their
initial investment is short. For some years now, for example, it
has been claimed that the reserves of Mantos Blancos are almost exhausted. Andina and Disputada, however, are known to have reserves
of 2 million and 1.4 million tons of copper content respectively, and
recently new reserves have been detected in the La Africana deposit.

_____

*It is, for example, essential to remedy the kind of inefficiency
that stems from the fact that Corfo as well as Codelco and Enami are
involved in the administration of various enterprises by virtue of their
control of part of their shares.

Finally, it should be mentioned that the development and operation of new medium-size mines have the added advantage that they do not require large capital investments and employ techniques that are within the capability of the country, especially for the production of cements and concentrates. No large-scale investment is required for housing and access roads since many of the deposits are near urban areas, thereby reducing investment and operating costs. Medium-scale mining has the additional advantage of being more labor-intensive than other competitive economics, thus generating more employment opportunities badly needed in the country. For these reasons, a new set of economic policies should be formulated to stimulate the small and medium copper enterprises, while protecting the general interests of the country.

NOTES

1. See W. A. Lewis, "Economic Development with Unlimited Supplies of Labour," Manchester School, May 1954; Anibal Pinto, "Naturaleza e Implicaciones de la Heterogeniedad Estructural de America Latina," El Trimestre Economico, vol. 37, no. 145; Oscar Munoz, "Teorfa Economica y Heterogeniedad Estructural," Estudios de Planification no. 39, Santiago, Ceplan, April 1974.

2. Decree no. 56 of the Ministry of Mining of May 1967, and Article 1, Act no. 10.270 of March 1952. Domestic currency was converted to dollars at the official exchange rate of 1967.

3. Figures based on an unpublished study of Enami's Planning Department, which was kindly made available to the author by Mrs. Luz Cereceda, professor at the Institute of Sociology of the Universidad Catolica de Chile.

4. This classification is adopted from the study by L. Cereceda, G. Wormwald, and L. Sosa, "Genesis y Evolucion de Trabajuena Mineria del Cobre en Chile," Documento de Trabajo (Santiago: Instituto de Sociologia de la Universidad Catolica, 1974).

5. See ibid. Enami itself has not published any official statistics on the subject.

6. L. Munoz, Estudio Ocupacional de la Mineria del Cobre (Santiago: Servicio Nacional del Empleo, 1971).

7. Servicio de Minas del Estado, Anuario de la Mineria (Santiago, 1971), p. 44.

8. Empresa Nacional de Mineria, "Resumen de la Labor Desarrollada en el Sexenio 1964-70" (mimeo.), 1971.

9. Acts nos. 10.270 of March 1952 and 11.127 of November 1952; and Servicio de Minas del Estado, op. cit.

10. Sources of the following data: re shareholders, the files of Bolsa de Comercio of Santiago; re ore-processing plants and their capacities, Servicio de Minas del Estada, op. cit.; re capital and reserves of the enterprises, balance sheets of each enterprise; re contributions of foreign capital, the publication by the Department of Foreign Investment of the State Development Corporation (Corfo), Analisis de las Inversiones Extranjeras en Chile Comparados por el Estatuto del Inversionista, no. 20, January 1972.

11. Information taken from Codelco, "Estadisticas de Adquisiciones" (mimeo.), 1967.

12. See Chapters 8, 9, below.

13. Corfo, op. cit., p. 111.

14. The rate of return of Disputada as far as its foreign shareholders were concerned was somewhat lower, averaging 15 percent for 1965-71 although it rose to 97 percent during the period 1969-70.

15. See Corfo, op. cit., pp. 111-12.

9

ECONOMIC PLANNING
IN THE NATIONALIZED
COPPER SECTOR
Ernesto Tironi

The purpose of this chapter is to discuss a state planning system for the nationalized copper companies that will maximize their contribution to the development of the nation as a whole.[1] It is obvious that nationalization imposes the necessity of defining a new national copper policy and planning system, to be implemented within a new framework of organization, regulation, and control of the copper enterprises that will fulfill the objectives the nation has assigned to this industry. These objectives can be reduced to one: that in the long-run the companies contribute to maximize the country's national product. (Concentrating on this objective does not imply ignoring other goals; it merely means we consider this one the most important for an economic activity like copper production.)

To attain this objective, in a country like Chile, the big mining companies should probably operate within a general planning framework normally known as a decentralized or indirect system of economic planning, with some autonomy of the firms with respect to microeconomic decisions.

The opposite alternative is represented in a form of organization and planning in which all decisions are adopted centrally, at the top of a bureaucratic pyramid, and are transmitted or implemented through direct administrative orders. Such a structure would combine the entire sector into one large state mining company. In the case of Chile, it would constitute the largest copper corporation in the world, formed by five big mines and smelters of different characteristics, several refineries, sulfuric acid plants, electric energy centers, ports, and so on. Their operation would require the employment of more than 25,000 workers and the supply of hundreds of thousands of different inputs. Running this gigantic complex would involve making millions of decisions, many of which should be delegated or decentralized, since in many instances the authorities would lack either the necessary information or the ability to control their efficient application.

Decentralized planning is also necessary because the copper-mining sector in Chile is composed not only of large companies but also of a number of middle-sized units that produce a significant fraction of the country's output. In several of these, the state has a majority interest in the capital stock, but the administration is independent, since these installations are scattered geographically and have different productive characteristics. Consequently, indirect control of them is much more advisable. In other words, mutatis mutandis, the organizational forms and economic policies suggested for the large mines will apply with the same or greater validity to the group of middle-sized enterprises.

For the nationalized companies to have a certain autonomy, and some basic criteria should be used for evaluating each one's operational success, so each company's administrators can adopt, on a common basis, the pertinent decisions for each unit. In order that these firms cooperate in increasing the country's national product to the maximum, the primary criterion should be their profit maximization but if and only if this goal is pursued within the frame of economic policies that reflect the general national interests as these are expressed in the development plans for the nation as a whole.

Such policies include determining exchange rates, interest and tariff rates, tax systems, and price control schemes that reflect the social priorities of the political economy in general. * All these variables are instruments that the authorities can manipulate deliberately in order that the companies, seeking their own benefit, fulfill the country's development plans.

For the profit motive to be a basic criterion for the evaluation of the companies, they need not be private or capitalist. The only requisites are that administrators or executives of each company adopt decisions accounting for their effect on company profits and that authorities evaluate the enterprises (and their administrators) on the

---

*Note that if the application of a correct exchange rate in Chile is included within the economic policies implemented by state authorities, the criterion of maximum profit necessarily implies the generation of maximum foreign exchange income, considering the social opportunity cost of obtaining it. This clarification is necessary because in the case of an export sector like copper some people think (erroneously) that the operational criterion ought to be exclusively the maximization of foreign exchange earnings, without taking into account the cost of generating the foreign currency.

basis of profits earned.[2] This criterion, consequently, can be used whether productive units are private, state owned, mixed, or even under a worker's self-management scheme.[3]

The purpose of this chapter is to analyze from a macroeconomic point of view the specific policies national economic authorities can use to achieve specific goals the nation has set for this industry. In other words, we shall analyze the new policies required to handle some particular problems faced in this area after the nationalization, many of which emanate from the previous foreign control over the companies. Besides analyzing theoretical aspects, we shall try to evaluate, even if crudely, the magnitude of the effects to be expected if the different policies suggested are implemented.

Some of the problems faced by the Chilean copper industry, and the specific instruments to deal with them, are as follows: (1) Chile's position in the world copper market after becoming an independent seller, separate from the multinational corporations that had previously controlled the sale of her copper; (2) the problem of the most efficient method of transferring to the rest of the economy the considerable surplus resulting from the operation of the mines; (3) the integration of the large mines into the rest of the nation's economy in such a way as to reduce progressively their traditional "enclave" character; and (4) the rationalization of production among the various companies and plants that comprise the large copper mines. This procedure implies determing which ones are relatively more (or less) efficient, in order to use fully the installed capacity in each firm, thus avoiding bottlenecks and increasing the overall efficiency of the sector.

Each of these problems will be analyzed in different sections of this chapter. It is obvious that they are of such a general character that they must be faced by an authority with global vision of the general problems and interests of the whole economy. Thus the need for an active planning role on the part of the national economic authorities.

CHILE'S PLACE IN THE COPPER MARKET
AFTER NATIONALIZATION

Chile obtains some general advantages from national control in the sale of her copper. The first is an increase in bargaining power with other nations in search of global national objectives. For instance, promotion of the copper-manufacturing industry, with a view to exporting manufactured articles, can be realized through conditioning the sale of minerals on the opening of markets for these manufactured products. Chile may employ this variable as a part of global strategy of foreign trade and economic development. Second,

Chile's control of her own sales also facilitates reaching agreements
with other copper-exporting countries in order to control prices.
The agency best qualified to discuss future agreements is, of course,
the Intergovernmental Council of Copper Exporting Countries (CIPEC).*
Finally, the separation of Chilean companies from the sales divisions
of multinational corporations like Kennecott and Anaconda grants Chile
greater freedom for penetrating into later stages of copper processing
formerly established mainly in developed countries. This demands
great caution, for, although at first glance, nationalization of Chilean
copper could seem to weaken these conglomerates, they can bounce
back by changing their production to substitutes for copper. Such a
danger obliges Chile to watch their actions and their tactics carefully.

A related aspect is, Does Chile's control of her own sales after
nationalization increase or diminish her monopoly power in international markets? More precisely, does nationalization per se strengthen
Chile's capacity to raise world prices on her own initiative and thus
increase copper revenues for the country?

Nationalization brought about an important change in the position
of Chilean producers in the relatively oligopolistic copper market.
In fact, before nationalization, the copper produced in Chile belonged
to different multinational corporations through which they commercialized the metal. Copper mines were not integrated with each other for
sales purposes; on the contrary, they were inserted into a corporate
structure and tied horizontally to subsidiaries in other countries,
with which they pooled to sell their copper. Consequently, the principal companies operating in Chile were competing in the market
among themselves and not with other producers located in other countries.

Table 9.1 shows that nationalization transformed Chile into the
single most important seller in the world market of primary copper:
This is a necessary but not sufficient condition to strengthen Chile's
monopoly power in the world market. One must consider other aspects too, such as the elasticity of the world demand and supply, and
make further assumptions in order to get some idea of the changes in
the monopoly power of Chile.

The existence of an oligopolistic market implies two major considerations. On the one hand, operating in it forces each company to
consider the possible reactions of all the other producers before de-

---

*These price agreements can be facilitated by the fact that the
governments of most of the CIPEC countries either have a majority
participation in the copper companies or have full control over them
through nationalization.

TABLE 9.1

Participation of Principal Producers of Raw Copper in World
Market
(percent)

	Before Chilean Nationalization (1969)	Estimated Situation after Nationalization
1. Kennecott	13.3	9.4
Chile	3.9	—
Rest of world	9.4	—
2. Anaconda	10.6	3.1
Chile	7.5	—
Rest of world	3.1	—
3. English-American Group (Zambia)	8.1	8.1
4. Union Mine (Congo)	7.6	7.6
5. Roan-AMC Group (Zambia)	6.9	6.9
6. Phelps Dodge	5.4	5.4
7. Other smaller producers	45.0	45.0
8. Chile	3.1*	14.5
Total	100.0	100.0

*This represents the production of small- and middle-sized
Chilean mines.

Note: Socialist countries are not considered, since they form
an almost self-sufficient trade bloc and prices have a much smaller
effect on their production in the long run.

Sources: American Bureau of Metal Statistics, 1969 Yearbook;
and Balanza de Pagos de Chile (Santiago: Banco Central, 1969).

termining any action on its own.* On the other hand, within an oligopolistic market we know the demand curve for Chilean copper will

---

*For example, if Chile increases her sales or lowers her prices,
probably such actions would reduce the demand faced by other producers. But should these competitors react by lowering their prices, the
initial advantage Chile gains will weaken, since demand for her copper

have a negative slope. Thus to increase its sales the country will have to lower its prices; and this price reduction will also affect its initial quantity sold.

It is not the same thing for a small producer to open a new mine of 100,000 tons annually or for Chile to do it. The increased production will lower world prices; therefore a large producer like Chile, initially selling, for example, 600,000 tons, will also receive less income for that previous production, which must be charged against the new product.

It could then be unfavorable for Chile to increase her production greatly, based only on demand consideration (that is, leaving aside the problem of costs of production or reserves). Such a tactic could penalize her by lowering the price considerably and reducing the copper revenue for the nation. But the opposite, too, could happen. In other words, Chile may not have a strong influence on world prices, and, consequently, increasing her sales may not force a reduction in prices (or force a very slight reduction) so the country's revenue would increase.

In order to determine what outcome is likely in practice, we need to know the magnitude by which prices would fall when the country increases its sales; in other words, we need to know the price elasticity of demand for copper faced by Chile after nationalization.

The problem is complex because, among other things, it requires some knowledge about the likely reactions of other world producers to changes in Chile's production or prices. The most reasonable approach in facing this issue seems to be to resort to historical data on the reactions of the principal copper producers when faced with price changes (which are reflected in demand and supply functions of those producers) and about which empirical estimates exist. The evidence shows that, compared to the experience of the multinational companies that commercialized Chilean copper, the nation's monopoly power increased after nationalization, since the long-run elasticity of demand for Chile fell from about 47 to 37. (The value in the long run of the elasticity of the world's demand for copper has been estimated at 0.5, if the "long run" is defined as that period in which producers have finally adapted themselves completely to the new market conditions. The elasticity faced by Kennecott, the largest copper producer until 1971, would be approximately 40 and 50 for Anaconda, which implies a weighted average of 47 for Chile before nationalization.)[4]

---

will fall. Chile in turn can react anew with consequent reactions from the others, and so on.

A value of 37 for the elasticity of demand for Chilean copper implies that if the country increases its production level by 50 percent, the world price will be reduced by less than 1.5 percent in the long run. Such a decline induces per se an increase of 8 percent in world demand. But, at the same time, if the other world producers accept "passively" the price drop, they will "reduce" their aggregate supply by about 8.5 percent. (Here, as well as in the rest of this chapter, the terms "increase" and "reduction" of production should not necessarily be understood in absolute terms. They may refer to a change in the historical tendency that the corresponding variable shows. Consequently, a reduction of 1 percent can be understood as an increase lower by one percentage point than the historical growth rate. The same may be true with respect to increments of a given variable.)

Although the monopoly power of Chile increases with nationalization, it is just not sufficient for advantageous price manipulation by the country alone. In fact, the elasticity of demand for Chile is considerably higher than one. In practical terms, this means that after having nationalized her copper, Chile would not increase her copper revenue by unilaterally raising the prices charged for copper, since that would reduce her sales considerably. Inversely, for that same reason, at the present levels of production, Chile need not fear that a given production and sales increase would bring about a considerable decline in the price at which she sells. Alternatively, even in the short run, reducing production to provoke a rise in prices, in the hope that ultimately this procedure would mean increasing total copper revenues for the country, would be an erroneous policy. If it were followed, Chile would not increase her revenues, even under the very strong assumption that the supply of the other world producers should remain constant in spite of the price rise that our country would induce. If there is any doubt about the danger of this option, it is worth pointing out that even if this policy was followed by all the nations that make up CIPEC, they would not increase the group's copper revenues in the long run because the elasticity of demand faced by the whole bloc is not less than one.

Summarizing, nationalization strengthens Chile's position in the world copper market. In order to use this greater power to achieve the different national development objectives, the country should concentrate all her sales in one entity, avoiding the competition among Chilean sellers. Given Chile's comparative advantage in copper production, or its low relative costs compared to other world producers, it is clear that, at the present, the most advantageous strategy for the country is production expansion. Moreover, given the lower Chilean costs, the slight fall in prices that the latter policy will cause is favorable for the country in the long run because it may expel the relatively high-cost producers from the market and discourage the development of new mines, including the exploitation of ocean manganese nodules.

METHODS TO EXTRACT SURPLUS
FROM THE MINING SECTOR

The high productivity of copper-mining in Chile generates tremendous surpluses that should be transferred to the rest of the national economy in order to finance the development plans of the nation. Thus it is important to study what percentage of those surpluses should be taken away from the companies and what method should be used for this purpose, in order to reach various objectives set up for the copper sector by the nation as a whole. (We will assume that the indicated agent to extract these surpluses is the state, which will distribute them for the benefit of the total economy.)

We must first define clearly what we mean by surpluses because these are not necessarily the same as the profits appearing in company books. We shall define the total surplus that can be extracted from a company as the difference between its total revenue and its total social costs. These costs ought to include the true depreciation of capital goods, the cost of inputs, imported and national, valued at prices that reflect the social cost of producing them, and the social cost of labor. If that definition is not used and, for example, labor cost is evaluated at excessively high wages and salaries, the surplus that can be extracted appears highly reduced. In such a case, it will be workers of the sector who will be capturing part of the real surplus, to the detriment of the rest of the economy, which will have fewer funds to invest in accordance with the nation's priorities.

With respect to what part and what method can be used to extract that surplus, the answer will depend on the organizational framework of the companies involved—that is, on who is responsible for making the different decisions, especially about investment and the criteria used for that purpose.

Traditional Policies

Traditionally, Chile has used three main devices to extract the surplus from the large-scale copper sector: profit taxes, an undervalued exchange rate, and government's favorable attitude toward raising wage and salaries of workers in the large mines. (This was not an explicit policy, but rather an implicit one.)

Between 1955 and 1970, the income tax policy collected approximately 56 percent of gross profits of the foreign companies.[5] These tax rates were really not so high if they are compared to the fact that the foreign companies' rate of net profits over the book value of their capital came to an average of 19 percent during this same period. These are explained basically by the high rent from the rich Chilean mines that the companies kept for themselves.

Regarding the discriminatory exchange policy, available studies show that per se it did not bring about an increase in government revenue from the large-scale copper sector. Therefore, in the final analysis, total foreign exchange income for the country did not increase when the exchange rate was used as an instrument to increase Chile's share of copper revenue. (However, that instrument is an efficient means of discouraging the use of foreign inputs and substituting for goods of national origin, as will be shown in the next section.)

On the other hand, the policy of supporting or allowing excessively high wages and salaries in the large-scale mining sector, together with special tax and tariff concessions that reduced capital costs, discouraged labor employment by the companies and the collection of more foreign exchange for the country by this means.

The traditional methods of extracting revenues from the large mines, notwithstanding the weakness of them, did generally discourage new investments to increase productive capacity. We must add, however, that this result was as much a consequence of the instability of these policies as of the policies themselves. Consider tax rates: Stricter policies for foreign enterprises, and consequently more favorable to the country, if applied in a proper and stable manner, can also prove more favorable to the enterprises (and encourage greater investment) than lower but unstable rates. These latter are doubly negative: They neither stimulate better conditions for investment within the country nor bring about a higher volume of government income.

Such an outcome, however, is to some extent unavoidable due to the permanent conflict of interests that foreign investment implies in the long run.[6] When one nation's policies become relatively unfavorable to a corporation, that enterprise will take the option of directing its investments someplace else.*

This conflict between the interests of the nation and those of foreign investors showed in the levels of copper production and revenue reached before nationalization took place. Both were considerably lower than they could and should have been had the mines been under

_____

*Tax rates on copper companies were higher in Chile than in any other country of the world. One must remember, though, that the Chilean deposits are considerably richer, too, a fact reflected in the volume of the net profit of the corporations.[7]

Chilean control. During the last two decades, copper production of the large-scale mining sector increased at a cumulative annual rate of 2 percent, compared to 4 percent in world production, and even 2.5 percent in the United States.[8]

This phenomenon is partially reflected, too, in the fact that the price elasticity of supply for Chilean copper has been relatively low, especially in the long run. Econometric studies estimate that parameter at around 0.4 for Chile, while for the world it reaches about 5; and even in the United States it is 1.7.[9] These figures imply that a 10 percent increase in world prices brought about a 48 percent increase in world supply but only a 14 percent increase in Chilean supply. On the other hand, one cannot logically conclude that technological limitations were the cause of the sluggish growth in production, since the foreign corporations operating in Chile had mastered the most modern techniques and were applying them in other countries.

Alternative Policies for the
Nationalized Companies

A policy for transferring the copper surplus to the rest of the economy cannot be decided without first defining the framework within which the companies operate and the degree of autonomy each will enjoy in deciding on the use of productive factors and investments.

This chapter assumes that the companies must have some autonomy over operation decisions and that they should adopt the latter with the criteria of maximizing profits, provided that they do it within a planning framework in which general interests of the country are safeguarded. Nevertheless, working on this same principle and considering the importance of the large copper companies in Chile's economy, the volume of resources they generate, and especially their repercussions on the other national economic sectors, it is necessary that all large investments (for instance, those referring to the development on new large mines) be decided on a global or macroeconomic level.[10]

Certain smaller investments, though, with high rates of return and productivity, can be evaluated, decided upon, and put into practice more efficiently by each company individually. In order to encourage these investments—small in size, but numerous, and hence ultimately quite important at an aggregate level—the companies should retain a reasonable fraction of their own resources to use for that purpose.

In brief, if one is interested in the extraction of the greatest possible surpluses from the copper sector in the long run, one should, on the one hand, define an organizational setup where (1) decisions

about large investments are decided at a macro-level and (2) enterprises have a predetermined margin of resources and autonomy to
make their own minor investments. On the other hand, the policies
to capture a given surplus margin from the copper sector should be
chosen to attain that objective without having undesired repercussions
on investment and other important variables.

The Exchange Rate Policy

During the first two years after nationalization, apparently
without adequately analyzing the problem, an undervalued exchange
rate was used as the principal policy to transfer profits from the large
mining companies to the state. (Indeed, the exchange rate for copper
toward the end of 1972 was 20 escudos to the dollar, compared to 30
escudos for manufactured exports and 40 escudos for imported spare
parts and machinery. During the previous 11 months this rate had
risen by 27 percent, while prices within the country had risen about
140 percent.)

Obviously, this procedure was not intended to oblige the companies to turn over a greater percentage of profits earned, as had occurred when the mines belonged to foreigners, since now they belonged
to the state. Instead, this policy allowed the Central Bank to receive
the bulk of the companies' revenues, because the bank received the
dollars from the copper exports and handed over "few" escudos in exchange. In this way the companies obtained lower accounting profits
in national currency, while the government revenues from profit taxes
also fell.* One of the reasons was that some municipalities and other
entities as well as the state received income from the taxes on the
copper companies. Faced with the difficulties of changing existing
laws, the executive branch chose the procedure described.

However, this policy had the same negative result as before
nationalization of cheapening the cost of imported inputs and encouraging their use. Imports clearly increased during the first two years
of nationalization, even though Codelco was administering the companies in a relatively centralized way. (In 1971 alone, imports of current inputs increased by 20 percent in real terms. During 1972 and
1973 no official statistics were published, but scattered information
available shows that the increase in those years was much higher.)

---

*For this reason it is incorrect to observe only the nominal prof-
its (in domestic currency) when the real exchange rate is undervalued
or varies considerably from one period to the next. Many public debates during the years 1971-73 on the losses of the nationalized mines
were based on this fallacy.

Codelco's lack of contact with the production operations nullified to a great extent the advantages of the bureaucratic apparatus set up to control imported inputs and thus handicapped the predominantly centralized planning of the operations.

Negative aspects of this policy could have been eliminated, however, if two exchange rates had been set up for the large-scale copper companies; one similar to that used for the rest of the economy would serve in acquiring inputs and capital goods outside the country; the other would apply for exports. A second undesirable consequence of using the exchange rate as an instrument to transfer the surplus from the copper sector is that, since that variable affects the costs of production and output, it is difficult to know, a priori, the exact margin of surplus that will finally be extracted with any given exchange rate.

Interest on Capital as a Surplus Extraction Policy

We have insisted that correct methods for extracting surplus not only ought to avoid distortions but should contribute positively toward fulfilling other goals important to the nation. Surely one such objective is saving as much capital as possible, since it is an extremel scarce resource in underdeveloped countries. Charging interest on each company's capital could result in better use and greater economy of this productive factor on the part of the companies operating in a decentralized fashion. The rate of interest should be the price charged by society as a whole, or the state, for "renting" capital to the companies, to cover the cost of leasing this capital to these companies. That cost should be at least equal to the potential increase of production in other national activities if the additional unit of that capital were employed there.

This policy could help bring about a better allocation of capital through the entire economy and among the different companies. If one of these is endowed with greater capital, a higher revenue should be demanded of it. The most efficient way to do this is by charging a fee proportional to the volume of capital it uses (that is, through an interest charge). If the company cannot generate sufficient returns, this obligation of paying an interest tax will oblige it to reduce costs by increasing its capital productivity or, eventually, force the company to reassign this capital to other activities that will be more productive—in other words, where it will contribute more to the country's national product.

This type of policy instrument attains the objective sought (surplus extraction) helping simultaneously to save capital efficiently. This is done by placing this decision on the company's administrators, who are better acquainted with the details of the productive processes.

They become responsible for deciding the kind of capital they can efficiently substitute without causing undesirable repercussions in production and total income or, at least, affecting these as little as possible. Such shifting of responsibilities is, in general, an attribute of indirect economic policies.

Application of an interest charge could help to correct distortions in the prices of productive factors within the large-scale copper sector, particularly the excessively low prices for capital goods in relation to labor costs. This situation comes about mainly because of an erroneous, traditional policy of allowing capital goods (almost all imported) to be purchased without having to pay tariffs on them. Thus the companies had the tendency to design relatively more capital-intensive than labor-intensive plants and processes, since they ended up being cheaper. This phenomenon had been reinforced by the high labor costs and caused the overmechanization of some productive processes.

Wages and salaries paid to the copper workers are higher than those paid to others in similar jobs in the rest of the economy. Prior to nationalization this situation was promoted to capture for the nation a greater return from the income generated by the foreign companies, and hence the copper workers enjoyed strong political support for their economic positions. The companies seemed to give in by increasing wages but defended themselves by increasing the mechanization of production and hiring fewer laborers than if wages had been lower. (However, it should be pointed out that this tendency toward excessive mechanization could show up mainly in the plans for new plants, since once they were built, the mining unions were able to block dismissal of workers, even though employing them could have become necessary because of advances in technology. This explains the paradox that, after all, at an aggregate level by 1970, the Chilean mines employed too many workers in comparison to other similar places in the rest of the world.)[11]

If such a distortion in the labor market is difficult to eliminate directly, via a decrease in wages, the second best solution is imposing an interest charge in order to increase the cost of capital goods relative to labor.* Using this policy, the Chilean companies may decide

---

*We have already mentioned a complementary form that will bring this about. It is increasing the rates of exchange used to import capital goods and the tariffs the companies have to pay to cover them. The greater costs derived this way do not represent necessarily greater costs for the country as a whole, provided that these exchange rates and the tariffs are fixed correctly. Actually these greater costs are either fiscal revenues or remunerations paid to new workers con-

autonomously to employ productively more workers in the copper industry.

To strengthen these arguments, we might add that modern multinational corporations usually charge an explicit or implicit fee for the capital they invest in their decentralized subsidiaries. The latter fee is reflected in the profit rate the parent corporation demanded or in the dividends claimed by stockholders on the Stock Exchange.[12]

A final reason for levying an interest tax is its potential effect on the attitude of the workers who demand relatively high wages based on the argument that the high profits of the copper companies are due exclusively to the productivity of themselves alone and to their enormous efforts. If one were to charge a rate proportionate to the capital of the nationalized companies (which belong to all Chileans), then the copper workers' argument would lose a great deal of its apparent validity and force. Expropriated capital is paid for by all Chileans through a certain indemnification, in paying off company debts, and, in general, through their sacrifice of imports, restriction of external credits, and through other reprisals that expropriation has partially brought about. If one discounts these costs from the gross revenue of the copper companies, the net income generated by the workers alone would really be much less than it appears to be. (This fact becomes yet more evident if the companies are charged a rent for the copper deposits they are exploiting, since these, more unquestionably than capital, do belong to all Chileans. They have never been the property of the foreign companies, nor can they become the property of the nationalized companies. Consequently, the income they generate cannot be appropriated exclusively by those working in the copper industry.)

From a practical point of view, the principal requirement for the application of charging interest on capital is a correct appraisal of the volume of the capital resources that each company uses. The minimum value of that capital is the book value of capital, as it appears in the accounting records of each company, which in the case of Chile is underestimated due to accelerated depreciations. Admitting these limitations, the book value of capital invested in the large copper mines up to December 31, 1970, reached approximately $650 million. (This excludes that part corresponding to rights over the copper ores.)[13] A 10 percent annual interest rate on capital would effectively capture almost $65 million from the companies' surplus, more than a third of the sum the country obtained that year from profit taxes.

---

tracted to do the job formerly handled by machinery and which proved too expensive for the country.

Charging Rent on the Copper Ores

A third means of transferring the surplus from the mining companies should be used to complement the other two. This one deals with a lease fee (or rent) for the copper deposits each company holds and which, according to Chilean law, belong to the state.

Rents should be a fee applied on each mineral deposit according to the quality of its ore (copper content, metallurgical properties, facility of transportation, water availability, and so on). Therefore, it should be variable for different deposits, but relatively fixed over long time periods.

A rent charge can correct differences in profit levels (and eventually in workers' income) among the different mining operations whose ore deposits vary in richness. It can be also important to extract surpluses and regulate the operations of small- and medium-scale mines. If rents are kept stable over long periods, they will produce no negative effects on either capacity or production levels of each company. Instead, a good use of this instrument would permit the state to exploit the richer deposits more intensely and reduce the wanton impoverishment of others. Eventually the rent charges may help regulate the optimum amount of mineral reserves.

Rent on the ore deposits can be figured at the difference between the "normal" revenues and cost of each mine, defined as those resulting from a "normal" long-run level of production and a copper price not affected by random changes in them. In practice they can be calculated as a moving average of those variables during a five- or seven-year period.[14]

Income and Excess-Price Tax Policies

In the context of a planning system, there is no reason why taxing of profits should be the only (or even the principal) method of transferring surplus from this sector to the rest of the economy. It is thus feasible to apply to the large mining companies the same income tax scales applied to other industries. Such a policy will maintain a principle of nondiscrimination and efficiency for the national economy as a whole. Given the very high profits of this sector, however, it would need to be complementary to other measures. Aside from the policies already mentioned in the previous sections, another such complementary measure is the taxation of the "excess price" of copper.

A tax on the excess price now exists legally. It consists of a 54-70 percent progressive tax rate on prices over a "base price" (which moves annually in relation to production cost increases). Its purpose is to capture the short-run increases in company earnings

above a certain limit arising from changes in world copper prices.
Thus, it stabilizes the net prices the companies receive. This enables enterprises to make plans considering real stable long-run
prices and encourages them to work constantly at cost reduction to
increase profits.

INTEGRATING THE COPPER COMPANIES
INTO THE NATIONAL ECONOMY

A third type of problem facing the copper mines concerns the
integration of this sector into the other areas of the national economy
—or, in other words, the need to reduce its "enclave" nature.

In a previous section, we referred to the need for increasing
the use of labor in the productive processes. In order to complete
our examination of the "backward" integration process, we shall examine the policies available for encouraging the use of national inputs
in the productive process.

Traditionally the copper industry imported a large number of
inputs that could have been acquired within the country if the correct
(shadow) prices would have been charged on them. Part of the explanation for this phenomenon stems from the integrated nature of
the multinational corporations that were exploiting the Chilean mines.
But another important part stems from certain national policies,
such as tariff exemptions and discriminatory exchange rates, that
reduced excessively the cost of imported in relation to national inputs.

A reversal of that policy occurred in 1965, when a system of
administrative control was imposed to increase the use of domestic
inputs. The administrative procedures, however—although quite successful—appear insufficient to reduce imports further. It seems advisable to complement these procedures with others destined to stimulate a greater use of national intermediate goods. A greater emphasis on an active price policy (raising exchange rates and tariffs on imported inputs) seems advisable, particularly when a point is reached
where the items that need to be substituted are numerous but small
ones. This policy has the advantage that it transfers the decision
about which and how many imported inputs to substitute to the specialists in each company. These specialists surely are better qualified
to do so than the state officials in the central offices of Codelco.
From the moment the companies (or the administrators) satisfy the
basic criterion of seeking maximum profits, the decisions they adopt
in this respect will tend to be the correct ones. They will minimize
costs by substituting for relatively cheaper inputs, without waiting to
receive orders "from above."

The control exercised by Codelco should not necessarily be
eliminated but should be changed. It can also complement the efforts

of the enterprises. Codelco can advise the companies about new forms of input substitution, based on the experience of other companies. It can also help the companies obtain the lowest prices possible, whether in the country or outside it, and promote a long-run program of development of a copper input fabricating sector.

In attempting to measure the impact of a price increase of imports on the use of imported and national inputs, it is necessary to have in mind also the effect on costs of production and output levels. To concentrate on the input changes here, however, we shall assume that the latter remains constant. This effect could be attained if other compensatory measures were applied.

Using this assumption, one may calculate the effect of a price rise of imported inputs in (1) variation in the cost of imported inputs and the volume of fiscal returns gained through tariffs, (2) increase in the consumption of national inputs, and (3) increase in total production cost. A rise in the price for imports implies an increase in the companies' total expenditures for those goods only if the amount purchased is reduced proportionately less than the rise in prices. In other words, if the demand elasticity is less than one. In the case of the large-scale copper mines, assuming a constant production, that elasticity has been estimated at approximately 0.33 at the average prices and quantities for the period 1955-70. (That elasticity of demand for inputs has been derived from the elasticity of substitution between them. The latter parameter reflects the degree in which inputs can be substituted, in this case national for imported by changing their relative prices. Ricardo Ffrench-Davis has estimated this value at 0.6.)[15] With a constant production level, a 10 percent price increase will imply a decrease in the quantity of imported inputs demanded of 3.3 percent. If the prices of imported inputs abroad (that is, in dollars) remain constant, the country will spend 3.3 percent less in goods bought abroad. If this policy had been applied in 1971, a mere 10 percent tariff increase would have saved $1.5 million in foreign currency. (Notice that these calculations give an approximate idea of the "potential" pressure for additional imports when the real exchange rate the companies pay for their imports is reduced. By December 1972 the nominal exchange rate, fixed a year before, had been reduced approximately 50 percent in real terms. Consequently, there was a potential pressure to spend approximately $7.5 million on additional imported inputs during 1971, or 165 percent more than in 1970. This figure is strikingly close to the 20 percent increase observed in practice during that year.)

Since the price elasticity of imported inputs is less than one, total company expenditures in these goods including tariffs will tend to increase by 7 percent. The difference between the companies' total import costs and the amount spent abroad by the country is the tariff revenue obtained by the state. The 10 percent increase in the price of imported inputs could increase those revenues by $3.5 million a year, or 17 percent.

On the other hand, the rise in the cost of imported inputs encourages the use of national inputs since, in relative terms, these become cheaper. Assuming a constant level of copper production, a 10 percent increase in the price of imported inputs would, in those conditions, increase the demand and expenditure in national inputs by an estimated 2.6 percent. In absolute terms, demand for national inputs could be expanded by $2.6 million dollars beyond the 1971 level of $100 million.

This policy does generate an increase of total costs of copper production for the companies but is of little relative significance since both imported and domestic inputs constitute only 37 percent of the total nominal costs. [16] A tariff rise of 10 percent as postulated here would increase total company costs by less than 1.5 percent.

This cost increase for the companies, however, is not prejudicial to the country as a whole. On the contrary, it allows a saving of foreign exchange, and the greater expenditures within the country will stimulate industrial development for the manufacture of inputs for copper and will increase at the same time fiscal revenues through the tariffs. In this way former distortions of the input market are corrected, while artificial reduction in the companies' surplus through unnecessary importations is avoided.

Another important aspect of this policy is its tendency to increase employment of labor in the industries producing domestic inputs. The companies may also be encouraged to expand their use of labor if this use of national inputs implies a more intensive use of labor than imported inputs. (For instance, importation of complete machinery can become less desirable if one charges the real [higher] cost of purchasing them abroad compared to repair and maintenance within the country, buying only spare parts. This latter procedure, relatively more labor-intensive, would become more convenient [cheaper] under the new prices.)

A warning is necessary. One of the most important simplifying suppositions used was that the prices of national inputs remain constant after the price rise of imported ones. Perhaps such an assumption is not very real, especially as it applies to homogeneous goods produced internally and externally in an environment of relative competition. It is probable that an internal price rise in the cost of an imported input may automatically raise the price of its equivalent produced internally, if the unit cost of producing those inputs is not constant. If this latter condition is not fulfilled—that is to say, if the supply of national inputs implies increasing unit costs—then the assumption that the prices of these inputs remain constant will not hold, and

one would have to modify the preceding results. However, even in
the most extreme case imaginable, when national input prices rise
as much as the prices of the imported ones, the use of the former
would increase anyway, if their elasticity of supply is greater than
zero. (It is likewise true that an eventual 10 percent increase of national and imported inputs already postulated does not raise total production costs more than 3.7 percent.)

Finally, it is worth pointing out that this process of input substitution takes time, and the effects mentioned will occur in the long
run. For this reason, the policies used to induce them need to be
well-defined and stable. The country should begin with substitution
of the easier, cheaper goods and move gradually into substitution of
the more sophisticated. This demands special attention to factors related to the quality of the national inputs, the continuity of their delivery, and the technical development they ought to incorporate.
There is indeed a clear need to formulate a long-range policy to develop a sector that will produce inputs and capital goods for the mining sector. This procedure should be stimulated at the same time by
a price policy like the one suggested here that will increase demand
for such goods on the part of the copper companies.

Recall also that the previous analysis and calculations refer to
current inputs and not to capital goods. In regard to the former, certain progress in substituting for imports has occurred in recent
years, but the lag in regard to capital goods is still considerable.
Since the production of many of these goods presents no great technological problems that impede their production in the country, they
represent an important potential market for developing Chilean industry.[17] That market can also be extended beyond the borders of the
country, particularly through agreements with other mining countries
within the Pacto Andino (like Peru and Bolivia) and with the CIPEC
countries.

RATIONALIZATION OF PRODUCTION
WITHIN THE MINING SECTOR

A fourth problem that shows up in the organization of the nationalized companies relates to coordination of the companies among
themselves in order to take greater advantage of the capacity of each
installation.

Before nationalization, the control of the copper companies by
different multinational corporations implied an artificial separation
among the companies that brought about rigidities and distortions
within the sector as a whole. A greater coordination among the companies, or rationalization of their productive processes, should in-

crease the efficiency of the sector and allow it to produce more with
the same volume of resources or factor endowment.

The deficiencies that existed at the time the firms were nationalized were due basically to the vertical integration of the Chilean
mines with foreign corporations, while each mine remained separate
from the others. For example, a company could decide to have excess refining capacity of its Chilean plant, which could be brought into
use when refineries in the United States were overloaded. Although
that excess capacity was idle during normal periods it could not be
used by another company operating in Chile, but belonged to a different multinational corporation.

From the national point of view, however, the optimum is the
equalization of marginal costs of each company, for which you need
to transfer factors of production and concentrate new investment into
the companies, mines, and plants with lower relative costs. To reach
this goal, an efficient planning system within the mining sector as
well as adequate macroeconomic policies are needed.

At an aggregate level, the inefficiency of this situation shows itself partially in the differentials of the rate of return on capital and
in the unit costs of production of each enterprise. During the entire
1955-67 period, the rate of profits before taxes for Braden averaged
129 percent, that of Chilex, 53 percent, and that of Andes, 8 percent. *
In the second place, data the companies themselves handed over to
Codelco show directly their nominal cost differences. During the period 1955-70, the average cost of producing a pound of copper in Chile
was 16.4 cents of a dollar for Braden, 20.4 for Chilex, and 20.5 for
Andes. The overall average for the group comes to 22.6 cents. **

If the differences of the marginal cost of production among the
companies are directly related to those between average costs, these

---

*Those differentials were also a consequence of different income
tax rates and depreciation rates each company was subject to. In the
case of Andes, this last item was particularly important. Its rate of
profits seems underrated (and costs overrated), because of the high
depreciation rate allowed on its books.

**This includes depreciation, but it excludes an imputation of cap-
ital costs. If this latter item were considered, however, the differences would be even more noticeable, precisely because Braden, in
relative terms, uses less capital to produce each pound of copper than
Chilex, which in turn uses even less than Andes. The exact cost data
should be interpreted cautiously, since the companies often use different definitions for some items, and depreciation figures are often
arbitrary. At any rate, differences in production costs are sufficiently appreciable.

figures suggest that El Teniente and Chuquicamata should be able to increase their profits if they raise production at least to the point at which their cost reaches a level similar to that of El Salvador (which was a profitable enterprise although its costs were 50 percent greater than Chile's). Since the revenues from each pound of copper are substantially greater than the costs of producing it in the former companies, the total surplus for Chile can rise considerably through a relatively more rapid expansion of those enterprises.

It is not pertinent that we discuss in depth here the economic policies that will improve this situation, since such an analysis refers more to a planning problem internal to the sector rather than to its relationship with the rest of the economy. That problem, too, is primarily an administrative and technological one, which can be met more efficiently through direct methods. Such a task involves detecting eventual bottlenecks and excess capacities in various phases of the production process (mine, refinery, and so on) within each company and the transfer and investment of necessary resources to overcome them.

In this sense, on a microeconomic level, adoption of an adequate price policy, planned within the frame of decentralized administration of productive operations, but which stimulates an efficient realization of those investments can be of inestimable usefulness. In general, such a policy should try to see that companies pay the social prices for all the inputs and factors of production they occupy and especially that they pay for the capital they are using.

CONCLUSIONS

Nationalizing the large-scale mine deposits calls for sketching out original forms of organizing and for regulating this economic activity; and it also demands the adoption of new criteria for evaluating results. The necessity of running this sector in the most efficient way possible makes it advisable to decentralize the administration of the firms with respect to current production decisions. This sector should, however, remain subject to central planning through state economic policies oriented toward fulfilling the national development objectives.

In this chapter, we have suggested specific policies to confront certain general problems that could sidetrack the companies in their attempt to reach these objectives. Such problems stem principally from the fact that the copper companies had previously belonged to multinational corporations. Therefore, once the companies are nationalized, it is even more important to set up a new planning system and new policies that will introduce those corrections necessary for their proper functioning.

Chile's monopoly power in the copper market necessarily requires centralizing sales into one single entity. Moreover, there existed a fear that nationalization would penalize the country in its sales activities. It was thought that Chile's separation from the sales system of the great multinational corporations would inevitably reduce the country's monopolistic power in the world market. Nevertheless, such fears were proved unjustified if some reasonable assumptions concerning the behavior of other producers are observed (no new and greater international conglomerates are formed, for example), and if all Chilean sales are centralized.

Another policy that should be revised deals with profit taxing or, more generally, with surplus extraction from the copper sector. That tax should be complemented by charging an interest on capital and a rent on the ore deposits themselves. These changes can be brought about without a reduction in the volume of resources the state receives from the industry, but, on the contrary, with an increase in them.

Foreign control of the copper mines and the domestic policies applied to participate in the revenues generated by them actually turned this industry into an enclave and caused a lack of integration to the rest of the national economy. As a consequence, the industry became involved in an overuse of imported inputs and the underuse of domestic inputs and labor. In this respect we have shown that a correction of this situation is possible partially by using as indirect economic policy such instruments as raising tariffs and exchange rates and imposing an interest tax on each company's capital.

Finally, another distortion that can be corrected is the noticeable efficiency gap between one company and the other. Those with lower production costs must expand their production (in relative terms) increasing in that way the overall benefits the country derives from the copper-mining sector.

NOTES

1. For failure to maximize national welfare in preceding decades, see Norman Girvan, "Las Empresas Multinacionales y el Cobre Chileno" in R. Ffrench-Davis and E. Tinori, eds., El Cobre en el Desarrolo Nacional (Santiago: Ediciones Nueva Universidad, Ceplan, 1974).

2. These are the criteria of Oskar Lange in his famous essay "On the Economic Theory of Socialism" (1938).

3. For use in socialist economy, see Yevsei Liberman, "The Plan, Profits and Bonuses" (1962); and Alexei Kosygin, "On Improving Administration in Industry" (speech of the Prime Minister of the USSR

at the plenary meeting of the Central Committee of the Government Part, September 7, 1966), where, among other things, he says, "in order to orientate enterprises towards increased efficiency, it would seem best to use the profit index, the revenue index. The bulk of profits or gains indicates in great measure the contribution made by an enterprise to the net income of the country which is used to expand production and raise the level of well-being of the population." Cited by Bornstein and Fusfeld, The Soviet Economy (Homewood, Ill.: Richard D. Irwin, 1970), p. 38.

4. Franklin Fisher, Paul Cootner, and Neil Bailey, "An Econometric Model of the World Copper Industry," The Bell Journal of Economic And Management Science, 3, no. 2 (August 1972).

5. See Chapter 6 above.

6. See Norman Girvan, Copper in Chile (Jamaica: ISER University of the West Indies, 1972); and E. Tironi, "The Historic Evolution of Foreign Investment in Latin America" (mimeo.; Santiago: Catholic University of Chile, Ceplan, 1974).

7. See Raymond Mikesell, Foreign Investment in the Petroleum and Mineral Industries (Baltimore: John Hopkins Press, 1970), p. 370.

8. Figures estimated from the American Bureau of Metal Statistics, Yearbook 1971 (New York); and Balanza de Pagos de Chile (Santiago: Banco Central, 1969).

9. Fisher, Cootner, and Bailey, "An Econometric Model of the World Copper Industry," op. cit.

10. As an example of such investment evaluation, see A. Foxley and D. Clark, "Social Returns from New Investments in Chilean Copper Production," Estudios de Planificacion No. 17 (Santiago: Catholic University of Chile, Ceplan, 1972).

11. Lecture by engineers Augusto Millan and Juan Pedrals (from Enami and Codelco in a seminar on copper sponsored by Ceplan, June 1973).

12. Compare "The View from Inside ITT," Business Week, November 3, 1973.

13. Company balances published in Diario Oficial, October 13, 1971 (which published the indemnity to be paid by the nationalized companies).

14. R. Ffrench-Davis, Politicas Economicas en Chile: 1952-1970 (Santiago: Ediciones Nueva Universidad, Ceplan, 1972).

15. See R. Ffrench-Davis, "Substitution of National for Imported Inputs in Copper Production: An Econometric Study" (mimeo.; Santiago: Ceplan, 1974).

16. Ibid.

17. Codelco has made several unpublished studies that confirm this statement.

III

INCREASING STATE
INTERVENTION IN
PERUVIAN MINES

10

LEGAL FRAME USED
FOR THE EXPANSION
OF BIG MINING
Diego Garcia-Sayan

NEW PERIOD OF ECONOMIC LIBERALITY

The military coup of 1948 that put Odria into the presidency of the Republic meant the assumption of political power by agro-exporting and landholders' sectors and, with it, the opening of a new period of economic liberality. These sectors did not take long in attaining the necessary modifications in the economic policy, eliminating the control of foreign exchange and prices. In this way, they turned from a timidly controllist structure into a structure as liberal as possible in the matter of exchange and commerce.

Economic liberalism, together with the weakness of national economic groups and the demand for raw materials by developed countries, created favorable grounds for the massive entrance of foreign capital into the mining sector. These results were made possible and assumed a specific form through legislation, which was soon recognized as the most liberal of that time in Latin America.[1] This legislation was mainly embodied in the Mining Code of 1950 (Decree Law 11357) and in the Oil Law (Law 11780).

THE MINING CODE: STEPS FOR
ITS CREATION

To outline the steps of creation of this transcendental legal regulation might appear as a useless exercise of enumeration of names and dates. Nevertheless, we can clearly see the prevailing interests from examination of these legal rules.

On August 22, 1949, by Supreme Resolution no. 76, a commission presided over by engineer Mario Semame Boggio was created,

to elaborate and present a preliminary outline of the new Mining Code
within 60 days. Such was the hurry of imperialist consortiums to invest and of agro-exporting sectors to safeguard the freedom of trade
and of exchange. But, "in order to meet our goal satisfactorily,"[2]
the commission obtained a 90-day extension. On January 21, they
presented the preliminary report to the minister of Development. Ob-
servations from national and foreign groups (among which was the
Northern Peru Mining and Smelting Company) and from lawyers and
engineers were considered by the commission for the final wording
of the definitive report, which was presented to the Ministry of Development and Public Works on April 15, 1950. Finally, a meeting
was held in the government's palace on May 11, 1950, between Odria
and representatives of mining institutions and enterprises, among
which were several foreign enterprises (including Northern Peru Mining and Smelting Company, Compagnie des Mines de Huaron, and San
Luis Gold Mines). Edgardo Portaro, president of the National Society
of Mining, president of the Directory of the Mining Bank of Peru, and
director-manager of the Atacocha Mining Company, speaking in the
name of the groups present, requested Odria to promulgate the Code.
The following day, May 12, it was promulgated by Decree Law no.
11357.

      In this way we see very clearly how the prevailing interests
were involved in the creation of this regulation. The authority in
charge of legislating (president of the military junta and the cabinet)
entrusts the elaboration of the project, first, and its conversion in
legal norm, afterward, to external investor groups. This legal instrument is dictated with the absolute approval of those whose conduct it
is supposed to regulate. Nevertheless, this is not a common characteristic in the generality of norms. A certain level of contradiction
is perceivable between the interests and demands of those that are to
be benefited by the norm and the content of the norm itself. The relative autonomy between the juridical and the economic levels apparently
cannot be perceived in the case of the creation of the legal norm on
which we are commenting. In this case, a mechanical correspondence
between the two levels, within the social structure, seems to be present. The process of application of this norm, shows, nevertheless,
that that mechanical correspondence is a mere appearance and that
there are secondary contradictions between its levels. Being the
right and the legal norms constituted not only by the text of the norm
and its creation but by the whole process of its application, this apparent initial mechanical correspondence will constitute only one stage in
the existence of the norm.

MAIN ASPECTS OF CREATION

It is not our intention to make a detailed analysis of the Mining Code. We are simply interested in analyzing its main characteristics.

In the first place, Article 1 establishes the principle that is more or less generalized in other legislation: Mineral substances are the property of the state. In this case, it followed Article 37 of the Constitution, which respects already acquired rights like those of Cerro de Pasco.

Exploitation by concessionaires is declared to be in the public interest.[3] This makes expropriations in favor of mining enterprises possible, in many cases to the detriment of Peasant Communities using Article 209 of the Constitution, which authorizes expropriation of communities' properties due to reasons of public interest. In 1967, for example, 20 hectares of the Peasant Community of Huayllay (Cerro de Pasco) were expropriated in favor of the French mining enterprise Compagnie des Mines de Huaron, with an "indemnization" of $10,000 (about U.S. $370 at that time).

Jose Rocha says of this concept, established in the Code of Mining, that

> The exploitation of the mines, declares the Code,
> is in the public interest because taking advantage
> of mineral deposits is of interest not only to the
> one who does it, but also the entire society; that
> is why the existence of a Special Right is applied,
> which tends to promote this industry through tute-
> lary dispositions of interest of the one who ex-
> ploits it as well as national interest. This special
> right is the Mining Right.[4]

This particular interpretation of this concept of the Mining Code is in contrast with that of Cuadros Villena, who, referring to that principle of social interest that assumes public interest, points out that,

> Truthfully, when analysing distribution of mining
> resources and verifying its majority ownership by
> foreign capitalist enterprises, which only use it as
> source of raw materials, without leaving either taxes
> or salaries for the countries that produce it, it is
> evident that the so-called social interest of mining
> is nothing but the protection of the interests of im-
> perialist enterprises operating in Latin America at
> the expense of the people's misery.[5]

Cuadros Villena emphasizes that this interpretation permits foreign firms to develop mines at the expense of the peasant population:

> Since the majority of our underground resources is
> in the hands of foreign capitalists, the so-called "so-
> cial interest" of the mining industry is nothing but
> the protection of foreign monopolistic consortiums,
> even to the detriment of the people's consumption,
> whom FAO [UN Food and Agriculture Organization]
> has described [as being] in "conditions of chronic
> hunger," requiring payment of preferential attention
> to the agricultural industry above any other. [6]

The tributary aspect is crucial to an understanding of the most important characteristics of the legal regulation:

1. The tax on exports was suppressed and replaced by the fiscal regime of rent (Article 50). This measure favored big mining enterprises dedicated to the export of gigantic tonnages of mineral, for it was less expensive than paying tax on profits.

2. The authorization of special contracts to be celebrated between the concessionaires and the Executive Power for the installation of power or processing plants, and for the exploitation of the so-called marginal deposits (Article 56). In the first case, a low rate of tax—ranging from 10 percent to 20 percent of profits for a set period of time—was given. In the second case, the right to apply the entire first profits in the mortgage of the invested capital is conceded. Code Article 56 allowed the Southern Peru Copper Corporation to start the exploitation of the rich copper deposits of Toquepala, under a contract reached with the state in November 1954.

It is important to glance at this contract, because of the importance of this enterprise in copper exploitation in Peru.

In paragraph (a) of the third clause, reference is made to Toquepala deposits as "deposits of great tonnage but of low standard of quality," meaning that these deposits were being qualified as marginal. Consequently, they could benefit from Article 56. Nevertheless, there have been disagreements about the character of these deposits, of which it has been said that

> The fraudulent aspect of the agreements rests on the
> fact that a deposit of the magnitude of Toquepala,
> with 1% average wealth, working "open cut" system,
> in no country has ever been considered marginal, es-
> pecially if we consider that during the first year of
> operations the average copper content was 1.7%, and

after 6 years it was 1.3%. The Brigham Young Mine, located at Salt Lake City (Utah), considered as the biggest of that country has only 0.8% of average copper content; Miami Copper works since 1936 a mine with 0.94% of average copper content.[7]

In relation to the use of first profits to mortgage capital (Article 56) the contract has been especially careful to stipulate that it was referred to as net profit, "after deducting depreciations, amortization of expenses and preparation of the mines, depletion and all taxes." With these specifications, the benefits of Article 56 grow.

Article 56 was modified in February 1968 through Law 16892. The law gives a more or less detailed specification of the benefits and guarantees of special contracts that that law authorized the executive power to make. Among other characteristics in the amending article, the use of marginality of deposits to obtain the benefits of special contracts is removed. Benefits that can be obtained through such contracts are limited to a period of 10 years. In the fiscal aspect, the maximum limit of the tax on profits established in the amended article is modified. The sum of the tax on profits and the complementary tax of set rates will not be less than 40 percent of taxable profit.

3. The "depletion allowance" (Article 54) permitted the concessionaire to reserve, free of tax, a percentage of 5 percent to 15 percent of its gross product, with a maximum of 50 percent of its profits. Because of this curious resolution, the concessionaire is exonerated of up to 50 percent of its taxes for depleting deposits owned by the state.

The data show the importance Article 54 had for big consortiums. For Southern Peru Copper Corporation, for example, in the year 1960-65, the amount exempted for depletion allowance was U.S. $45,000 million, over a total gross income in the same period of U.S. $448,871,682.[8] In the 1953-66 period, the total amount exempted for depletion allowance of the Marcona Mining Company was U.S. $32,700,000 out of a gross income of U.S. $349 million.[9] It would seem that the tendency carries the depletion allowance to around 10 percent of net income which is considerable.

4. For the simple payment of territorial duty and tax on profits, the concessionaires were exonerated from any other obligation or tax created within 25 years from the date of promulgation of the Code (Article 53). In this way, the state was expressly renouncing the exercise of its sovereignty and was tying its hands in respect of what, at that time, was its main source of income from the mining sector.

5. Freedom from the taxes on the import of machinery, equipment, spare parts, and materials for specific use in mining (Article 242).

The concessionaires, be they national or foreign, have the right to exploit the deposits for an indefinite length of time. The only causes for termination were voluntary abandonment or lack of payment of the superficial duties for two consecutive years (Articles 3 and 27). In this way, mining enterprises are allowed to keep areas unexploited as they wish, turning them into reserves according to their convenience. The formation of these private reserves, the consequence of laissez-faire in law, derives directly from some juridical acts.

Let us examine the Southern Peru Copper Corporation program for the exploitation of the Toquepala and Quellaveco deposits that was divided in two stages. The first refers to the necessary work for the exploitation of the mines of Toquepala, "which should start before the period of 18 months counted from the date it became public writing and its signing by the contracting parties" (21st clause). The second stage referred to the exploitation of the Quellaveco deposits, "which the Company will undertake when, according to its judgment, economic and technical conditions of exploitation justify it." For almost 20 years, economic and technical conditions of exploitation do not seem to have justified the exploitation of Quellaveco. The deposit has remained a reserve of the foreign consortium, waiting for its all-powerful decision.

Juridical ideology has masked reality of cases such as this one, maintaining the analysis at a purely formal level. Jose Rocha has stated that[10]

> The Code of Mining, which is a State Law, does not authorize the Government to reserve mining zones, since this would limit the right to unhindered appropriation that according to Article 3 all natural or juridical national persons have, as well as foreigners of private right, which so request.

In the same way, juridical ideology, in which doctrine plays the main role, has been responsible for determining the real nature of the declaration.

The prevailing doctrine, following a profoundly formalist route, is elaborated and profuse in support of this view. Fernando Barco is emphatic in pointing out that,[11]

> Through a concession, particulars acquire real estates over these goods of the State, which, nevertheless, remain in their possession, since the concession, let us repeat it, does not entail transference of ownership of the mineral deposit.

The real right given to the concessionary, according to the same author,[12]

> is founded on the possibility of appropriation of the substance and it is made effective once the substance, through the work of the miner, is extracted from the deposit.

He also notes that,[13]

> In this sense the area to which Article 24 makes reference (Art. 24 of the Mining Code) is only the terrenal corps over which right is exercised, which, at the same time has as an object the mineral deposit within the granted space.

Rocha emphasizes that,[14]

> When we talk about property in Mining Right, we are referring to a right over the concession, not to a right of property over the mining deposits; these are two very different things.

In stating the nature of the concession, he observes,[15]

> The mining concession is not a constituent act of property, but a legitimator of a preceding juridical situation. The mine appears when, over the material fact of the mining deposit, concession exists, that is the title, the official garment, both elements linked by human activity, which backed on this situation, acts over the demarcated ground. That is why excavation without having the concession does not originate the concept of mine, juridically speaking.

He concludes that,[16]

> If we approach the theme based on the regulations of our Civil Code, mines are immovables and belong to the State, which gives them in concession to individual persons, and results are received when extracted.

With a similar idea in mind, Cesar Polack states that, "What is conceded is the right to exploit the minerals, that is to make them evident, to extract them."[17]

The right of property over the deposit is formally different from the right over the concession. Nevertheless, this formally valid reasoning shows its weakness when confronted with certain material proofs that make us wonder if the state continues to be owner of the resource conceded by the system of concessions.

In reality, the large number of jurisdictions that alienation in its various forms allows (in hypothecary guarantee, contribution to a corporation or society, contribution in contract of bargain and sale, inheritance, and so on) makes us realize that we confront a kind of property. It seems that the Mining Code, as well as the prevailing doctrine, are both oriented toward a privatization of mining resources. It is formally defensible—as some of the writers argue—that there is no transference of property to particular parties, but the facilities transmitted, "with irrevocable and indefinite title," as Article 27 of the Mining Code says, makes us think that in a material sense privatization does occur. In this way, the legal norm as well as the juridical doctrine forms an organic and coherent whole that, in practice, contests control by the state of the mining resources, as established in Article 37 of the Constitution, Article 822, Clause 4 of the Civil Code, and in the Mining Code itself.

A general term of special prescription is set to give operation and security to the mining title (Article 93). This title cannot be discussed administratively once it is inscribed in the Concessions and Mining Rights Registry. Neither can it be discussed in the Courts after five years of being inscribed.

Finally, it is worthwhile pointing out, as an outstanding characteristic of the Mining Code, that it constitutes an organic and coherent whole, as is typical of capitalist legal norms. It is applicable to all natural or corporate persons, national or foreign (with the only restriction of being of private right if it is a foreigner, as Article 3 specifies) that participate in the mining business as concessionaires or pretend to do so. There is formal and abstract liberty for anyone to make use of the same legal mechanisms under which, at a formal and abstract level, everyone is assumed to be equal. As one of the lawyers commenting on the preliminary report of the Samame Commission points out,

> Within a healthy orientation, we want to establish
> the bases for investment by big and small capital
> in this very important sector of our economy. We
> don't want monopolization and exclusion of nationals.
> The doors remain open for nationals as well as for
> outsiders.

This formal equality, however, conceals the facts. Liberty for and equality between "big and small," "national and outsiders," enables those who are more powerful to use the benefits of the Mining Code, to the exclusion of others. We can obviously infer, and practice has confirmed this, that those benefited will be the "big" and the "outsiders."

Another factor of the general structure of capitalist law present in this particular norm is the factor of calculability, of predictability. Capitalist rationality, and, in this case, the large investments to be made, created in mining enterprises the need to have security as the dominant element in their juridical relations, and risk as subordinate and secondary. From the brief descriptions of the main characteristics of the Mining Code above, we can conclude that in this norm the level of predictability is very high and the margin of the juridical-political level much reduced.

CONCLUSIONS

The legal norms established created a favorable climate for the expansion of enterprises dedicated to big mining exploitation and the possibility of expropriation of land even when it belonged to Peasant Communities. Magnificent tax incentives were also established, like the suppression of the tax on exports and its replacement by a tax on rent; the ability to make special contracts benefiting from the (also special) incentives of Article 56; the possibility of using the depletion allowance as a means to greatly reduce the taxable amount; tax stability for 25 years; freedom from taxes on imports; and so on.

By paying only a surface duty, concessionaires could continue indefinitely to possess their concession, with the consequent privatization of the mining resources.

These advantages, in the context of a norm with structural factors typical of capitalist law, such as liberty and equality at a formal and abstract level and the predictability factor, were an essential part of the intensification of the imperialist dependence of our country through the expansion of big mining in the hands of big consortiums in the 1950s and 1960s.

NOTES

1. Official letter dated October 6, 1949, sent by Mario Samame Boggio, president of the Commission, to the Ministry of Development.

2. Here, for example, Mario Samame Boggio says that among the factors that favor the investment of foreign capital in Peru is "an

excellent Mining Code, which gives the country, from this viewpoint, the most liberal and progressive [industry] in the Americas and, perhaps, in the world." Article published in the weekly Peruvian Times, January 22, 1960.

3. Letter dated April 15, 1950 sent by the commission to the minister of Development and Public Works. Included in the official edition of the Code of Mining of 1950.

4. Jose Rocha, "Apreciaciones sobre la Aplicacion del Codigo de Mineria y la Naturaleza Juridica de las Minas y Minerales," Revista de Derecha Minero, June-December 1965, pp. 8-9.

5. Carlos Cuadros Vilena, "Principios de la Reforma Minera Latinoamericana," Revista del Foro, no. 3 (1970), p. 149.

6. Ibid.

7. Carlos Malpica, Los Duenos del Peru (Lima: Peisa Editorial, 1970), p. 171.

8. Judgment of the Bicameral Multipartidary Commission in charge of revising the agreement subscribed between the Peruvian government and the Southern Peru Copper Corporation. Annex 23 by ibid., p. 174.

9. Sven Linduvist, Slagskugga, free translation to Spanish in Documentos de Trabajo sobre Mineria (Lima: Federated Center of Social Sciences of the Catholic University).

10. Rocha, op. cit.

11. Fernando Barco Saravia, "El Control del Estado sobre los Yacimientos Minerales," Revista de Derecho Minero, no. 26 (December 1969), pp. 16-17.

12. Ibid., p. 17.

13. Ibid., p. 18.

14. Rocha, op. cit., p. 4.

15. Ibid., p. 10.

16. Ibid., p. 12.

17. Cesar Polack, "Consideraciones sobre el Desarrollo Economico y Social en Ocasion del Dia del Minero," Revista del Derecho Minero, no. 25 (June 1969), p. 22.

11

THE ANATOMY
OF IMPERIALISM:
THE CASE OF THE
MULTINATIONAL MINING
CORPORATIONS IN PERU
Claes Brundenius

This chapter aims not only to show the methods of exploitation of the modern multinational corporation but also to demonstrate the intimate links between the "national" bourgeoisie in Peru and foreign capital.

THE EXPANSION OF MULTINATIONAL MINING
CORPORATIONS IN PERU AFTER
WORLD WAR II

The end of World War II had immediate repercussions for U.S. monopoly capital. Surplus capital that had accumulated during the war boom now had to look for markets abroad. Between 1950 and 1953 the book value of U.S. direct investments in mining and smelting in Peru increased from $55 million to over $155 million, primarily due to iron mining investments by the Marcona Mining Company, a subsidiary of Utah Construction Company and Cyprus Mining Company. Until then, Cerro de Pasco had been the single giant in mining.*

This essay was financed by the Nordic Cooperation Committee for Internal Politics including Conflict and Peace Research, Stockholm. It is reprinted in an abridged form with permission from the Journal of Peace Research, no. 3 (1972).

*From 1907 to 1931 the Vanadium Corporation of America operated the sole vanadium mine in Peru. During this time, this mine supplied approximately 44 percent of the world's total consumption of vanadium (a refractory metal used in the fabrication of catalysts for nuclear reactors). Output has since then diminished each year; in

Marcona expanded production rapidly: By 1968 over 9 million tons were produced. During the 1950s the output of lead and zinc, primarily from properties belonging to Cerro de Pasco, also increased by 86 percent and 50 percent, respectively. Due to the stagnating world market price on both lead and zinc in the 1960s, these metals are of much less importance to the mining companies than is copper. Thus, in 1968 over 38 percent of the value of metallic mineral production was accounted for by copper, while iron represented 13 percent, precious metals (gold and silver) 14 percent, zinc 8 percent, lead 6 percent.[1]

By the end of the 1950s, U.S. mining investments in Peru were steadily increasing, this time due to the investment by Southern Peru Copper Corporation (SPCC) in the Toquepala mine. Toquepala is the largest copper mine opened anywhere in the world since World War II and has the cheapest production costs available, just over 10 cents per pound.[2] SPCC is a so-called joint-venture project, a modern and more sophisticated version of the earlier cartel agreements where Cerro Corporation in New York participates with 22.3 percent of the equity capital, American Smelting and Refining Company (Asarco) with 51.5 percent, Newmont Mining Corporation with 10.2 percent, and Phelps Dodge Corporation with 16 percent.

With the aid of the U.S. government—a $115 million loan from the Eximbank—SPCC has installed South America's third largest smelter in Ilo, with a yearly capacity of 130,000 tons of blister.[3] When the Toquepala mine started producing in 1960, Peruvian copper output rose to 184,000 tons per year—over three times the 1959 level. Since then, however, copper production has increased only moderately, with a record output of 215,000 tons in 1970. The rapid increase of the value of mineral production during the 1960s is primarily the result of the abnormal increase of copper prices in 1963-70, when the price of blister copper on the world market more than doubled.

Essentially due to SPCC investments in Toquepala, Peru during the 1960s increased its share of world copper output—but only in mine and blister production. Refining capacity has not expanded at the same rate. In 1953 all of smelter (blister) production went on for refining (electrolytic copper). In 1968 only 21 percent of the smelter production was refined in Peru. The reason is that the multinational, mostly U.S. corporations that totally dominate the mining sector of Peru for economic and political reasons prefer to process the metals in the United States, Japan, or Europe. Since blister copper contains many valuable by-products—gold, silver, molybdenum, arsenic, selenium—

1968 no production was reported, presumably due to exhaustion of the mines.

separated only during the refining process, the value of which can be estimated only roughly when leaving the country, the corporations can effectively circumvent state control on commercialization of the mineral production. In addition, this of course represents a hindrance for efficient industrialization of the country.

In 1968, less than a fifth of Peru's output was refined. Less than 2 percent was consumed in a semiprocessed form in the country. Of the total copper exports, over half went to the United States, a fourth to the EEC countries, 12 percent to Japan, and 6 percent to the rest of the world.[4]

THE OWNERSHIP STRUCTURE OF THE PERUVIAN MINING INDUSTRY

The mining sector in Peru is completely dominated by the giant U.S. mining corporations. In the field of iron, U.S. domination is total. In 1968 Marcona Mining Company accounted for 98.3 percent of iron production, while the remaining 1.7 percent was produced by Pan American Commodities (Pamecsa), a company linked to Cerro de Pasco (see below).

In the field of nonferrous metals, which account for 69 percent of total mining output in Peru (including nonmetallic minerals), the situation is more complex. There 36 percent of the equity capital belongs to SPCC, and 33 percent to Cerro de Pasco. These two U.S.-based companies represented over 75 percent of the sales value of the sector in 1969 and more than 83 percent of the copper production in 1968 (see Table 11.1).

Northern Peru Mining Company, a subsidiary wholly owned by Asarco, represents only some 2 percent of the sales and between 3 and 4 percent of copper output. The Japanese firm Mitsui and Nippon Mining, with four subsidiaries, represents as yet a rather insignificant share of output, as is the case with the French company Huaron. The British property in the Lampa mine was sold some years ago to the W. R. Grace Company, which in 1971 sold it to a Peruvian fishmeal magnate.

The Natomas Company, a California-based U.S. company linked with the influential Prado family, is of special interest, although it represents only 0.2 percent of the sales in this sector. Since 1961 this company has been exploiting one of the few remaining gold mines in Peru near the Bolivian border in violation of the Peruvian constitution.[5] In 1968 the gold production of the Natomas Company accounted for no less than 20 percent of the total gold output in Peru. Although SPCC is by far the biggest copper producer in Peru (see Table 11.1) it is nevertheless Cerro de Pasco that is the most influential company.

TABLE 11.1

Ownership Structure in Peruvian Mining Industry

Company	Equity Capital, 1969 (million soles)	(per-cent)	Sales Value, 1969 (million soles)	(per-cent)	Mineral Production, 1968 Copper (thousand tons)	(per-cent)	Zinc (thousand tons)	(per-cent)	Lead (thousand tons)	(per-cent)	Iron (thousand tons)	(per-cent)	Gold (kilos)	(per-cent)	Silver (thousand kilos)	(per-cent)
Southern Peru (S.P.)	1,828	35.8	6,530	31.1	133	62.5	—	—	—	—	—	—	*	—	*	—
Cerro de Pasco (C.P.)	1,660	32.5	6,447	30.7	33	15.5	141	49.4	52	34.6	—	—	261	10.2	298	28.6
Companies directly controlled by C.P.	457	8.9	1,055	5.0	1	0.5	45	15.8	31	20.7	—	—	218	8.5	167	16.1
Companies indirectly controlled by C.P.	429	8.4	1,855	8.8	11	5.2	54	18.7	22	14.6	214	1.7	755	29.4	320	30.8
Cerro de Pasco Total	2,546	49.8	9,357	44.5	45	21.2	240	83.7	105	69.9	214	1.7	1,234	48.1	785	75.5
Marcona Mining	114	2.2	2,757	13.1	—	—	6	2.1	1	0.7	8,802	98.3	—	—	—	—
Northern Peru	202	4.0	410	1.9	6	2.8	—	—	—	—	—	—	146	5.7	28	2.7
Japanese Companies	12	0.2	441	2.1	3	1.4	—	—	—	—	—	—	32	1.2	—	—
Huaron (French)	38	0.7	372	1.8	3	1.4	12	4.2	7	4.7	—	—	18	0.7	42	4.1
Natomas (United States)	5	0.1	41	0.2	—	—	—	—	—	—	—	—	503	19.6	—	—
Other companies	366	7.3	1,121	5.3	22	10.7	28	9.8	37	2.7	—	—	636	24.7	184	17.7
Total	5,111	100.0	21,029	100.0	213	100.0	286	100.0	150	100.0	9,015	100.0	2,569	100.0	1,039	100.0

*No production recorded.

Sources: The annual declarations of the various companies for 1969; Anuario Minero del Peru, 1968; H. Espinoza, Concentracion del Poder Economica en el Sector Minero (Lima: Universidad Federico Villareal, 1970).

228

Cerro de Pasco owns all three existing refineries in the country and two of its four smelters. Through an intricate interlocking system, it also controls more than a hundred smaller mining companies. In addition, Cerro de Pasco owns 60 percent of the shares of Compania Minera Raura, and 33.6 percent of the shares of Compania de Minas Buenaventura. The latter company was recently described as an example of a successful Peruvian company. As a matter of fact, not only does Cerro de Pasco own a third of the shares of Buenaventura, but its president is Alberto Benavides de la Quintana, president of the board of directors of Cerro de Pasco (see Table 11.2).

Cerro de Pasco also owns or controls practically the entire Peruvian metal industry. Thus, the company owns 32.4 percent of the shares in the ammunition firm Explosivos S.A. with its wholly owned subsidiary Electrodos Oerlikon (welding rods), 28.54 percent in Metalurgica Peruana (steel casting), 42 percent in Refractarios Peruanos (refractory bricks), 50 percent in Fundicion de Metales Bera (lead alloys), and 49 percent in Metales Industriales del Peru (extruded products).[6]

In addition, the mother company in New York, Cerro Corporation, owns directly Compania Industrial del Centro (with two semifabricating plants and one sulfuric acid plant in La Oroya). It also owns 76 percent of the shares in Industrias del Cobre (a wire and cable mill in Lima) and is the sole owner of Cerro de Pasco Petroleum Corporation (still at the exploration stage).[7] Cerro de Pasco is also the owner of the railway between Cerro de Pasco and La Orroya and was until January 1969 the owner of 270,000 hectares of pastures.

(Because of poisonous gases from the smelter in La Oroya, which destroyed hundreds of thousands of hectares of crops and pastures, Cerro de Pasco was in 1925 obliged to buy 27,000 hectares of land for a sum of 342,000 Peruvian punds, about $1.5 million in today's equivalent values. In March 1968, following procedures initiated in 1965 under the Agrarian Reform of the government of Fernando Belaunde, the authorities expropriated one of Cerro's haciendas. In January 1969 a further 18 haciendas belonging to Cerro de Pasco were expropriated. The military junta paid Cerro in cash a negotiated price of $2 million for the livestock, the land to be paid for in 18-year, high-interest-bearing, tax-exempt government bonds and the installations in cash, after a value fixed by appraisers and approved by a court.)[8]

There are about 220 operating mining companies in Peru. Of these, however, about 100 are controlled directly or indirectly by Cerro de Pasco. Members of the board of directors of Cerro de Pasco are managing directors of no less than 14 medium-sized mining companies (see Table 11.2). Through these companies Cerro de Pasco controls some 80 or 90 other smaller companies, in collaboration with the most influential Peruvian families (see Figure 11.1, Table 11.3).

TABLE 11.2

Cerro de Pasco Representation on Board of Directors of "Peruvian"
Mining Companies

Mining Companies	Members of Board of Directors of Cerro						Group Brazzini-Gildemeister-Mola					Group Rizzo Patron		Group Hochschild		Group Beltran
	1	2	3	4	5	6	7	8	9	10	11	12	13	14	15	16
Cerro de Pasco	PB	D	D	MD												
Buenaventura	MD	D														
Raura	MD				D	D										
Culquipurro	MD	D											D			
Minera Malin	MD				D		PB								D	
Minera Mantaro	MD	D		A								D				
Santa Rita	MD															
San Vicente		MD										D	D	D		
Minera Pepin		MD														
Venturosa			MD				D	D	D	D	D					
Volcan Mines				MD												PB
Chabuca Gold Mines				MD												
El Palomo				MD												PB
Milpo				A												A
Rio Pallanga				A								D	D	·		
Puquiococha					MD	D										
Pativilca					MD	D	D							RE	D	
Cercapuquio					MD			D	D	D						
Ganades					MD							D				

Legend:
1. Alberto Benavides de la Quintana
2. Harold A. Gardener
3. Michael A. Kurila
4. Aurelio Garcia Sayan
5. Joaquin Schwalb and J. F. Miculicich
6. Alberto Brazzini Walde
7. Angelica de Osme Gildemeister
8. Alfonso Loret de Mola
9. Carlos Loret de Mola
10. Hernan de Aguila
11. Aurelio Miranda Villanueva
12. Alfonso Rizzo Patron Remy
13. Antenor Rizzo Patron Remy
14. Joseph Pospisil
15. Luis Hochschild
16. Felipe Beltran Espantoso

PB: president of board of directors
MD: managing director
D: director
A: stock owner
RE: representative

Source: H. Espinoza, Concentracion del Poder Economico en el Sector Minero
(Lima: Universidad Federico Villareal, 1970).

FIGURE 11.1

Peruvian Mining Companies under Cerro Corporation Control

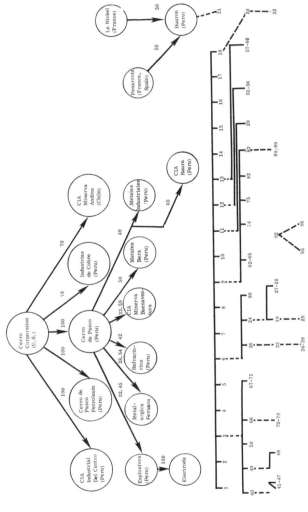

Numbers over arrows represent percentage of shares owned; each number below line represents a company.

Note: For number key, see Table 11.3.

Source: H. Espinoza, Concentracion del Poder Economico en el Sector Minero (Lima, 1970).

231

TABLE 11.3

Peruvian Mining Companies Directly or Indirectly Controlled by Cerro

Direct Influence:	
1. Cia Minera Malin	48. Cia Minera Korani
2. Cia Minera Mantaro	49. Cia Minera Palca
3. Cia Minera Colquipucru	50. Arenera Jicamarca
4. Cia Minera Santa Rita	51. Cobre Asia
5. Cia Minera Pepin	52. Cia Minera San Florencio
6. Minas de Venturosa	53. Cia Minera Castrovirroyna
7. Alberto Gubbins V.	54. Sindicato Minero Pacococha
8. Volcan Mines Company	55. Corporacion Minera Patara
9. Chabuca Gold Mines S.A.	56. Minsur S.A.
10. San Luis Gold Mines Company	57. Minas de Yarabamba
11. Sindicato Minero Rio Pallanga	58. Corporacion Minera Castrovirreyna
12. Cia Minera Milpo	59. Castrovirreyna Metal Mines Company
13. Cia Minera El Palomo	60. Cia Minera Huamanrauca
14. Sociedad Minera Puquiococha	61. Cia Minera de Canta
15. Cia Minera Pativilca	62. Cia Minera Sayapullo
16. Cia Minera Ganades	63. Cia Minera Alianza
17. Cia Minera Algamarca	64. Cia Minera La Virgen
18. Cia Minera San Vicente	65. Cia Minera Condor
	66. Chavin Mines Corporation
Indirect Influence:	67. Cia Minera Toronto S.A.
19. Minas de Cercapuquio	68. Equipos Mineros
20. Minas de Arcata	69. Exploraciones Peruanas S.A.
21. Cobre Acari S.A.	70. Latin American Mines Ltd.
22. Sociedad Minera Yauli	71. Peruvian Oils & Minerals Ltd.
23. Cia Explotadora Millotingo	72. Cia Minera Parcoy
24. Cia Minera Huampar S.A.	73. Sindicato Minero Parcoy
25. Cia Minera Cerro Noroeste	74. Empresa Minera La Suerte
26. Minas de Millococha	75. Cia Minerales Santander
27. Cia Minera Las Lomas	76. Capitana Gold Mines Company
28. Consorcio Metalurgico	77. Cia Explotadora de Huaranguillo
29. Cia Minera Santa Rosa de Comas	78. Cia Explotadora de Minas del Peru
30. Cia Minera Chanchamina	79. Eugenia Gold Mines Company
31. Cia Minera San Juan de Lucanas	80. Tangana Mines Company
32. Cia Minera Explotadora de Huallanca	81. Cia Administradora de Minas
33. Cia Minera Turmalina	82. Andaray Gold Mines Company
34. Cobre Caraveli	83. Minera Metalurgica del Centro
35. Concentradora Sinchao	84. Cia Minera Cochas
36. Sociedad de Inv., Com. y Mineras	85. Cia Minera San Fernando
37. Cia Minera Minas del Peru	86. Arias Davila S.A.
38. Sociedad Minera Beatita de Humay	87. Cia Minera Azulcocha
39. Cia Minera Condoroma	88. Cia Minera San Ignacio de Morococha
40. Panamerican Commodities (PACSA)	89. Cia Minera Canta
41. Cia Minera Bella Union	90. Consorcio
42. San Antonio de San Juan del Peru	91. Cia Aurifera de Buldibuyo
43. Cobre San Juan	92. Sociedad Minera El Brocal
44. Cia de Explotaciones Mineras	93. Corporacion Minera Sacracancha
45. Hierro Sur S.A.	94. Cia Minera San Jorge
46. Minerales del Peru	95. Sociedad Minera Huaraz
47. Cia Panamericana de Inv. Mineras	96. Neg. Min. Fernandini Clotet Hnos.

Source: H. Espinoza, Concentracion del Poder Economico en el Sector Minero (Lima, 1970).

The much celebrated "national" bourgeoisie is thus mostly non-existent as far as the mining sector is concerned. Pacsa is a typical example of the traditional links between the Peruvian bourgeoisie and foreign capital. Pacsa was created in 1952 by U.S. citizens together with members of the Dasso Drago family. In 1957 the company received a large loan from the American Overseas Finance Company. After some years, Pacsa was in default with respect to amortizations, so a new loan was arranged with the Chase Manhattan Bank (belonging to the Rockefeller empire) in 1965. General Jose Benavides, the son of a former president of Peru, and until June 1969 the agricultural minister of the present military junta has been a member of the board of directors of Pacsa since 1964.[9]

THE MULTINATIONAL MINING CORPORATIONS
AND THE "NATIONAL BOURGEOISIE"

Pacsa is but one example of integration between foreign and national capital. As already mentioned, there are some 80 to 90 smaller mining companies over which Cerro de Pasco has indirect control through board representation by board members of Cerro de Pasco or by its directly controlled companies.

That so many companies have come under the control of Cerro de Pasco has an explanation. Most of these smaller companies have no concentrating plants of their own and are hence completely dependent on the demand for copper-, zinc-, and lead ores by the leading producers. In 1968 there was in Peru a total concentrating capacity of 107,000 tons of ore per day. Of this capacity no less than 81,000 tons—or 75 percent—were controlled by Cerro de Pasco and SPCC.[10] This shows a so-called monopsonic relationship between most of the smaller mining companies and the giant multinational mining corporations: In practice there is only one buyer for the ores of the smaller companies.

Pressed financially, the smaller mining companies have to sell their ores at low prices to Cerro de Pasco. In order to finance the expansion of mineral production, the companies receive loans from Cerro de Pasco and Hochschild Limited. The latter firm is owned by the notorious usurer Mauricio Hochschild, who was one of "tin baron" Patino's associates in Bolivia until he was thrown out after the revolution of 1952.

The late Bolivian historian Sergio Almarez Paz has described Hochschild's activities in Bolivia. According to him, Hochschild found a way of acquiring smaller mining companies in Bolivia through purchases of their production, protected by obsolete legislation. In practice he converted the smaller mining proprietors into his own employees through a system of advance payments, commissioning, and price manipulations.[11]

In 1969, the smaller mining companies in Peru borrowed 153 million soles (about $4 million). Of this amount, $77 million were lent by Hochschild and $21 million by Cerro de Pasco.[12] Together these two companies thus accounted for 64 percent of the total lending. Gradually, as the companies have been unable to repay the loans, Cerro do Pasco and Hochschild have nominated representatives to the boards of the declining companies.

The situation in Peru today is thus that a growing number of the smaller mining companies are being incorporated into the Cerro de Pasco empire, partly with the generous aid of Hochschild. Through this incorporation the great majority of the Peruvian mining bourgeosie is also integrated with international monopoly capital at large.

Figure 11.2 shows the interlocking relationship of the major copper producers as of 1968. The mining corporations represented in the figure accounted for about 83 percent of world copper output in 1968. It seems that American Metal Climax (Amax) enjoys a kind of key position, which is not surprising since Chase Manhattan Bank (as mentioned, one of the flagships of the Rockefeller empire) has a strong influence over Amax. Some years ago, Chase Manhattan acquired Societe Belge in Brussels[13] through which it got a hold on the former Union Miniere du Haut-Katanga, the Zairois affiliate that has been "nationalized" by the Congolese state.[14]

The internationalization of the capitalist production process by the emergence of the multinational corporations has greatly limited the possibilities of a "nationalist bourgeoisie" to emerge in the so-called underdeveloped world. The only means for the native bourgeoisie and the state to take part in the productive process in these countries is in close alliance ("joint ventures") with the multinational corporations. The small independent native entrepreneur is confronted with the alternative of either being absorbed by international monopoly capital or being outcompeted and joining the ranks of the proletariat. This certainly seems the case in the Peruvian mining sector.

ACCUMULATION AND TRANSFER OF CAPITAL

According to official U.S. sources, North American mining corporations repatriated in the period 1950-60 about $790 million in profits from their subsidiaries in Peru, of which $669 million was repatriated during the 1960s. In the same period U.S. mining corporations invested $284 million in their Peruvian subsidiaries. Thus, on a net investment of $284 million the U.S. mining corporations repatriated almost three times as much profits. Yet these direct net investments

FIGURE 11.2

Major Copper Producers, 1968

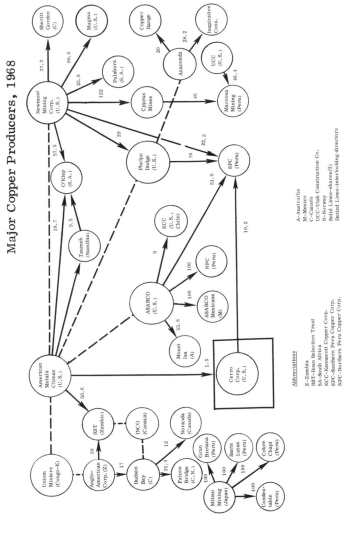

Abbreviations

Z—Zambia
RST—Roan Selection Trust
SA—South Africa
KCC—Kennecott Copper Corp.
SPC—Southern Peru Copper Corp.
NPC—Northern Peru Copper Corp.

A—Australia
M—Mexico
C—Canada
UCC—Utah Construction Co.
N—Norway
Solid Lines–shares(%)
Dotted Lines–interlocking directors

Numbers over arrows represent percentage share of shares held.

Note: See Figure 11.1.

Source: H. Espinoza, Concentracion del Poder Economico en el Sector Minero (Lima, 1970).

235

TABLE 11.4

Official Capital Transfers Between U.S. Mining Corporations and
Their Peruvian Subsidiaries
(millions of dollars)

	Net Capital from Mother Company to Subsidiaries	Repatriated Profits from Subsidiaries to Mother Company	Reinvested Profits in Subsidiaries
1950–54 (average)	16	12	1
1955–59 (average)	19	13	*
1960–64 (average)	3	40	*
1965	21	66	*
1966	–9	92	1
1967	43	72	1
1968	22	83	1
1969	38	102	2
1970	–21	54	—
1950–70	284	790	26

*Less than $500,000.

Sources: Survey of Current Business (U.S. Department of Commerce), October issues of 1968, 1969, 1970, and 1971, September issues of 1965, 1966, and 1967, August issues of 1961, 1962, 1963, and 1964; and U.S. Balance of Payments 1870-1961 (Washington, D.C.: Department of Commerce, 1963).

are considered part of the "aid flow" to the "underdeveloped countries," according to Organization for Economic Cooperation and Development (OECD) definitions! Of course, the repatriated profits are not deducted: If they were, the "aid flow" would turn out to be negative.

It is thus one of the myths of bourgeois economists that economic development in a country like Peru would result from foreign capital. The expansion of mineral output in Peru in the postwar period has primarily been the result of a ruthless exploitation of the Peruvian mining capital through which increasing amounts of surplus have been transferred to the metropolises of the capitalist world.

How big are the profits appropriated by the multinational corporations operating in the mining sector in Peru? Everything depends on how the accounting is made. Lindqvist has estimated that Marcona's profits in relation to invested capital from 1953 to 1968 vary from 29.7 percent to 132.6 percent, all depending on the accounting methods used.[15] With the aid of annual declarations submitted to the Peruvian authorities by the 64 most important mining companies operating in Peru, we have reconstructed the cost structure for the Peruvian mining industry for 1967 and 1969. The details of this analysis are shown in Table 11.5. But the official Peruvian figures for the repatriation of profits abroad do not agree with the official U.S. figures (Table 11.4). According to the declarations to the Peruvian authorities, the repatriation of profits totaled 2.3 billion soles (about $60 million) in 1969, whereas according to U.S. sources no less than $102 million were repatriated in that year.

How have some $42 million disappeared? Some kind of double accounting must be the explanation. What is scandalous is that this double accounting is obviously known to the Peruvian authorities. An investigating commission set up by the Peruvian parliament in 1966 made the authorities aware of the strange double accounting methods used by SPCC. The commission studied the income and expenditure accounts of SPCC for 1960-65. To the Peruvian authorities, SPCC had declared a net profit of $69.1 million in this period, but according to the U.S. Securities and Exchange Commission, the corresponding net profits amounted to $135.1 million—a difference of $66 million.[16]

Actually, it is not so strange that the Peruvian authorities close their eyes. The Peruvian parliament has itself adopted laws that legalize double accounting.

The importance of the depletion allowance is shown in Table 11.5, which shows the cost structure of the mining companies in 1967 and 1969. Thus, in 1969 the depletion allowance made up 7.9 percent of the sales of the sector as a whole, and 8.5 percent of the sales of the three largest U.S.-owned corporations. The depletion allowance can vary from case to case. It was 4.9 percent in the case of Cerro de Pasco, and as high as 13.5 percent in the case of SPCC. The reason is that SPCC is still allowed to make depletion reductions, in addition to generous depreciations, as a result of its investments in the Toquepala mine.

If the depletion allowance and the depreciation charges are instead calculated as a share of gross profits (or "cash flow" as the multinational companies so adequately put it),[17] these deductible "costs" represented 74 percent for Cerro de Pasco, 47 percent for SPCC, and 82 percent for Marcona. By calculating this cash flow as "costs, between half and three-fourths of the real profits are manipulated away,

TABLE 11.5

Cost Structure of Peruvian Mining Companies

	1967				1969			
	Three Largest Companies*		Other Companies		Three Largest Companies		Other Companies	
	(millions of soles)	(per-cent)	(millions of soles)	(per-cent)	(millions of soles)	(per-cent)	(millions of soles)	(per-cent)
Wages	839	8.3	503	15.2	1,070	6.8	715	13.5
Salaries	627	6.2	205	6.2	1,109	7.0	308	5.8
Energy	757	7.5	207	6.3	1,283	8.2	303	5.7
Peruvian taxes	613	6.1	201	6.1	2,504	15.9	750	14.2
Other costs	3,473	34.5	1,341	40.6	4,722	30.0	1,654	31.2
A. Total costs	6,309	62.6	2,457	74.4	10,688	67.9	3,730	70.4
Other deductible items								
Depreciation	1,180	11.7	242	8.4	1,665	10.6	641	12.1
Depletion	476	4.7	184	6.4	1,340	8.5	328	6.2
B. Total deductible items	1,656	16.4	426	14.8	3,005	19.1	969	18.3
C. Total (A + B)	7,965	79.0	2,883	89.2	13,693	87.0	4,699	88.7
D. Total sales	10,066	100.0	3,307	100.0	15,734	100.0	5,295	100.0
E. Gross profit (cash flow) = D - A	3,757	37.4	850	25.6	5,046	32.1	1,565	29.6
F. Net profit after tax = D - C	2,101	21.0	374	10.8	2,041	13.0	596	11.3
G. Repatriation of profits abroad	2,613	26.0	11	0.4	2,233	14.2	67	1.3

*Cerro de Pasco, Southern Peru Copper Corporation, Marcona Mining Company.

Source: The Annual Declarations of the Mining Companies to the Peruvian Authorities for the years 1967 and 1969 (unpublished).

and thus exempt from Peruvian taxes, which are calculated as a percentage of net profits.

COVERT TRANSFER OF PROFITS

The officially recognized repatriations of profits are so large that one might wonder why the mining subsidiaries receive any capital at all from abroad. According to Table 11.4, U.S. mining corporations operating in Peru invested $284 million from funds supplied by the mother companies in the United States, in 1950-70, while in the same period the subsidiaries repatriated, to the same mother companies, $790 million of profits. The reason behind the capital transfers from the United States to subsidiaries in Peru could not possibly be the need of hard currency (dollars) since the revenues of the subsidiaries are received in convertible currencies.

The real reason is quite different. The net capital flows (sometimes negative) from the United States to subsidiaries usually concern loans from corporations in the United States to their subsidiaries. (With loans from a bank or a state institution such as Eximbank, the transaction does not show up at all in the direct investment statistics but is accounted for as a "portfolio investment" in the balance of payments statistics.)

Is it really necessary for multinational corporations to lend money to their subsidiaries if the latter are able to repatriate annually three or four times the money borrowed? The reason is simply that this is one of many ways of covert profit transfers from the subsidiary to the mother company, because the subsidiary has to cancel the loan with interest. According to the annual declaration of Cerro de Pasco for 1969, the company had in that year an outstanding debt of $28 million to the mother company (Cerro Corporation in New York). In that same year, $1.6 million were accounted as interest on this loan.

One might also ask why these powerful multinational corporations that contribute to tidying up the U.S. balance of payments statistics have to seek loans from public institutions. Nevertheless both SPCC and Marcona had, according to their own statements, large debts to Eximbank. In 1969, SPCC paid more than $500,000 and Marcona no less than $3.4 million, as interest to Eximbank.

Still more curious, it seems that subsidiaries of multinational corporations find it necessary to resort to the national capital market for funds. But in 1969 Northern Peru (a subsidiary of Asarco) had a debt of 207 million soles (about $560,000) to the state-owned Banco Minero del Peru (a bank set up by the Peruvian state to help the smaller Peruvian mining companies). What interest could foreigners have in borrowing from a Peruvian bank that does not lend dollars but soles? One explanation might be that it is advantageous to have loans (debts) in countries that suffer from a chronic inflation. A devaluation of the national currency, for instance, leads to an automatic increase in the price level.

On September 4, 1967, the Peruvian currency was devalued from 26.8 soles per dollar to 38.6 soles per dollar. For the multinational mining corporations operating in Peru this meant that their costs paid in national currency, primarily wages, decreased by about 30 percent. One could say that a devaluation in this way constitutes a loan from the Peruvian state to the export industries in the country, a loan that never has to be repaid. The "costs" of the devaluation have to be repaid by the state through an increasing external indebtedness. Foreign companies may also take advantage of the devaluation in another way, for instance if they have debts to Peruvian banks before the devaluation occurs. During 1968, the price level in Peru increased by 20 percent, primarily as a result of the devaluation the year before. Companies that amortized their debts to Peruvian banks in 1968 could thus make a significant extra profit through this transaction.

In addition to the above-mentioned transactions, there are still other ways of repatriating covert profits, legally and illegally. These could be summarized as follows.

Illegal Exports of By-Products

We have already mentioned several reasons for the mining companies to export blister instead of refined copper. Blister contains so-called impurities separated only at the refining stage. These "impurities" contain such valuable by-products as gold, silver, molybdenum, the value of which can only be roughly estimated by customs officers. Of course, this question concerns illegal exports whose value is very difficult to estimate. But, in Chile, Chilean authorities now estimate that, after the nationalization of the U.S.-owned mining companies of Kennecott and Anaconda, these companies have illegally exported by-products in this manner for an average value of $27 million a year.[18] With respect to Peru, a North American author has said that "in that way a copper exporting concern can hide considerable gold in the ingots or ore which are shipped to the United States and elsewhere for refining."[19]

Double Invoicing with Respect to Imports

The huge multinational mining corporations are integrated not only horizontally but also, to a steadily increasing degree, vertically.

Most of the equipment and material used in the mining industry in Peru is imported. In 1969, the three leading U.S. mining comp nies in Peru bought goods for a value of 2, 889 million soles (about $75 million), of which foreign fabricated goods constitued 55 percent.[20] When machinery and equipment are from firms owned or controlled by the multinational corporation (or for instance, a used machine bought from the mother company) a deliberately overestimated price may be charged for the goods. In this way there is an increase in the "costs" of the subsidiary in order to reduce the taxable income.

Double Invoicing with Respect to Exports

Since the mineral output in Peru is largely exported from sub-sidiaries to refineries abroad that belong to the same corporation, it is very difficult for the Peruvian authorities to check that the exports on a certain day have actually been effectuated at the prevailing quota-tions on the world market (that is, the U.S. Producers' Price quoted in New York and the London Metal Exchange Price). According to an investigating commission set up in Peru in 1964 in order to study the accounting of Marcona Mining Company, the declared value of the iron export was much below the value on the world market. It has been estimated that the revenues of Marcona 1953-66 should be adjusted by about 23 percent to 31 percent to correspond to the real sales value.[21]

Technological Fees

One cost item for the subsidiaries of foreign corporations is "transfer of technology and know-how" from the mother company to the subsidiary. These payments include such diverse items as li-censes, royalties, and management fees and are reported annually under the heading "Direct Investment Receipts of Royalties and Fees" in Survey of Current Business, published by the U.S. Department of Commerce. Detailed branch and nation data are not available, nor are these data available in the companies' annual declarations, but it is obvious that there is no relationship whatsoever between these "technological payments" and the actual transfer of technology. In 1970 alone, $321 million were transferred from Latin America to the United States (from subsidiaries to mother companies), to cover such "transfer of technology."[22] But as one economist working in Santiago has recently stated, the much celebrated "transfer of technology" is one of the many myths justifying foreign capital. It is rather a ques-tion of the "commercialization of technology," as he says.[23] And even a Rand report concedes "it is conceivable that royalties and fees represent, in part, a covert repatriation of capital in addition as serving as a reflection of the transfer of intangible capital."[24]

EXPLOITATION MYTH?

For more than four centuries, enormous amounts of wealth have been appropriated by foreigners in Peru. The accumulated wealth has been transferred to the metropolises of the capitalist world, contributing to the expansion of the productive forces in the industrialized capitalist world.

Still many economists doubt that the continuing domination of foreign capital in the so-called underdeveloped world would be tantamount to exploitation in the Marxist sense. Many bourgeois economists and politicians hang on to old myths such as "employment effects" and "external effects." Since bourgeois economists frequently write on the explicit request of private industry this should perhaps not be surprising.

The real world is different, and the mining industry is a good illustration. The share of the workers in the value added of the mining sector as a whole (including petroleum) decreased from 24 percent in 1950 to 17 percent in 1970. In the same period, the share of net profits, depreciation, and depletion allowance has increased from 37 percent to 54 percent. It is true that the nominal wage in the mining sector has steadily increased since the war; but because of accompanying increases in the cost of living, the real wage has been stagnating. By 1970 it was 14 percent below the 1964 level (Figure 11.3).

The large multinational mining companies are not exploiting the Peruvian workers any less than the "national" companies. As shown in Table 11.5, in 1967 the average share of the workers in the turnover (sales) in the three largest multinational companies was about half of the average share of the other companies. And in all companies the workers' share was lower in 1969 than in 1967.

It is true that Southern Peru pays wages that are about 70 percent higher than the average wage in the mining industry, but this means that while the average wage per hour was about 19 soles (about 43 cents), the maximum wage, including social benefits, was 32 soles (about 73 cents) at Southern Peru (1969 figures).[25]

The miners at Southern Peru would, in the opinion of some economists, constitute the so-called labor aristocracy in the underdeveloped world. One author, for instance, claims that the disadvantage of the multinational companies investing in Latin America is not that they pay low wages but that they, on the contrary, tend to pay too high wages, "which might undermine their international competitiveness; in most cases they will be forced to close down, with a decline in em-

FIGURE 11.3

Wage Trend for Mining Proletariat in Peru
(index 1963 = 100)

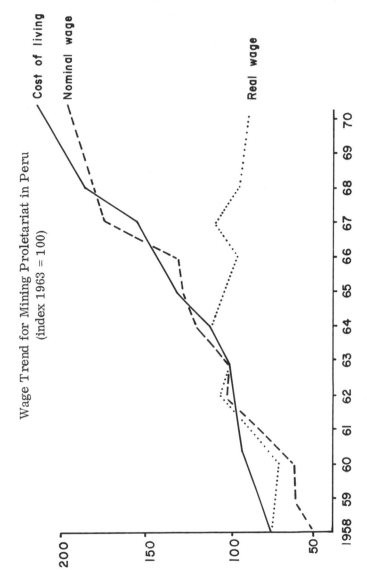

Source: Boletin del Banco Central de Reserva del Peru, March
1971, unpublished data from Banco Central de Reserva.

ployment as a result."[26] Such a reasoning seems to be based on the
tacit assumption that the wage level is too high in relation to labor
productivity. But is it really true that labor productivity is lower in
plants operated by subsidiaries in "underdeveloped" countries than in
plants belonging to the mother company in the metropolis?

At least according to data supplied by Southern Peru, this does
not seem to be the case (see Table 11.6). According to the annual
declaration of the company in 1969, there were 1,168 workers employed at the Toquepala mine in that year. These workers worked altogether 321,880 days that year. If they worked a legal maximum of
8 hours a day this would make 2.6 million hours worked. According
to the 1970 annual report of Asarco, 64,301 tons of copper were extracted from the Toquepala mine in 1969. This means a labor productivity of 25 kilos per man-hour. According to U.S. statistics, average labor productivity in U.S. copper mines (1963) was 19.4 kilos per
man-hour. With respect to smelting and refining, the average productivity in the United States appears to be higher than the productivity
at the SPCC installations at Ilo, but the latter is still considerably
higher than the average of such industrially advanced countries as
Japan, France, and Norway. It could thus be maintained that the low
wages at the U.S.-owned mining properties in Peru are justified by
low labor productivity is in fact an imperialist propaganda myth.

What about the "employment effect?" Do not the multinational
mining companies operating in Peru, after all, help maintain the level
of employment in Peru? The fact is that in 1969 Southern Peru, Cerro
de Pasco, and Marcona together employed only 10,237 mining workers,
or 15 percent of the total number of mining workers in Peru, while
these same companies produced 75 percent of the output. In the same
year, smelters and refineries employed 4,960 workers, of which the
earlier mentioned companies employed 100 percent—because these companies own all smelters and refineries in the mining sector (excluding
petroleum refineries and the state-owned steel-works in Chimbote).

Thus the real employment effect seems to have nothing to do
with the "employment effect" suggested by bourgeois economists. The
matter is rather that the multinational corporations through their imported capital-intensive technology (SPCC is a good case) contribute to
the worsening of the employment crisis. The alternative to foreign
direct investments is not, of course, no investments at all, which
seems to be the underlying assumption of those defending foreign direct investments, but nationally planned investments that contribute to
increasing employment and welfare in the country at large.

The exploitation of the proletariat is reflected by more than
these stagnating real wages and unemployment. Workers in the Peruvian mines are daily exposed to accidents and working conditions frequently leading to permanent invalidity or even death. In 1969 alone

TABLE 11.6

Cost Structure of Largest Multinational Mining Companies Operating in Peru, 1969

	Cerro de Pasco Corporation		Southern Peru Copper		Marcona Mining Company	
	(millions of soles)	(per-cent)	(millions of soles)	(per-cent)	(millions of soles)	(per-cent)
Wages	620	9.6	232	3.6	218	7.9
Salaries	475	7.4	257	3.8	383	13.9
Material and equipment	594	9.2	790	12.2	133	4.8
Energy	885	13.7	213	3.3	185	6.7
Peruvian taxes	499	7.7	1,841	28.2	164	5.9
Royalties	4	0.1	—	—	329	11.9
Amortizations and interests	63	1.0	134	2.0	131	4.8
Sales costs	132	2.0	125	1.9	146	5.3
Other costs	1,670	25.9	97	1.5	374	13.5
A. Total costs	4,942	76.7	3,683	56.4	2,063	74.8
Other deductible items						
Depreciation	539	8.4	444	6.8	427	15.5
Reserves	255	3.9	—	—	—	—
Depletion	315	4.9	881	13.5	144	5.2
B. Total deductible items	1,109	17.2	1,325	20.3	571	20.7
C. Total (A + B)	6,051	93.8	5,008	76.7	2,634	95.5
D. Total sales	6,447	100.0	6,530	100.0	2,757	100.0
E. Gross profit (cash flow) = D - A	1,505	23.3	2,847	43.6	694	25.2
F. Net profit after tax = D - C	396	6.2	1,522	23.3	123	4.5
G. Repatriation of profits abroad	588	9.1	1,522	23.3	123	4.5

Source: The Annual Declarations of the Mining Companies to the Peruvian Authorities for 1969 (unpublished).

there were 3,421 serious accidents in the Peruvian mines, smelters, and refineries, of which 65 were fatal. According to crass evaluations of the mining companies, these accidents corresponded to "the loss of 373,249 working days" for the companies.[27] But these are "losses" that the companies would find too embarrassing to show in their public propaganda, although they are eager to denounce the "losses" to the companies as a result of workers' strikes. These are also "losses" that are in fact not very cumbersome for the companies, since they can easily find new workers in the huge industrial reserve army of unemployed workers in Peru. This also applies to all those workers who never appear in the accident statistics: for instance, the hundreds who yearly die a "natural" death as a result of silicious injuries or tuberculosis. One seldom finds workers above 35 years of age in the mining districts. If they have survived to that age, they emigrate with incurable pulmonary diseases to the shanty towns down on the coast.

NOTES

1. Anuario Minero del Peru 1968 (Lima, 1970).

2. J. C. Carey, Peru and the United States, 1900-62 (Notre Dame, Ind.: University of Notre Dame Press, 1964).

3. Ibid., p. 171.

4. Anuario Minero del Peru.

5. Oiga, weekly magazine, January 15, 1971, p. 16.

6. Cerro Corporation, Annual Report (New York, 1970).

7. Ibid.

8. Carlos Malpica, Los Duenos del Peru (Lima, 1967); Cerro Corporation, Annual Report, 1969.

9. Ibid.

10. H. Espinoza, Concentracion del Poder Economica en el Sector Minero (Lima, 1970).

11. S. Almarez Paz, El Poder y la Caida (La Paz, 1967).

12. The Annual Declarations of the Peruvian Mining Companies to the Peruvian Authorities (unpublished) for 1969.

13. C. P. Kindleberger, Six Lectures on Direct Investments (New Haven, Conn.: Yale University Press, 1969).

14. K. Nkrumah, Neo-Colonialism: The Last Stage of Imperialism (London, 1965).

15. S. Lindqvist, Slaskuggan (Stockholm: Aldus, 1969).

16. Malpica, op. cit., pp. 174-81.

17. See the 1970 Annual Report of the Cerro Corporation.

18. "Anaconda se Robo los Sub-Productos del Cobre," Punto Fina August 17, 1971.

19. Carey, op. cit., p. 175.

20. Annual Declarations of the Mining Companies, op. cit.

21. Lindqvist, op. cit., p. 210.

22. Survey of Current Business, October 1971.

23. J. Vaitzos, "The Commercialization of Technology in the Andean Market," paper prepared for the Preparatory Meeting of the Group of 77 in Lima, October 1971.

24. L. Johnson, "The Course of U.S. Private Investments in Latin America since the Rise of Castro." Rand Corporation, prepared for the office of the Assistant Secretary of Defense, International Security Affairs.

25. Annual declaration, Southern Peru.

26. A. Kung, "For och emot utlandska investeringar i Latinamerika," Ekonomisk Revy, November 1971, Stockholm.

27. The Annual Declarations of the Mining Companies.

12

CERRO DE PASCO CORPORATION IS NOW CENTROMIN-PERU

Manuel Cabieses Barera

In October 1968 the Peruvian Armed Forces, headed by General Juan Velasco Alvarado, now president of Peru, engineered a coup d'etat, removed Fernando Belaunde's government, and began a reformist-nationalist process, which continues to show vitality.

We think we can rightly qualify the political process that began in 1968 in Peru as reformist-nationalist. On the one hand, the Peruvian government carries on transformations that imply important changes in the productive, distributive, and ownership structure of the country. Those who are affected by these changes are the ones who traditionally controlled the Peruvian society: foreign imperialist capital and the Peruvian intermediary capitalists and landowners. However, on the other hand, the government maintains the basic relationships of capitalist systems and the conditions necessary for the penetration of foreign capital.

At the end of 1969, the government signed a contract with the Southern Peru Copper Corporation (SPCC) to exploit the copper mine of Cuajone, an open mine, like Toquepala, and quite close to it. The contract was very advantageous to the SPCC and was much criticized. The government justified it by saying there was no other alternative; Peru could only get the necessary capital under these conditions. The Soviet Union, it said, had refused to help or take part in this deal. The following new measures, it announced, would be taken:

1. A new mining law would put an end to the lenient conditions of the previous law;

2. The state would recover all the mines given in concession to foreign companies, but not exploited;

3. A state mining company, Minero-Peru, would be created to work them; and

4. A state monopoly of trading and refining Minero-Peru's minerals would be established.

These measures were gradually carried out, but not completely. It cannot be said that the new mining law corresponds fully to national interests. It is, however, an important step in comparison with the previous conditions. Minero-Peru has not carried out all its plans and has made a lot of mistakes. However it is an important step forward.

This is the Peruvian military's way: bourgeois and nationalist, strong enough to create problems and irritate imperialism but not to defeat it thoroughly and expel it from the country. It creates hopes and illusions in different sectors of the population, but not enough to obtain an enthusiastic and militant support from the majority of the population.

The government is basically sustained by the institutions that gave it birth: the armed forces. All initiatives come from them and are discussed and approved by them. Inside the armed forces, there are clashes between those officers who think that the process has gone too far and those who think that it has not gone far enough and that more should be done.

On January 1, 1974, the Peruvian Military Government expropriated and nationalized Cerro de Pasco Corporation (CPC). The measure was taken in the midst of intensive negotiations with the U.S. government. After the nationalization of the International Petroleum Company in 1968-69, the Cerro de Pasco Corporation remained the major imperialist company in Peru.

In this chapter we try to show the meaning of the measures that changed Cerro de Pasco Corporation into Centromin-Peru.

WHY CERRO CORPORATION OFFERED TO SELL CPC TO THE PERUVIAN GOVERNMENT

The year 1971, when prices for minerals were very low, was a bad one for mining companies. But, for special reasons, it was worse for CPC. In 1970 the worker's unions had gone on strike, marched to Lima, and obtained important wage increases, granted by the government to obtain their support. In 1971, the unions became stronger and organized a new strike involving the 15 unions of each of CPC's working centers (mines, foundry, railways, and so on). One of the demands of the Workers' Federation was the nationalization of the CPC.

At the end of 1971, the CPC published its balance sheet, declaring its losses, and officially communicated to Peruvian govern-

ment that it wanted to sell its mining business. The government then
created a committee to study the situation of the company and the conditions in which Peru would eventually buy it. This committee worked
secretly for a year and a half, until June 1973.

Why did Cerro Corporation want to sell its mining business in
the central part of Peru?

Most of CPC's ores are worked by underground tunnels. This
system does not allow the use of machines but requires many workers. Therefore, capital productivity is necessarily low. In the mine
of Cobriza, the newest of its seven mines, exploited since the end of
the 1960s, CPC attempted an underground exploitation with a high organic composition of capital. This attempt failed, and CPC became
convinced that Peru offered better economic possibilities through the
exploitation of open mines.

Moreover, CPC had in mind the experience of Toquepala. This
is shown by Table 12.1. SPCC obtained four times more earnings than
CPC, using only 25 percent as many workers and only $25 million ex-
tra in assets. The U.S.-owned Cerro Corporation faced the following situation: In 1969 it had obtained U.S. $14 million earnings
through its 22.5 percent ownership of SPCC, while it obtained only
U.S. $10 million in earnings from its wholly owned CPC mines in
1971. CPC's initial investment in SPCC was lower than the total
profits it obtained from that source in 1969.[1]

When, in 1969, SPCC signed a contract with the Peruvian government for the exploitation of Cuajone, Cerro Corporation actually
arranged the agreement (through its 22.5 percent ownership of SPCC),
and from then on the government pressured strongly for financing the
necessary $500 million or more to get the operation started. Cerro
Corporation, at its level, had to respond to these pressures. Toward
the end of the 1960s, CPC reduced its investments in the mining industry in the center of Peru from 530 million soles in 1967, to only 240

TABLE 12.1

Comparative Data of Three Largest Mining Companies, Peru, 1971
(millions of dollars)

Company	Fixed Assets	Wages	Gross Earnings	Workers
Cerro de Pasco Corporation	192	20.3	10.2	13,170
Southern Peru (SPCC)	218	8.7	44.3	3,030
Marcona Mining Company	108	77.5	11.1	2,325

Source: Balance sheets of companies.

million in 1971.[2] Meanwhile in the United States, Cerro Corporation made new investments of more than $100 million.

The reason why Cerro Corporation wanted to sell is therefore very clear. The problem was at what price, under what conditions, and so on.

THE PERUVIAN GOVERNMENT EXPROPRIATES
AND NATIONALIZES CPC

In August 1973, as the Peruvian government was living through its second big crisis, and as the army and the navy were openly confronting each other on the future of the big press in the hands of the reactionary factions opposing the government, Cerro Corporation announced in New York that the Peruvian government wanted to confiscate CPC, paying only $12 million in compensation. It was the first public expression showing that Cerro Corporation and the Peruvian government were conducting conversations concerning the purchase and sale of the company. The government adopted a defensive position, informing the public that a committee was studying the operation.

At the end of September 1973, after the reactionary coup d'etat of the Chilean Armed Forces that overthrew the legal Government of Chile and murdered President Salvador Allende and while in Peru all governmental circles were under the shock of this news and in a defensive position, Cerro Corporation made a statement in New York. This time it announced that it was taking back its sale offer and interrupting negotiations with the Peruvian government, which, it said, wanted to confiscate the company, paying compensation the company could not accept.

The Peruvian government answered more strongly, calling CPC an imperialist company, responsible for the misery of its workers, and for the pollution of lakes and rivers in central Peru. It denounced CPC as stepping on workers' rights and so on. Finally, President Velasco himself declared that before the end of the year CPC would be transferred into the hands of the Peruvian state. This second public show of conflict between the government and CPC opened the way to direct conversations at government level between Peru and the United States.

These negotiations included not only the CPC but also a number of other North American countries that had been expropriated or were soon to be expropriated under terms that were the object of disagreement between the parties. The companies involved were as follows: the Peruvian branch of Cerro Corporation, incorporated in the state of Delaware; Sociedad Paramonga Ltd. SA (industrial complex: paper, plastics, chemicals); Compania Papelera Trujilo (paper); Cartavio SA (rum); Envases San Martin SA (prinnia) Pesquera de Ceishco SA, Compare Meilan SA, Gloucester Peruvian SA, and Gold Kist SA (fishmeal); Refineria Conchan-Chevron SA (oil refinery); and Morrison-Knudsen Co. Inc. and Zachary International Inc. (road builders).

In the midst of these negotiations, President Velasco announced, on December 31, 1973, the expropriation and nationalization of CPC, by decree of the government. No information was given on the price that would be paid or on the form this payment would take. Nothing was said either, as to whether these terms had been established in agreement with the U.S. government, or otherwise.

The disagreements between Peru and the United States on the subject of nationalizations were already public knowledge.

CPC had declared that the book value of its mining business in central Peru was $176 million. The Peruvian government had determined that the total assets value was 4.986 million soles—that is, some $124 million. Cerro had announced that the government wanted to pay it only $12 million. At some stage in the negotiations, the government mentioned the need of deducting allowances for pollution damages, for housing that should have been built by CPC and was not, and that the values of minerals in Peruvian soil could not be computed as CPC property. But at no time were the concrete sums mentioned. Finally, the agreement came into effect through a complicated arrangement of decrees by the Peruvian government and an International Agreement.

The agreement specifies the following:

1. The payment of $76 million for the whole lot of companies mentioned, including CPC. No sums were established for individual companies, the distribution of this sum being left to the U.S. government, dealing directly with the companies.

2. An additional payment of $74 million, of which $67 million was to go to the CPC. This sum of $67 million allegedly represented certain assets, like money in banks, earnings not distributed, and so on, that had not been expropriated because they were considered to represent extra assets not necessary for normal company activities.

3. A loan of $80 million to Peru by a consortium of U.S. banks, headed by the Morgan Guaranty Trust to enable Peru to pay the $76 million in compensation. The remaining $74 million were paid with the foreign currency reserves of Peru's Central Bank.

As a result of this agreement, the governments of Peru and the United States declared that "the Agreement seals definitively all controversy between both countries relative to the activities of the above mentioned transnational companies . . . [the Peruvian government wishes to develop] the best relations with the United States of America on the basis of equality, mutual respect and loyal cooperation."[3]

It is clear that this agreement (aside from formal friendly political statements) constitutes a mutual concession, after the deterioration of Peruvian-U.S. relations that can only be compared to what
followed the nationalization of another traditional imperialist company
in Peru: the International Petroleum Company. Although on this concrete issue, the main conflicts have been solved, important contradictions remain between a government that attempts to establish new relations with U.S. imperialism and the U.S. government, which are
trying to limit Peru's elbow room.

CONCLUSIONS

The nationalization of CPC and its transformation into Centromin-
Peru means a partial reduction in the presence of foreign capital (especially U.S.) in Peruvian mining. The Peruvian mining production
structure is highly monopolistic and the foreign capital appears not
only in the form of wholly owned branch companies but also in the form
of "joint ventures" in which native capital is absolutely dependent.
Furthermore, the government, pursuing a rapid increase in copper
production, has signed new contracts with foreign companies, including the one with SPCC for the exploitation of the rich mine of Cuajone.
We find here a feature that appears throughout the Peruvian
economy and that marks the terms and limits of the contradiction with
imperialism: The influence of monopolistic companies and imperialism
have been reduced, but at the same time, there appears a new relationship, more structured, which could be even more difficult to break.
With the nationalization of the CPC, the role of the state as a direct agent in mining production is strengthened, if we consider the
strategic position that CPC, which now belongs to Centromin-Peru,
occupies in the mining sector. This productive activity will be able to
complement and strengthen the relative control that the state exerts
on commercialization. On the other hand, middle-term plans for the
sector indicate an important increase in copper production (it is expected to increase from 250 billion tons in 1973 to 500 billion tons in
1975). Thus, the development strategy of the Peruvian government,
in which the state plays a most important part, is consolidated by the
nationalization of CPC.

NOTES

1. Cerro Corporation Report, 1971.
2. Ibid.
3. Prime Minister Mercado, statement.

13

THE REFORMS OF
COPPER MINING AND
THEIR ENFORCEMENT
Luis Pasara

From the very beginning of the political process commanded by the Peruvian armed forces, in October 1968, the theme of mining reform has been stated repeatedly. President Juan Velasco Alvarado pointed out in July 1969,

> The mining field is of an enormous interest for the
> country, because national development is intimately
> linked with the development of mining. The exter-
> nal sector of our economy will depend primarily
> upon the export of metals with the highest degree of
> processing our industrial development allows. . . .
> The Revolutionary Government has often expressed
> its decision to support the investor who comes to
> our country and works respecting our laws. Within
> this plan, there are numerous requests, for invest-
> ment coming from big enterprises with whose repre-
> sentatives we are discussing the conditions of opera-
> tion, negotiations that are still going on in an effort
> to reach agreements that can satisfy legitimate na-
> tional aspirations. In this situation, it would have
> been very easy to announce to the Nation the signing
> of several contracts, if these had been written in
> conditions similar to those accepted by preceding
> Governments; but we could not have acted that way,
> because it would have meant loss of national wealth
> belonging to future generations.[1]

Since then, several legal and administrative measures have been enacted: The objectives have been to change the rules of the game and directly enforce the new policy affecting certain interests. New state agencies, the most important of which are a new Ministry of Energy and Mining and a new state enterprise, Minero-Peru, have been created.

The president of the republic synthesized this legal order when he said,

> The General Mining Law consecrates the permanent right of the State over all mining resources of the country. For the first time, it takes effective control of the country's abundant mining resources, and reserves for the State the refining of metals and the commercialization of products. In the same way, the law establishes that the mining resources cannot remain unexploited indefinitely, using the traditional system of payment for a simple territorial rent. Now, mining enterprises are obliged to invest in the stages of exploration, preparation and initiation of the mines' production; stages for which the Law sets very definite and precise limits in relation with the volume of minerals of each deposit.
>
> The Law establishes that the enterprise activity of the State is an essential factor in mining industry development, a principle that supports the creation of Minero-Peru, the big state mining enterprise, as well as the establishment of rules and regulations that should guide the enterprises in which the state will participate together with the private sector.[2]

Our approach here is to emphasize two main aspects of the new legal order. On the one hand, our "law in action" approach tries to contrast the purposes of the new policy with the effective, practical results achieved, at least in the most important aspects. On the other, we want to suggest some general conclusions drawn from the Peruvian experience concerning (1) the overall significance of copper policy in maximizing benefits to the nation and (2) the problems and limits of implementing mining reforms in a dependent country.

We will divide the analysis into four main themes. The first three refer to the different stages of the copper industry: mining, refining, and marketing. The fourth is worker participation, a declared objective of the Peruvian government.

MINING

A first important element to be taken into consideration is the fact that in Peru the majority of the mineral resources are not yet being exploited. Contrary to the case of Chile, for instance, the copper deposits that are being exploited in Peru constitute a relatively small part of the total amount of the available resources. Mining policy has had to contemplate two completely different aspects: the unexploited and the exploited deposits.

The general policy has been to concentrate in the state responsibility for exploiting large-scale mines. Most present and potential copper mines fall into this category.

Reversions

The first measure of the new mining policy is aimed at establishing the juridical principle called, in the mining doctrine, "protection for labor"—that is, a concession is only legitimate if it produces or heads toward production.

The new measures required private enterprises to submit for approval of state authorities timetables of work on their concessions. Enforcement of these rules has meant that numerous copper concessions reverted to the state, which has rapidly reserved them to be directly exploited by Minero-Peru on its own or in joint ventures with foreign enterprises—for example, Antamina, a very important copper deposit formerly held as a concession by a private firm and now to be exploited by a mixed enterprise formed by the Romanian and the Peruvian governments, to which we will shortly refer.

State Exploitation

Acting directly or through mixed enterprises, the Peruvian state is developing a series of mining projects. Cerro Verde is among the biggest, with a reserve of 800 million tons of copper that require an estimated investment of U.S. $105 million.[3]

Financial requirements and technological needs are given as justifications of Minero-Peru's joint ventures with foreign enterprises. Two formulas are used: mixed enterprise (with 51 percent of the stocks belonging to the Peruvian state) and management contracts. In the first case, the participation of foreign capital takes over the risk of exploitation and commits resources directly, foreign functionaries being in charge of technical decisions. A contract of this type is the one signed with the Romanian government to exploit the Antamina deposit. The Antamina Special Mining enterprise has foreseen an investment of $70 million for an annual production of 40,000 tons of copper (and the same amount of zinc). Minero-Peru owns 51 percent of the capital of the enterprise. The rights of exploitation of the deposit have been determined from the outset, a situation that constitutes a novelty in the regime of these contracts.[4]

Under the other formula of management contracts, used in the case of Cerro Verde, Minero-Peru is the only stockholder of the enterprise but has contracted for the provision of financial services and management by a foreign enterprise. British, Swedish, and Japanese enterprises have participated in this model. It seems that this intervention is informally suggested by the financial sources, which demand a "sound administration" to administer their loans. As a consequence of these demands, the state is very seldom able to carry on the exploitation on its own, without resorting to associations with foreign enterprises, be it by way of mixed enterprises or management contracts.

One of the few cases of exclusive state exploitation is that of Tintaya. This is probably explained by the smaller size of the deposit: 50 million tons of reserves and an investment of U.S. $48 million.[5]

The lack of capital and the need of acquiring technology combine to demand from the state a foreign administration vested with a state garment. It is important to determine the orientation of this foreign presence—that is, whether or not it has a transitory character. Available information is not adequate enough to answer the question. There probably exists an important field of maneuver for the Peruvian state, which has not been sufficiently exploited, but we do not have the available verified information to prove that other formulas could be used without foreign participation. Obviously, international financial sources are directly linked to the interests of multinational mining enterprises; thus, the former exercises pressure for the incorporation of the latter. Hence, finance, technology, and foreign administration arrive in the same package, even though they might appear, formally, through different institutions. This picture shows one of the factors limiting the maximization of benefits through state exploitation as substitute to exploitation by foreign enterprise.

State Control

It is obvious, though it has never been explicitly established as an objective of mining policy, that it is the purpose of the Peruvian government to have all big mining in the hands of the state. For this purpose, it is not enough for Minero-Peru to exploit the new deposits.

All present mining exploitations should pass to the control of the state exclusively, or under the mode of mixed enterprise (legally called "special mining enterprise"). The two biggest enterprises that produce copper in Peru have been Cerro de Pasco Corporation and Southern Peru Copper Corporation. The first of these was acquired by the state, and a new state enterprise was created (Centromin-Peru).

The opposite case seems to be represented by the Southern company, which not only keeps its rights in Toquepala, acquired under the preceding legal regime, but also received in December 1969 a new contract to exploit the rich deposits of Cuajone, before the government approved the new rules of the game under which the state reserves refining and the marketing for itself. This contract has been considered by leftist critics as a sign of the government's conciliatory policy toward imperialism.[6] Nevertheless, everything seems to indicate that the contract of Cuajone, with all its importance, does not reflect the general policy of the regime. It is too early to anticipate whether a new nationalization could be decided in the future or whether the state would associate with Southern in a joint venture like the one it is negotiating with the third biggest mining enterprise, the Marcona Mining Company, for the production of steel. What is very clear is that the government signed the Cuajone contract at a time when financial resources and private investment in general were hard to attract to Peru. It is true that this is not the model operation of the regime, but the government will probably have to respect it, given the desire to exploit the deposit as soon as possible and the financial interests already committed to it. The $550 million needed for this operation is provided by a consortium of banks, headed by the Chase Manhattan, the Eximbank, and a group of purchasers, which will be paid with the product of sales of the first 15 years.[7] In these terms, it seems difficult to anticipate a nationalization.

For the time being, what the government has tried to do with enterprises like Southern and Cerro, until the latter was nationalized, is to cut off the mechanisms that allowed them to take the profits outside the country. The depletion allowance has been eliminated from the tax benefits, and financial controls over these enterprises are rather severe. Nevertheless, the presence of foreign enterprise creates certain mechanisms that have already been analyzed. Brundenius[8] lists imports of material and of machinery from the mother firms and the effect of loans from those same enterprises, and the payment of royalties and other deductions, as the main forms of exportation of profits, which are inherent to foreign enterprise. A different factor is marketing, which in the hands of foreign enterprise meant an additional way of extracting profits by way of the undervaluation of exports; later on, we will talk of the reform introduced in this stage.

REFINING

A little less than one-fourth of the total production of copper is refined in the country in two refineries, owned by the biggest enterprises Cerro (today Centromin-Peru) and Southern. The rest is processed in refineries outside the country.

The consequences of this situation are obvious, and a solution is being sought. A state refinery is being installed in Ilo by Japanese contractors and should start operating by the end of 1975. [9] Even with this new refinery, the majority of the copper to be produced in 1977, an anticipated 565,000 tons, [10] will still be exported unrefined. On the other hand, it should be noted that the Southern refinery constitutes an "acquired right" and consequently is an exception to the general principle that refining should be restored to the state. In the case of Cuajone the minerals extracted may be freely disposed of by the enterprise, thus permitting the foreign firm to make decisions concerning refining. *

We do not mention use of refined copper in national industry, because the supply already largely surpasses the needs: This is a characteristic of being an underdeveloped country. The aim of building an enormous industrial complex around the future state refinery of Ilo seeks to overcome this aspect of the problem, using copper that is refined there and looking at the Andean and Lafta markets of the finished output. Even this, because of the reduced national markets, resulting from historical internal domination in our countries, is not large enough to realize the economics of a scale essential for a major industrial complex. [11]

MARKETING

As has been pointed out previously, the new legal order aims to govern the marketing by the state of all the minerals the country exports (except for the ones extracted from Cuajone). Such an expecta-

---

*All the preceding shows that the stage of refining is highly vulnerable for the national maximization of benefits coming from copper exploitation. On the one hand, the aggregated value is much less in the case of exportation of concentrates and blister. On the other hand, the capacity of the Peruvian government to maneuver and manipulate—in the world market—over its production is less when an essential part of the productive process does not depend upon decisions controllable by the state itself but by enterprises outside the country.

tion appears exaggerated if one examines the concrete modes of commercialization being used by the state firm Minero-Peru.[12]

Let us examine the three forms of marketing used by the state. The first of them corresponds to the contracts already existing when the new legislation was enacted. The state enterprise has intervened in the contractual relation between producer and purchaser, with the objective of supervising the terms of the contract and of watching over its fulfillment, so as to control possible evasions through undervalued exportation. In reality, Minero-Peru does not itself handle marketing but assumes control of the preexisting contractual relationship. The first form is called "three parties contract," making reference to the state's intervention, by which the state enterprise charges a commission of 0.5 percent on the agreed price. This cannot be compared to a commission, since Minero-Peru does not play the role of marketing agent, but it is equivalent to an additional tax. In many cases, contracts have been made many years ago; this means, for example, that all the production of the state company Centromin-Peru (formerly, Cerro de Pasco Corporation) is being actually sold by a marketing agent, the Cerro Sales Company.

Due to the persistence of many contracts previous to the present legal regime, this form covered 40 percent of the total value of sales of state enterprise in 1973. (Copper constituted 73.7 percent of the total value of the sales of minerals of Minero-Peru that year.) The second form of state intervention is the "sale to Minero-Peru through agents." In this case, the state firm allocates the copper production through "brokers" on the world copper market, paying commissions that vary between 0.5 percent and 2 percent. The state receipts here are bigger than in the preceding form (third-party contract), but the marketing mechanisms are identical. The only difference lies in the fact that Minero-Peru has the initiative of choosing the intermediary agent, thus adding itself to the relation that the agent established directly with the producer before. This form comprises 23 percent of the value of international sales by state enterprises in 1973.

The third form of intervention used by Minero-Peru is direct selling. In effect, government-to-government contracts are made that do not go through the usual mechanisms of world market. This type of intervention has served, especially, for the sale of large amounts of copper to socialist countries. The sale enterprise subtracts a commission for the sale and pays the rest to the producer. In 1973, the proportion of this form increased to 30 percent of total sales, although the percentage of copper seems to have been lower.

Table 13.0 shows the incidence of commissions of intermediary agents and of Minero-Peru on the total amount of foreign sales made in 1973.

TABLE 13.1

Destiny of Product of Sales to Exterior, 1973

	Value (in soles)	Percentage
Sales to the exterior (total value)	17,839,309	100.00
Commissions received by agents	44,220	0.25
Price paid to producer	17,454,076	97.85
Total income Minero–Peru	341,033	1.90

Source: Statement of gains and losses of Minero–Peru, 1973, unpublished.

No doubt, the tendency is toward direct marketing by the state enterprise. Nevertheless, internally there seem to be several different possible forms of carrying out this responsibility. In 1976, the old contracts in which Minero–Peru has introduced the "third-party" modality will disappear. The most radical approach is that the state enterprise should sell directly to the consumer, which implies operating without agents or intermediaries in the world market. A second approach is for the state enterprise to act as the sole purchaser from Peruvian producers, but to contracting with agents for sale of the minerals abroad. And a third proposal is that state enterprise assume the role of monopolistic intermediary between national producers and a selling agent associated with the state enterprise on the world market.

The first possibility implies entering the world market in terms of monopoly of Peruvian foreign trade of copper. This position also implies that Peru play an active role among copper-producing countries. It is obvious that this last would seriously suffer if Minero–Peru continued acting through agents in the world market, where the possibility of speculation within the limits of offer and demand really lies.

There is an additional element we would like to introduce in the discussion of the marketing problem and the role that might be played by the state mining enterprise—the fixing of prices. The policy followed through 1974 has been to take the world market selling price effectively obtained and pass it on to the producer, minus a commission for Minero–Peru (and the other selling agent, if there was one) for its intervention in the copper trade. This policy is different from that followed by the Peruvian state in other productive sectors in which it monopolizes marketing—for instance in the agrarian sector, which includes export products. The state sets the price of purchase to the producer as a function of its costs, plus a reasonable profit; the difference obtained by the state in its marketing activity (which is influenced by a series of speculative factors, including the force of monopolization of supply by the state) is its own profit. The fact that the opposite way is being used in the mining sector, and particularly with the marketing of copper, means that the state limits itself to being an intermediary between producer and purchaser in return for a commission but does not really exercise monopoly for its own benefit.

All this leads us to three different levels of the problem of state marketing. A first form of state performance, which is the predominant one in the activity of Minero-Peru right now, limits itself to mediating between producer and selling agents of the world market; by the old contracts, this performance only implies direct control over possible machinations on the contracts of sale to the exterior and secures to the state's small margin of additional benefit, as intermediary, which is equivalent to an additional tax.

A second level is the one stated by some officials of Minero-Peru demanding that the state enterprise act on its own in the market, without using agents and, together with the rest of the countries of CIPEC, be able to impose prices on the world market by exercising substantial monopoly power. At this level, the state would not be a simple intermediary but would try to maximize benefits, speculating with its package of production of copper and challenging the world market to the extent possible.

A third level, which we have not found in discussion within Minero-Peru, is the possibility that the state would set its purchase prices as a function of costs of production and would subtract the rest of the returns from sale at the world price to contribute to national accumulation for the process of development.

This is extremely significant because private enterprise continues to be a producer of copper in the country, and there are signs that make us predict it will continue to be so in a certain proportion. Furthermore, in the mixed enterprise, there is presence of foreign capital. If the state, in its marketing activity, remains satisfied with only charging a commission as intermediary, the monopoly of marketing directly benefits the producer—partly foreign—and not fundamentally the state. This would render illusory the conception of state marketing as a mechanism to guarantee national maximization of benefits.

WORKER PARTICIPATION

One of the most frequent themes in the political terminology of the Peruvian Military Government is economic, social, and political

participation.[13] At the level of enterprises, an institution to put this
social goal into practice has been constituted, gathering all the workers (stable full-time workers in the enterprise, Decree Law no.
18880, Article 277) with the goal of their participating in the ownership of the enterprise and, consequently, in profits and management.
The participation in ownership is progressive and is a function of a
percentage quota of the annual profits of the enterprise.

Eighty percent of the enterprise receipts, in both levels of participation (for cash distribution and for enterprise shares), is transferred to a compensation community that gathers all the entrepreneurial base communities (Law no. 18880, Articles 282 and 287). This
compensation community operates as a mechanism of redistribution
between workers of the different enterprises seeking to reduce worker
income differentials attributable to differences in profitability or productivity. Nevertheless, this principle operates only partially within
each enterprise, since half of the net profit is distributed among the
workers proportionally to their salaries (Decree Law no. 18880,
Article 283)—that is, the one who receives more salary receives a
higher share, a principle that worsens and stresses the existing differences of remuneration. In the case of the mining sector, the community participates with 4 percent of the net profit of the enterprise
as direct cash distribution to workers and 6 percent for participation
in ownership (Decree Law no. 18880, Article 281). This means that
4 percent is shared among workers as an added salary, whereas the
other 6 percent is dedicated to reinvestments of the enterprise, or
if these are absent, to the acquisition of shares of the enterprise that
are collectively owned by the workers. This mechanism of participation in the net rent of the enterprise, according to law, will allow the
workers to acquire up to 50 percent of the social capital of the enterprise.

As has been pointed out, the mining community, to the extent
that it owns share capital of the enterprise, will participate in its
management through the board of directors to which it designates its
representatives (Decree Law 18880, Article 296). In the case of
state or mixed enterprises, the community does not receive shares of
the enterprise, but bonds or shares in the state development bank
(Cofide). Consequently, its participation in the board of directors is
fixed at two representatives (Decree Law no. 18880, Article 295).

This last point can be of special importance when we try to
analyze the implications of the creation of the mining community. In
the context of the mining workers' movement, the mining community
is an unrequested participation—that is, an "undemanded demand" of
the workers who, undoubtedly, are among the most organized of the
Peruvian proletariat. This may explain the relative skepticism with
which this new institution has been received. Several trade unions

have denounced it as an instrument of unacceptable reconciliation with imperialism.

The arguments used against the mining community may be summarized as follows. First, the high capital of mining enterprises and their plans for reinvestment definitely postpone the opportunity for the community to reach significant share of ownership. Second, and as we have noted in other parts of this chapter, the bookkeeping of enterprises allows management to deduct as costs such items as royalties and interests on loans, so the net profit that appears in the balance and in which the community may participate is nothing but a part of the effective profit. Third, the leaders of the communities can become competitive with trade unionists and, in the long run, weaken the workers' movement. Fourth, the community has not been demanded by the workers, who, instead, have called for nationalization of big mining and its conversion into state enterprises.

Some of these arguments have proved valid in practice. In two years of operation of the new system of participation, the mining communities have acquired only tiny percentages of participation in the capital of their respective enterprises, which symbolize the participation of workers' delegates in the decision-making bodies of the enterprises. Marcona Mining Company and the Cerro de Pasco Corporation declared losses, not profits, after the law was approved, which strengthens well-founded doubts about the truth of their balance sheets and bookkeeping methods. On the other hand, if leaders of communities have not openly adopted an anti-trade-union attitude, their political behavior seems not to have the combative tone of trade unions. The leaders of communities seem to be in the process of being coopted by political agents of the government and the responsible bureaucrats of the sector. In certain political situations, like the nationalization of the Cerro de Pasco Corporation, the mining community has played an important and active role within the tactical action of the government, while the trade union presented their own alternatives, demanding nationalization without payment, which was considered as a "provocative demand" by the government and the leaders of the mining communities.

In relation to the redistributive effects of the mining community, the figures for the first 18 months of practice of the new system (1971-72) indicate that the total amount of net participation (4 percent of the net profit of enterprises) was 180 million soles.[14] Since this was an amount shared by the 50,441 workers of the mining sector under the regime of community[15] we find a participation of 3,600 soles (U.S. $82) a year for each mining worker. This amount is only an average, since half of the sum is to be distributed in direct proportion to the salaries, which is equivalent to saying that technicians, professionals, and white-collar workers received much more than that sum and the blue-collar workers received much less.

TABLE 13.2

Net Profits of Enterprises of Sector
(in million soles)

Year	1967	1968	1969	1970	1972
Net profit	4,021	4,943	7,481	7,411	4,500

Source: Daniel Rodriguez Hoyle, La Mineria Metalica Peruana
(Lima: Sociedad Nacional de Mineria, 1971).

The 180 million soles of net participation represents 4 percent
of the profit of the enterprises in the sector; we can project the total
amount of the net profit as 4.5 billion soles in 1972. In Table 13.2,
we compare profits of the latest years with this last figures.

We can verify an obvious coincidence between the beginning of
functioning of the community, through which workers should participate in profits, and a clear reduction in the net profit. Adding local
information permits doubt about the declared returns of the enterprises in 1972. In 1970, the social capital of the mining sector was
14,673 million soles.[16] If we project this figure, using the rate of
growth of capital over the last four years, we may moderately estimate that the social capital of the sector in 1972 was of 18 billion
soles, which, compared with the 4,500 million soles of net profit of
the sector that same year, indicates a profit rate of 25 percent. If
we compare this figure and those corresponding to the last years, we
find the results outlined in Table 13.3.

TABLE 13.3

Rate of Profit of Mining Sector
(percent)

	1967	1968	1969	1970	1972
Rate	57.31	46.54	67.7	50.51	25.0

Source: Daniel Rodriguez Hoyle, La Mineria Metalica Peruana
(Lima: Sociedad Nacional de Mineria, 1971).

This profitability, suddenly reduced after the mining community came into existence, implies participation of the community in 2.5 percent of the capital of the sector. One percent is distributed in cash, and only 1.5 percent represents the participation of community in the capital of private enterprises in the mining sector.

CONCLUSIONS

We would like to summarize some of the elements of our evaluation of the "Peruvian model" of copper exploitation, trying to indicate the more general issues that relate to the problem of copper exploitation in underdeveloped countries.

The reforms that may be introduced in the regime of exploitation of copper have to be placed in the context of external dependence. This situation, which implies a domination of our economies from the hegemonic centers of world capitalism, goes beyond political will or the autonomous decision process. Dependence penetrates our economies, makes us exporters to a world market controlled by imperialism. In this way, finance and technology are managed by the multinational enterprises that impose their rules over copper exporting in the country and control it, whether the level of production is formally controlled by nationals or not.

As a consequence, copper-exporting countries cannot, at this stage of their development, do anything but negotiate with imperialism to try to maximize benefits for the nation and accumulate surplus and hence finance its own process of development and liberation. It becomes an obvious requirement for this renegotiation to be done in agreement with the other underdeveloped countries in CIPEC.

Even the margin of the possible does not seem to be employed in the best way by the "Peruvian model." This is clear in the prolonged or renovated foreign presence in the stage of exploitation, in the intermediate conception of marketing under the state, and in the timid or obscuring mechanisms of workers' participation. Political objectives, even the new principles of the mining policy, shaped in legislation, are "translated" into regressive ones through the concrete measures of implementation.

In the outline of legal mechanisms, as well as in the policies implemented, we can find the persistence of mystical abstract nationalistic patterns, which are very inadequate in relation to the concrete need of effectively maximizing national benefits. This is illustrated by the role of the state. The discretionary power given to the state, if managed by a technobureaucracy over which there is no popular control, as in the case in Peru, does not maximize national benefits. The state's intervention in marketing clearly shows how, under the garment of the state, a mechanism is established that benefits the producers, including powerful multinational enterprises.

Summarizing, the myth of nationalization and transfer of ownership to the state should be revised to incorporate the concrete meaning of class interests that the state apparatus manifests and serves. This point, no doubt, drives us to the political analysis of the society as a whole, in order to identify the specific meaning of formal legal measures in the field of mining policy.

NOTES

1. Velasco, La Voz de la Revolucion, Table 1, p. 72.

2. Velasco, op. cit., "Mensaje a la Nacion del 28 de Julio de 1971," Table 2, p. 131.

3. Peruvian Times 33, no. 1719 (December 14, 1973): 3.

4. El Peruano, September 19, 1973.

5. Peruvian Times 33, no. 1711 (October 19, 1973): 8.

6. Anibal Quijano, "Nationalism and Capitalism in Peru: A Study in Neo Imperialism, Monthly Review Press, 1972.

7. Peruvian Times 24, no. 1736 (April 18, 1974): 3.

8. See C. Brundenius, Chapter 11.

9. Peruvian Times 33, no. 1706 (September 14, 1973).

10. El Peruano, September 19, 1973.

11. Peruvian Times 33, no. 1706 (September 14, 1973).

12. The following information has been obtained through personal interviews with Minero-Peru officials and through unpublished documents of the state enterprise. Obviously, we have put quotes aside.

13. Compare Velasco, ed., Participacion (Lima, 1972), 2 volumes.

14. El Peruano, September 19, 1973.

15. Dela, Oficina de Comunidades Laborales del Ministerio de Energia y Minas, no. 1, 1972, pp. 14-15.

16. Daniel Rodriguez Hoyle, La Mineria Metalica Peruana (Lima: Sociedad Nacional de Mineria, 1971).

14

THE STRUCTURE OF MULTINATIONAL CORPORATIONS IN ZAIRE

Mulumba Lukoji

The purpose of this chapter is to analyze the forms of investment chosen by the foreign companies that have recently joined Gecamines in the exploitation of Shaba's copper mines. The chapter is concerned partly with the mining companies of Tenke-Fungurume and partly with the Company for the Industrial Development and Mining of Zaire (Sodimiza).

THE MINING COMPANIES OF TENKE-FUNGURUME

In a mining contract signed September 19, 1970, Zaire gave a five-company consortium of American, British, Japanese, and French interests the rights of exploration for and exploitation of mines in the Tenke and Fungurume regions in eastern and southern Shaba; the companies were Amoco Mineral Company (United States), Charter Consolidated Limited (Great Britain), Mitsui and Company (Japan), Leon Tempelman and Son (United States), and Le Bureau de Recherches Geologiques and Minieres (France).

With the participation of the Government of Zaire, this consortium formed two mining companies, La Societe International des Mines du Zaire (SIMZ) and La Societe Miniere du Tenke-Fungurume(SMTF).

Since the purpose of SIMZ was only to effect geological studies for the consortium, SMFT was to exploit the mines already discovered. Of 10 zones for exploration (totaling 30,692 square kilometers) granted in 1970 to SIMZ, half were returned to the state in March 1974.[1] By 1975, the total expenses of exploration will be about 500,000 Zaires ($2 equals 1 Z).

SMFT had exclusive rights of exploration in a zone of 1,425 square kilometers, which have returned to the domain of Gecamines.

271

The construction of the mine began this year; exploitation should begin in 1976 or 1977. The annual production will initially be about 100,000 tons. It should return on the company's initial investment of 2 million Z about 150 million Z.

Although there are some minor differences, the agreements about the association between the Zairois state and the consortium of foreign financial interests are the same for both mining companies. In effect, the identity is absolute concerning general financial conditions and the promises the state has given the companies. Now we will examine the contract and the terms of the financial and fiscal institutions.

The corporate capital of SIMZ and SMTF are owned by Zaire and the foreign consortium in the following proportions:

| | SIMZ | | SMTF | |
	Shares in Zaires	Percent of Total Shares	Shares in Zaires	Percent of Total
Zaire	75,000 free shares	20.00	400,000	20.0
Amoco	106,363	28.36	560,000	28.0
Charter	106,364	28.36	560,000	28.0
Mitsui	53,182	14.18	280,000	14.0
Tempelman	7,500	2.00	60,000	3.0
BRGM	26,591	7.09	70,000	3.5
Omnium de Mines (France)	—	—	70,000	3.5
Total	375,000	100.00	2,000,000	100.0

The new corporations receive certain tax privileges. First, in Article VIII of the contract made September 19, 1970, it is agreed that from their commencement of operations until 20 years after the mine has opened, the SMTF and the SIMZ will be exempt from "all proportional and/or fixed tariffs, taxes, duties and laws of export and import which now exist or which may be created during this period." The same holds true for personal taxes on dividends and interest on the loans. The companies are free from compensatory taxes accruing to the institution and to the renewal of mining rights and from all contracted duties. Then, from the date of signing the contract until the end of the fifth year following the start of mining, the companies are exempt from all taxation. Finally, from the sixth year to the end of the 20th following the opening of each mine, the companies will not have to pay contracted taxes exceeding 50 percent of the net taxable profits; 40 percent of the vocational tax on profits; and 10 percent of the special tax on profits.

These important financial and fiscal concessions made in the contract signed September 19, 1970, were doubtless claimed because of the high cost of geological research and preparations for mining. One might also mention the financial difficulties of the copper industry because of the steady drop in the world price of copper.

In the following section, which concerns Sodimiza, we have tried to understand the constraints that the evolution of the price of copper imposes on the economic feasibility of the mining activities of Sodimiza.

SODIMIZA

October 2, 1972, two months before the agreed time, the smelting plant of the mining company Sodimiza was inaugurated. Here we will examine (1) the legal structure of Sodimiza; (2) its technology; and (3) the conditions necessary for its economic survival—all of which are needed to compare Sodimiza with the other mining companies of Zaire—for example, Gecamines, the consortium of Tenke-Fungurume and the Miba.

On November 10, 1967, a contract was made between the Democratic Republic of the Congo, which became Zaire and the Nippon Mining Company Limited (Nikko), a Japanese company. This contract, which came into effect December 18, 1967, completed the general permission for geological research granted the Japanese Company January 1, 1967. Nikko associated with other Japanese financial groups to form the Company of Mining Development in Zaire, the Sodimiza. All rights and obligations given to Nippon Mining by the contract of December 18, 1967, and its annex were delegated to Sodimiza. On April 17, 1969, Sodimiza was founded. The Sodimiza is a Zairois company with limited liability. It was constituted for 17 years, renewable.

The corporate capital was initially fixed at 100,000 Z in 10,000 shares, 10 zaires each, and later increased. After December 30, 1971, the capital was 3 million Z, apportioned in the following manner:

	Shares
Republic of Zaire	45,000
Compagnie de Developpement Miniere du Zaire	102,000
The Nippon Mining Co., Ltd.	87,210
Sumitano Metal Mining Co., Ltd.	15,300
Toho Zinc Co., Ltd.	15,300

Furakawa Mining Co., Ltd.	12,240
Mitsui Mining & Smelting Co., Ltd.	15,300
Missho-Iwai Co., Ltd.	7,650
Total	3,000,000

The Zairois shares may not fall below 15 percent even when capital is added in material or specie. The Japanese companies may buy shares of Sodimiza equal to 50 percent of the company capital; this right is theirs from the company's inauguration to the fifth year of exploiting the mine.

In Sodimiza, the Zairois state forms a mining consortium with the Japanese companies; the foreign firms form a majority. In this, Sodimiza differs from Gecamines, which is owned 100 percent by the state. Also, the association between the state and these Japanese interests is based simply on the structure of the corporate capital, the most classic form of international consortium.

In the case of Gecamines, by contrast, the association is based in the daily management and commercial operation of the mines. Which type of association is more advantageous for the Zairois state and which for the foreign investor?

The problem of control of the company is raised here. Thus, although participation in the structure of the corporate capital gives the right to dividends when the association is ended, it does not necessarily give the state control of the means of production—that is, of the strategy of expansion, commercialization, and investment in the mining enterprise. Such control is above all a function of control of the world market in minerals, of the direction and daily management of the enterprise, of the technological mastery, and of the overall fiscal structure of the country. This is a complex problem that we can only touch on incidentally in these pages.[2]

Since the Japanese financial groups have a majority participation, control of the enterprise goes to them. First, the company representatives of the Japanese consortium are the "managers" who organize the day-to-day plan for exploitation of the mine complex and determine the mines' economic conditions. Article 11 of the contract gives Sodimiza and Nikko the right to establish progams for development of the enterprise and to decide on the amount of funds necessary to put them into practice.

The mining contract has in Article 15 given the exclusive right to Nikko to furnish all technological equipment and materials, and the personnel needed for research, prospecting, and development of the mines. In fact, Article 15 of the contract only approved the contract for technical cooperation entered on October 27, 1969, by Sodimiza and Codemiza, the Japanese mining consortium. Finally, the commercialization of the mining production is entirely left to Nikko. The 1970 annual report of the Department of Mines on the mining industry in Zaire shows that the area of the concession accorded to Sodimiza consists of eight exclusive zones of research that total 36,232 square kilometers.

In 1969, the company developed a five-year plan for prospecting and exploitation. It is estimated that the company wants to increase its annual copper production from 44,000 tons to about 200,000 tons. The company was able to obtain numerous fiscal pledges and financial conditions. In the first place, during the research stage, Sodimiza has the right to export/import permits for all material needed for its research and for the mineral samples sent outside the country for final analysis and metallurgical studies. To this is added an exemption from all taxes and from state, local, and customs duties. The only exceptions are remunatory taxes to the institution and the renewal of mining rights.

The next phase is during the first five years of actual mining: As in the preceding stage, the company will be exempt from state, local, and customs imports and duties, save for remunatory taxes due to mining rights. Sodimiza will benefit from permits for the material and supplies imported for the realization of the investments used in the exploitation phase. The company is equally exempt from laws dealing with the export of concentrated ores and metal products. The profits from this exemption are to be used, however, for the expenses made or to be made in the framework of the development of the company.

The third phase extends from the 6th to the 20th year following the opening of the mines, a total of 15 years. A long, stable period was agreed on in the contract for the benefit of Sodimiza, fixing the imports, duties, and laws in effect from the moment of signing the contract. The mining company will be free to opt for the application of a more favorable general fiscal or customs regime promulgated during this period.

During this same time, the mining contract provides for special export taxes on minerals, mineral concentrates, and metal products: between the 6th and 10th years following the start of exploitation, the taxes will be 25 percent; between the 11th and the 15th years, the tax will be 50 percent; between the 16th and 20th years, the tax will be 75 percent.

The contract also provides for certain financial advantages for Sodimiza. In effect, until the 20th year following the start of exploitation, Zaire will not take measures that affect Sodimiza and its affiliates. Zaire agrees to permit the following:

1. The export from Zaire of sums owed by Sodimiza to Nikko, suppliers, carriers, and, in general, the export of sums to anyone by the company and in view of the realization of its company object and of the financing of its programs of development.

2. The transfer of wages of foreign personnel working for Sodimiza and affiliates.

3. The transfer of dividends and all other profits originating with the Japanese participants to the capital of Sodimiza and its affiliates, or loans granted by Nikko to Sodimiza.

4. The placing of foreign currency earned by production of exports at the disposal of Sodimiza and affiliates; these currencies must be used for providing the company with materials, machines, equipment, replacements for the above, and in general, everything necessary for the functioning of the technical side of the company.

5. In the case of the Japanese companies' selling out, the transfer of the proceeds from the sale. This principle applies also to the repatriation of the foreign partners' share in the company assets on the dissolution of Sodimiza or its affiliates, excepting, however, the limitations imposed by Article 30 of the Mining Law of 1967.

Altogether, these financial and tax measures form an ideal framework for the start of Sodimiza. After this beginning phase, it will be possible to evaluate at least the modern methods of "forecasting" the economic effect on the state and on the company.

To compensate for these fiscal and financial guarantees accorded to the company, Zaire obtained from Nikko and Sodimiza the following promises: The realization before 1973 of investments needed for the extraction and treatment of crude ores, with a view to an annual copper production of 40,000 tons in the first stage of mining; the construction of smelters of a sufficient capacity to transform 50 percent or more of concentrates into 95.5 percent pure electrolytic copper. This expansion should exist by the time Sodimiza's annual ore production passes 60,000 tons, on the condition, however, that the company use sufficient energy resources, 25,400,000 kilowatt hours (kwh) per month.

In its mining concession, Sodimiza has discovered important deposits, notably at Musoshi and at Tshinsenda, localities on the plain of South Shaba, toward the Zambian border. At Musoshi, the exploitable ores amount to 30 million tons averaging 3 percent copper. The deposit at Tshinsenda is also evaluated at 30 million tons of copper ore, but its average copper content is higher than 4 percent. The veins of this deposit are varied and complex, which situation necessitates prospecting for inclined galleries.

Actually, only the Musoshi deposit is exploited. A total of $90 million—45 million Z—have been needed to work the Musoshi mine, notably for the installation in no. 1 pit of shoring 60 meters high, and winches of 1,300 kw and 2,400 kw for the transport of personnel and material and for the evacuation of ores, and in pit no. 2, the installation of a winch of 450 cv to assure the transport of materials during the construction of a car for crushed ore and the installation of four compressors and six water pumps.

The cost of the smelter was about $8.6 million (4.3 million Z). This includes installations to store 6,000 tons of ore, a reservoir for the crusher of 5,000 tons; two crushers with bars using 500 kw; five crushers with balls of 350 kw; two machines for flotation; four groups for decantation; a filter and a dryer. The motor force for all the installations of the smelter is on the order of 13,000 kw. Its capacity is 40,000 tons of ore a month.

In the smelter, copper concentration in the ore is rasied from 3 percent to 36 percent. This concentrated ore is exported to Japan and refined there while waiting for the construction of a refinery in Zaire by Sodimiza. More than 3,500 people work for Sodimiza at present; 3,000 are Zairois. It is estimated that exploitation of the Tshinsenda deposits will give employment to about 3,000 workers.

As Sodimiza envisions increasing its annual copper production to about 20,000 metric tons in 1980, it is foreseen that, in consequence, its technical capacities must increase. In what measure can this expansion guarantee the feasibility of the entire enterprise? Two categories of factors, internal and external, present difficulties for Sodimiza.[3] The internal factors involve constraints on mine production in the Shaba.

First, it is necessary to note the nature of the mines. The high copper content of the ores of Shaba are, to some extent, counterbalanced by the difficulties and the cost of extraction. It is estimated that Musoshi ores are only 6 percent copper—94 percent is useless.* Second, there is not enough energy. Gecamines itself is unable to satisfy its needs without Zambian aid. Sodimiza has had to negotiate with the Zambian Electric Supply Company for 4 million kw for its expansion program. While waiting for Inga to supply the Shaba, when conditions should improve, the use of Zambian electricity by Sodimiza constitutes an additional cost for production. Third, internal transport expenses are considerable. There are many exportation routes for mining companies situated in the Shaba. Beyond the national routes, there are Beira, Lobite, and Dar-es-Salaam. Compared

---

*Editor's Note: This appears to be a high grade of ore, compared to new open-pit mines being developed elsewhere with ores containing less than 1 percent copper.

with the costs of external transport (which includes the costs of transport itself from the port of embarkation, freight charges, transport agency, handling, insurance, and so on), internal costs (which cover comparatively short distances) are very high. The 1970 Report of the Department of Mines indicates that for Gecamines' copper, internal expenses (Zaires per ton) for the metal were 5,312 Z, as against 1,352 Z for external transport.[4]

The external factors are basically concerned with the world consumption of the copper and the evolution of its price. The factors are always related. If 1971 was catastrophic for copper, 1972 showed, if timidly, some encouraging signs. Unless the member countries of CIPEC (Chile, Peru, Zambia, Zaire) moderate their programs of expansion, it is feared that despite the growth of world copper consumption, we will not see a meaningful recovery, much less a boom comparable to that of the 1960s.

In conclusion, we can show various provisional findings:

1. Sodimiza's output will immediately augment Zaire's capacity for copper production. If the constraints on energy are removed, if the gain in world consumption is confirmed as a trend, and if the price of copper recovers significantly, the establishment of Sodimiza will be a real advance that may contribute to government revenue. The fiscal and financial guarantees made to Sodimiza, however, are enormous; and if the present situation of the copper market changes, it will be vital to review these guarantees.

2. The relationship between foreign private capital and public interests will not work unless a climate of real collaboration between the state and the Japanese interests is ensured.

3. Sodimiza's investment will reinforce the importance of Shaba region to the country's economy. Its mining industry must be the motivating force for the development not only of the Lubumbashi subregion but also of the subregion of North Shaba and the neighboring regions of the two Kasai.

4. Finally, the study of the structures of the Sodimiza and of the mining companies of the Tenke-Fungurume reveal the state's approach to association with foreign private capital as a minority shareholder in the exploitation of the country's mineral resources.

NOTES

1. Minister of Mines, Mining Industry of the Democratic Republic of the Congo, Annual Report, 1970, p. 60.

2. The author, who is consultant for the United Nations on negotiations with mining consortiums, is researching these problems.

3. For discussion, see <u>Le Mineur</u>, Journal d'Enterprise Sodimiza, special of September 1972.

4. Ministry of Mines Report, 1970, on the mining industry of
the Democratic Republic of the Congo, p. 48.

CHAPTER 15

THE GENERAL COMPANY OF
QUARRIES AND MINES
OF ZAIRE
Mulumba Lukoji

On January 1, 1967, Zaire took over all the goods, movable and immovable, belonging to the Union Miniere du Haut-Katanga (UMHK) and its affiliates.

A law passed on January 2, 1967, authorized the constitution of the Congress General Company of Minerals (Gecomin), which since 1971 is the General Company of Quarries and Mines of Zaire, abbreviated Gecamines, and which is 100 percent state owned.

The aims of this chaper are to describe the political context of the creation of Gecamines; analyze its economic activity and feasibility; consider the constraints imposed on the expansion of its activities; and detail the mode of association between Gecamines and the General Company of Minerals (SGM), and suggest alternative paths for the sale and transformation of copper products.

POLITICAL CONTEXT OF THE DEVELOPMENT
OF GECAMINES

To affirm its economic independence, Zaire passed two laws in 1966: The first ordered the transfer to Zaire of the registered and administrative offices of companies having their centers of exploitation there.[1] The second, called "Bakajika law," canceled all the concessions and transfers accorded before the country gained independence.

This study was made with the collaboration of Ilunga Ilunkamba, assistant in Economic and Financial Sciences at UNAZA.

MEANS OF PRODUCTION AND FEASIBILITY
OF GECAMINES

The Mines and Quarries

The mining domain that Gecamines inherited is located in Haut-Shaba; its limits are determined by the decree of October 30, 1906.[2] They correspond to the basins of the upper Lualaba, the Lufira, and the upper Luapula, an area of 34,000 square kilometers.

Part of UMHK's concession was taken from Gecamines and given to the Company for the Industrial Development and Mining of Zaire (Sodimiza) and the Mining Company of Tenke-Fungurume (SMTF).[3]

Gecamines' concessions are above all rich in copper, cobalt, zinc, cadmium, germanium, and silver. Their recovery is through quarries and subterranean mines, thus justifying the company's name. The open-pit exploitation (quarries), of which Kamoto, Musonoi, and Ruwe are the most important, require the use of giant mining machinery. The products are excavated with electric shovels, whose scoops can lift a volume of 4 to 12m^3 at a single time. Giant buckets, of from 45 to 100 tons, assure the transport of minerals to the smelters. Such material and much more call for considerable investments. For example, the price of a shovel lifting 12m^3 is 45,000 Z, and a bucket for 100 tons costs 131,000 Z (1 Z equals $2).[4]

An underground mine is justified by the depth of the mineral vein. Thus, the uncovering work that would be necessary to start with makes a quarry-type exploitation less feasible. The methods used here are not the same as those in open-pit exploitation. They are primarily dictated by the form of the mass to be exploited, the shoring (lining of tunnels), the richness of the minerals, and the possibilities of mechanization. The methods of exploitation of underground mines have evolved with new technological acquisitions by the company and the mining industry in general. Table 15.1 shows what mines and quarries Gecamines was exploiting as of December 31, 1972.

The ores from the various mines and quarries are taken for treatment to the many smelters and concentration plants established at Kipushi, Kambove, Kolwezi, Kamoto, Mutoshi, Lubumbashi, Panda, Shituru, and Luilu.

For the functioning of all the industrial installations, as of December 31, 1972, the company employed (administrative personnel included) 26,046 laborers and 2,204 management. European management increased greatly in 1968, growing from 1,212 to 1,581, then decreased until it reached 1,384 in 1972. The Zairois management grew from 379 on December 31, 1967, to 820 at the end of 1973, rep-

TABLE 15.1

Mines and Quarries Exploited by Gecamines, as of
December 31, 1972

Quarries	Displaced Earth (m^3)	Valorizable Ores Extracted (in tons of copper)
	Western Group	
Musonoi	1,901,250	106,815
Kamoto	10,800,311	256,502
Mupine	5,110,775	66,721
Mutoshi	4,818,425	52,772
Mines	Extracted Ores	
Kamoto	1,626,561	63,346
	Central Group	
Quarries	Displaced Earth (m^3)	
M'sesa	1,334,412	59,242 (1971)
Kazibizi	931,755	20,198
Kakanda	3,559,288	34,422
Mines	Extracted Ores (tons)	
Kambove-West	938,344	41,153
	Southern Group	
Mines	Extracted Ores	
Kipushi	1,066,967	44,943 (cu)
		120,759 (zinc)

Source: Office of Gecamines, Lubumbashi, Zaire.

resenting 37.2 percent. Until the recent replacement of M. de Merre
by Citizen Umba, the posts of director general and assistant director
were filled by Europeans. One must remember, however, that Zairois occupy important directorships, including those for the Directory
of Mines and Quarries, the Administrative Directory, the Directory
of the Treasury, and the Commercial Directory.

It is possible that in the near future all aspects of the large state mining complex in the Shaba will be controlled by competent and sufficiently trained Zairois managers. Africanization of management is expected to bring to more than 2,000 the number of Zairois management employees in 1980.[5]

EXPANSION PROGRAMS AND FEASIBILITY PROBLEMS

In 1970, the company decided to intensify exploitation of its vast mineral resources through two five-year programs. The first, which planned to raise the production of copper in 1974 to 460,000 tons, was approved by the government. In the second the company hopes to raise production of copper to at least 600,000 tons per year by 1980.[6] This last five-year plan is still being studied.

All the stages of the first five-year plan have been progressively realized up to now. In 1970, the company undertook the exploitation of the underground mines at Kamoto; production started in a new section of cleaning at the Ruwe cleaning factory. The concentrator at Kambove was enlarged and the electrical foundry at Panda was put back into use. An extension was added to the new installations for electrolytic cobalt in the Luilu factories. Finally, a new limestone crusher in Kakontwe and various developments on numerous mines were made to increase the company's productivity.

In 1972, Gecamines put into service at Kolwezi a production unit for 55,000 tons of copper per year. In 1974 in Lubumbashi, an oxygen factory was inaugurated that makes it possible to increase production capacity for raw copper to 30,000 tons per year, thus taking the production of the Lubumbashi works to 160,000 tons per year. For this process, it is necessary to feed the water-jacket fusion furnaces and converters with pressurized air enriched to 30–35 percent oxygen. This permits, moreover, a saving on coke and the activation of the fusion, as well as better use of the calories not used during the oxidation of the bath; from this the direct fusion of rich sulfureted concentrates becomes possible.[7]

The final stage of the first five-year plan foresees the coming into production of a new pit at the Kipushi mines, which, thanks to its enormous diameter, will permit the recovery of ores located at a depth of 1,150 meters. The total cost of the first five-year program, 1970-74, has been valued at about 51 million. Because of the fall in the price of metals that began at the end of 1970, the company preferred to finance its investments through external borrowing. Ordinance law no. 70/021 of March 26, 1970, authorized Gecamines to borrow up to 17 million Zaires with a state guarantee, for the realization of the first five-year plan.

By December 31, 1972, the company had obtained the following loans:

1. $5 million (2.5 million Z) with an annual interest of 6 percent from the U.S. Export-Import Bank, to finance the buying of U.S. equipment. This credit is to be paid back after November 15, 1972 in 10 half-years, at the rate of U.S. $75,000.

2. Credit of 160 million units (which is 8,800,000 Z), granted by the European Investment Bank, at 8.5 percent per year, and repayable from March 5, 1974, in 11 half-years.

3. A loan of $3 million (1.5 million Z), given in 1972 by the Morgan Guaranty Trust Company at a rate of interest equal to Eurodollars, at 12 and 18 months after the London interbank market plus 1.75 percent. The first part of this loan is to be repaid after 12 months, the other at 18 months.

To facilitate the repayment of these loans, the Finance Department of the government arranged, by ministerial decree no. 51 of July 2, 1970, a temporary tax exemption for Gecamines. The company is free, during the period of realization of the first five-year plan, of all duties and taxes on production after 360,000 tons of copper and 11,000 tons of cobalt. The exemption also extends to equipment for implementing the expansion program. The decree says, finally, that the profits exempted must be used, until the expiration of the exemption period, only to finance the investments planned for Gecamines' development program and to pay back the loans contracted for the realization of this program.

The expansion of production projected by Gecamines indicates a yearly rate of growth of around 5 percent. This rate is in line with the growth of consumption of copper worldwide. Production in 1972 represented an 8.6 percent increase over 1971, due primarily to the recovery of the economies of Japan (15 percent) and the United States (11 percent). The expansion in copper production is accompanied by a significant increase in the production of cobalt. From 10,600 tons in 1969, production grew in 1974 to 14,400 tons. Taking into account the taxation on Gecamines and admitting a rate of realization of investments of 10 percent, the rate of profitability of the expansion programs grows to 40 percent for a copper price of 500 Zaires per ton and to 48 percent with a copper price of 600 Zaires per ton. This rate of profitability might be higher if the tax charges supported by Gecamines were not so heavy. In effect, besides the customs duties on imported merchandise, the professional taxes and duties levied on the occasion or at the time of production, the companies subjected to the tax on profits, to the tax on exports and the exist duty, to which is sometimes added a complementary exit duty.[8]

In 1968, these duties represented respectively 12, 77, and 11 percent of total tax charges. And all the fiscal charges represented more than half the net turnover of Gecamines.[9] In 1972, the duties paid by Gecamines were valued at 88,220,993 Z, against 86,482,066 Z in 1961. Gecamines' share represented 85 percent of all contributions paid by mining companies to the state in 1971.[10] And its contribution to the ordinary receipts of the state usually exceeds 40 percent.[11]

Gecamines' total turnover was 259.1 million Z in 1972, against 246.4 million Z in 1971. Initially assessed at 104,700,000 Z, the corporate capital recently grew to 259,600,000 Z through the incorporation of reserves.[12]

CONSTRAINTS ON EXPANSION

Essentially five problems confront Gecamines, partially due to the very heavy taxation. These are the ambiguity of the objectives assigned it; the management and direction of the company; the energetic development of new mines and factories as foreseen in the five-year plan; the high transport costs on mining products destined for export; and, finally, the marketing of mining products.

The Nature of the Objectives Assigned to Gecamines

Conforming to Article 3 of the company regulations, its goal is "the research and exploitation of mineral ores; the treatment of mineral substances from these ores; the sale of these substances and other connected operations."[13]

Above all, the management of Gecamines rests on considerations of commercial viability.

On the other hand, having been constituted a state company (all the shares are owned by Zaire), immediate profitability may be sacrificed to considerations of overall development. In this framework, the president of the republic, in his speech of November 30, 1973, decided that by 1980, all copper produced in Zaire would be refined there. He also called for investments in the agricultural sector from the mining companies. This orientation may, perhaps, diminish the present profitability of the company but is justified by the option for economic independence taken by the country. It goes without saying that the company often finds itself torn between considerations of immediate profitability and its efforts to satisfy the new orientation at the risk of compromising the execution of its expansion programs.

The Management and Direction of Gecamines

The problem posed here is that of the present importance of foreign personnel in the service of the company. Since the creation of Gecomin on January 1, 1967, the proportion of nationals reached 37.2 percent of the management personnel by December 31, 1972. However, on the qualitative level, it is necessary to recognize that the results of Africanization are not very encouraging. If the rhythm of Africanization in the administrative services is very rapid, that in the technical services is slower, above all in that which concerns directory personnel—that is to say, at the university level.

The percentage of Africanization is lowest among the engineers, the engineer technicians, and the doctorates; for the first two categories, it is on the order of 15.24 percent, and it is 17.5 percent for the doctorates, against 75 percent for the other disciplines.

In the second place, for some disciplines, the effort of Africanization has been checked by the increase in recruitment of expatriate personnel. This is especially true for the doctorates. While the number of nationals has been constant, that of expatriates has grown from 26 in 1969 to 33 in 1972. The very small proportion of technical management is the result, in our opinion, of the disproportion established at the level of the university campus between the student population in technical faculties and that of other faculties. This disequilibrium is also reinforced by the too high percentage of failures in the technical faculties. To assure a rapid replacement of foreign technical management by locals, it is necessary to orient more students toward the technical faculties at the level of the university and the higher institutes and to assure that the assignment methods will guarantee the "production" of enough competent managers.

Concerning the other university-level management, especially the geologists and doctorates, it must be hoped that the requisition affecting the finalists of the university and the higher institutes will be flexible enough to permit recruitment by the company. Finally, it is necessary to remember that even in the areas where the results of Africanization are on the order of more than 75 percent, the effort of Africanization has not been total.

Often an increase in Zairois university-level management corresponds to the increase in recruitment of new expatriates. This is the case for graduates in economics and for those under the heading "miscellaneous." Finally, sometimes when the number of national university-level management is increased, that of the expatriates remains constant. This is the case for doctorates and law graduates.

With these reservations, it can be admitted that the general results of Africanization are positive. If it is established that radical change in the rhythm of Africanization can hinder the normal functioning of the company, it is best that the company maintain its present rhythm and accelerate it where possible, especially in the effective administrative posts.

The Energy Constraint

In two earlier studies on Sodimiza, we have underlined the need for energy in the Shaba.[14] In effect, to be able to continue activities, Sodimiza had to negotiate for 4 million kilowatts with the Zambian Electric Supply Company. The mines and factories of Gecamines obtain their energy from four main sources, producing about 2.5 million kilowatt hours annually. These are the Mwadingusha (ex-Francqui) Station on the Lufira at Mwadingusha Falls (ex-Cornet) with an installed power of 77,100 kilovolt-amperes (kva), which, until November 30, 1973 belonged to Sogeform; Koni Station (ex-Bia), 7 kilometers before the Mwadingusha Falls, with a power of 46,800 kva; Nzilo (ex-Delcommune) Station, with 120,000 kva and Seke (ex-Le Marinel) Station, with 276,000 kva of installed power, both constructed on the Lualaba. The last three stations belong to Gecamines, but Sogeform operates them. For treatment of its ores, Gecamines needs 2,350 kwh for each ton of electrolytic copper, 6,600 kwh per ton of electrolytic cobalt, 4,500 kwh per ton of electrolytic zinc, and 1,800 kwh per ton of electrolytic cadmium.

These figures are sufficient to illustrate the importance of Gecamines' energy requirements. In 1972, its consumption reached 1,821.6 gwh,* an increase of 8.88 percent over 1971.[15]

For 1972, the realization of the stages of the expansion program necessitated a total electricity production of around 2,850 gwh. The future needs of the Shaba, following these calculations, may reach 3,809 gwh by 1980;[16] especially after the realization of Gecamines' second five-year program, the intensification of exploitation of the Musoshi and Tshinsenda mines by Sodimiza, and mineral exploitation by SMTF. To make up the energy deficit in 1973, Gecamines expected to import energy from Zambia.

To realize its programs of expansion, the company decided to construct, downstream from the Seke power station, the Busanga station, which would have almost the same capacity as Seke, about 276,000 kva. However, because of the government's decision to realize the Inga-Shaba high-tension line, this project has apparently been abandoned. Moreover, a fall in the level of rain in the Zambezi and Lualaba basins has confirmed earlier studies by the company, which

---

*One gwh equals 1 million kwh.

emphasized the limited hydroelectric possibilities of the Shaba region.[17] The realization of this line thus constitutes a condition sine qua non for the execution of the second five-year plan; Gecamines can only count with difficulty on Zambian energy, for, like Zaire, Zambia has embarked on a large program of mining expansion.

Transport Costs on Mining Products

In our studies of Sodimiza, we have already indicated that the expense of transport for mining products destined for export are considerable, especially the cost of internal transport. Internal transport, in 1970, cost 5,312 Z for one ton per km for Gecamines, against 1,352 Z for external transport.[18] An excessive increase in these costs translates for the company as a real increase in the costs of production, which could compromise the feasibility of expansion plans.

The Marketing of Mining Products

Almost all the mining production of Gecamines is exported to European consumers. Local consumption, especially in copper, is not more than 10 percent of the production of Gecamines. This is mainly by Latreca (rolling mills and wire and cable makers) and Cabelcom companies.[19] At the start of the execution of the expansion programs, world consumption of copper showed encouraging signs compared to 1969, especially in Europe (2,441,000 tons, against 2,328,000 tons). But it fell considerably in 1971 (2,353,000 for Europe; America, 2,346,000 in 1969, then 2,247,000 in 1970, and 2,236,000 in 1971; Japan: 806,000 in 1969, then 815,000 in 1970, and 820,000 in 1971), a total reduction of 17 percent compared to 1970.[20]

This was followed by a catastrophic fall in the price of copper.[21] In the fourth trimester of that year, 1971, the average price of a ton of copper sold in Brussels fell to 415 Z. As indicated above, it is imperative that the price of copper should be at least 500 Z per metric ton to guarantee the profitability realization of the expansion programs. The 1973 tendencies of world consumption of copper and its price are very encouraging. The expectations of 500 Z and 600 Z expressed in the expansion programs have mostly been surpassed. If the world demand for copper remains high enough, the feasibility of the expansion programs will be certain, as well as the company's capacity to repay the external debt borrowed for the realization of these programs.

However, it is feared that the present petroleum crisis, and above all the rise in price it has provoked, will entail a reduction in economic expansion begun by the industrialized countries of Europe,

the United States, and Japan, the principal consumers of Zairois copper. This reduction is translated in effect by a decrease in world
consumption of copper, accompanied no doubt by a fall in price.[22]

If Gecamines is able to overcome these internal and external
constraints, it will succeed in realizing the objective to which it is assigned.

THE ASSOCIATION BETWEEN GECAMINES
AND SGM

After the "Congolization" of UMHK and the creation of Gecamines, Zaire decided to give the marketing of the mineral products of
Gecamines to SGM. This last, an affiliate of the General Company of
Belgium, served before 1967 as the marketing agents of UMHK. The
terms of cooperation between Gecamines and SGM were detailed by
the Convention of February 15, 1967[23] and the additional Accord of
September 24, 1969.[24] The convention seems to be a response to
pragmatic considerations: The country did not, in 1967, have enough
national high-level manpower to operate the large mining complex of
Shaba, nor could it assure the commercialization of mining products,
the principal source of its budgetary receipts. The convention gave
SGM the following responsibilities, for an undetermined duration:

> The study of general programs required for the
> functioning of the means of production and their de-
> velopment, for the employment of personnel as
> well as for the mining, industrial and commercial
> management of the enterprise;
> The execution of programs undertaken by the
> Administrative Council;
> The recruitment and placement at Gecamines'
> disposal of non-African technical personnel neces-
> sary for the functioning of the enterprise;
> Refining, transformation into saleable products,
> and marketing (1st Article).

To this are added buying services and provisioning, studies of
civil engineering, metallurgy, promotion of metals, and so on.

SGM also assumed the direction of all the industrial and commercial operations (Article 2c). SGM recruits non-African personnel
for Gecamines. For each of these employees, Gecamines gives SGM
2 percent of the total currency repatriated, until 600,000 FB (Belgian
franc) per non-African employee is reached, under a technical assistance contract. This total corresponds to a rate of 130 on the official

index of the cost of living in Belgium. Its adjustment is foreseen; it produces, to the profit of Gecamines, an equal interest at the common rate of banking deposit for three months in Belgium.[25] At the expiration of the convention, the balance of this deposit will be returned to Gecamines.

Finally, SGM sells the mine's output for Gecamines. The products themselves are sent freight on board (FOB) to the African port of embarkment—that is, Matadi, Lobito, Beira, or Dar es Salaam. If at the time of delivery of its merchandise, the price owed by SGM is not known, "SGM effects to the profit of Gecamines at the beginning of each year, a provisional payment at the value of the price of products bought in the course of the preceding years." An adjustment is made so that the price due is collected.

In deciding on November 30, 1973, that the products would be marketed cargo, insurance, and freight (CIF), the president of the republic wanted Gecamines to support directly all the expenses tied to the transport of products until bought from now on. This solution may seem onerous, but its advantage is that it lifts from SGM the power it had to declare to Gecamines and make it pay all the expenses of marketing it had incurred. The presidential decision has thus changed the procedure of repayment of the expenses of commercialization foreseen by Article 6 of the convention.

It is also possible to interpret the creation in Lubumbashi of a commercial management as a modification of the 1967 Convention, which recognized SGM as the only commercial agent. It cannot be changed unless the Lubumbashi management effectively becomes solely responsible for the marketing of mining products. This does not seem to be the case. The Lubumbashi management seems rather to be a registering division, for the control needs of Gecamines, of the movements of mining products from the Shaba to Belgium. For the provision of these diverse services, SGM has a right to commissions. The 1967 Convention provided first for a commission for technical assistance equivalent to 4.5 percent of the value of the company's mining production and then 2 percent for "engagements associated to technical aid," for a total of 6.5 percent.

The additional Accord of 1969 reduced this total to 6 percent. This accord was to be maintained for 25 years of association between Gecamines and SGM. It detailed, moreover, that the commission will be 6 percent for the first period of 15 years; the period that followed would see the commission reduced to 1 percent.

The arrangement between Gecamines and SGM resembles the "management" and "consultancy" agreements that exist between Nchange Consolidated Copper Mines (NCCM) and Anglo-American Corporation, on the one hand, and, on the other, Roan Consolidated Mines (RCM) and Roan Selection Trust Company in Zambia and Swaziland.

In the limited framework of this study, it is not possible to analyze in depth all the aspects of this association. One must be content to give the principles of its functioning. By the light of these principles, and taking into account other more highly developed aspects of Gecamines, one can at present try to establish some developments. This is also an occasion to evaluate the impact of these last measures on the activities and functioning of Gecamines.

Gecamines plays an overwhelming role in the Zairois economy. It constitutes the principal base of government resources, as much in fiscal receipts as in currency it procures for the country. (See Table 15.2.)

The expansion programs undertaken by Gecamines, as well as the investment by Sodimiza and SMFT, reinforce increasingly the degree to which the Zairois economy relies upon the mining sector. When the price of copper is sufficiently elevated, the feasibility of all these investments will be guaranteed and the public Treasury regularly supplied. But to prevent lean periods, it is necessary to use the revenues of this sector for feasible activities that can equip overall economic development. The mining complex of the Shaba can thus serve, like the German Ruhr, as a real motor unit for development. This is, moreover, the orientation decided upon by the president's speech of November 30, 1973, in which he asked that the mining sector invest in other sectors of the national economy.

In deciding to give the direction of the company to a Zairois, the authorities of the country call for at least two things: on the one hand, real control by the nation of this important industrial complex, and, on the other, effective management to facilitate its real integration into the national economy. The refining and transformation of copper in Zaire especially is one of the forms of the integration of this industry into the country's economy. As for the realization of other expansion programs by Gecamines, energy is still an important constraint, if the refineries and transformation industries are to be constructed in the Shaba. If, however, they are established near Inga, this constraint will be lifted, as well as the difficulties of transport and export of refined products as long as the internal consumer market is reduced. One way to encourage this market, as suggested by the research group "Wajengaji" (that is, builders), at the second seminar on mines organized in May 1973 in Lubumbashi, would be to give the fabrication of high-tension lines for Inga-Shaba to local companies such as Latreca and Cablecom.

Concerning the marketing of Gecamines' mining products, it is certainly possible to envisage, even now, the eventual participation of Zaire in SGM. But we are of the opinion that the country is going to organize little by little its own system of marketing. The deliveries

TABLE 15.2

Evolution of Taxes, Duties, and Profits of Gecamines
(in Zaires)

Year	Taxes and Duties	Net Profits
1967	69,827,606.03	9,298,718.78
1968	92,153,403.00	12,868,967.00
1969	129,738,403.00	23,866,000.00
1970	132,922,565.00	14,507,955.00
1971	86,482,066.00	17,354,039.00
1972	88,220,993.00	8,869,822.00

Sources: Table calculated with the help of Reports by Gecamines (1967-72); the Department of Mines (1967-71); and the Bulletins Trimestriels of the Bank of Zaire.

of the SGM limit is essentially to traditional consumers, that is, to Europeans; it is indicated that the state will change its deliveries to new customers, such as China, Algeria, and Japan. The state can abandon to this national company of mining marketing the ownership of 10 percent of the production of each mining company given it by the mining law.

Zaire would thus join Zambia, which is putting an end to the role of marketing agent played by the Anglo-American Company and Roan Selection Trust in deciding to create the Metal Marketing Corporation of Zambia, Ltd.[26]

Finally, in deciding to raise its participation to 50 percent in all mining companies, the state attempts primarily to reinforce its power of control over the important mining complex of the country. Gecamines is not concerned in this measure. The company belongs to the state 100 percent. Differing again from the other mining companies, it does not belong to the category of traditional international consortiums—that is, those in which the association of groups of financiers operates at the level of the structure of corporate capital. Because of its association with SGM, Gecamines offers a new mode of association, limited to the level of management of the company.

NOTES

1. Mining Code 1966, 1st Part, p. 523.

2. Compare Belgian Ordinance 1906, p. 434 and ss; Belgian Ordinance 1908, p. 446 and ss; Nouvelles, 71, p. 626 and ss.

3. Convention of December 11, 1967 between RDC and Nippon Mining Company Ltd. for Sodimiza. Convention of September 19, 1970 between RDC and a consortium of American, British, Japanese, and French interests for SMTF.

4. Gecomin, Monographie, p. 35.

5. Gecamines, Rapport Annuel, 1970, p. 6.

6. Ibid., 1972, p. 7.

7. Gecamines, Mwana Shaba, industry journal, no. 211.

8. Compare notably Ordinance Law no. 69/018 of May 14, 1969.

9. See the declarations of MM. De Merre and Jorion at the December 1972 seminar at Lubumbashi on the mining industry, in Industrie Miniere et Developpement au Zaire (Kinshasa: University of Zaire Press, 1973), pp. 53, 55, and 129.

10. Ministry of Mines, Industrie Miniere du Zaire, 1971, p. 58.

11. See Ilunga Ilunkamba, "Industrie Miniere des Finances Publiques," in Industrie Miniere et Developpement au Zaire, op. cit., p. 98.

12. Compare Ordinance no. 72/291 of September 25, 1973: "The decision of the director general of Gecamines to increase the capital of this organization, which is presently at 104,700,000 Zaires from a sum of 154,900,000 Zaires, by incorporating to the apparent capital the sum of 154,900,000 Zaires, to 259,600,000 Zaires, raised until 122,506,000 Zaires on the special reevaluation reserve of immovables and to 32,393,135 Zaires on the provisional funds, is approved."

13. Ordinance Law no. 72/050, February 14, 1972, JORZ (provisional version of the Supreme Court), no. 21, December 12, 1972.

14. Ibid.

15. Gecamines, Rapport Annuel, 1972, p. 30; and Ministry of Mines, op. cit., p. 63.

16. W. K. Malu, Transport du Courant vers le Shaba et Son Impact Economique, paper no. 1 at the 2d Seminar on the Mining Industry of Zaire, Lubumbashi, May 16-23, 1973, p. 20.

17. Statement by M. Saquet at the 2d Seminar on the Mining Industry, commenting on the paper by Professor Malu.

18. Ministry of Mines, op. cit., p. 41.

19. Wa Sanga, Possibilites de Transformation des Produits Cuivreux au Shaba (example de LATRECA), Paper no. 17 at the 2d Seminar on the Mining Industry in Zaire.

20. Mining Annual Review, 1972, p. 35.

21. Idem.

22. See especially the pessimistic predictions of the National Economic Development Council of Great Britain, in Mining Journal, December 7, 1973, p. 469.

23. Crisp, Congo 1967, pp. 481-86.

24. Press Release, Belgo-American Development Corporation, New York, October 16, 1969.

25. Article 7 of the Protocol to the convention.

26. These accords were recently renounced by Zambia. See "Zambia Economic Survey," African Development, October 1973, p. 17; and Mining Journal, October 1973, p. 282.

16

THE INVESTMENT CODE
AND THE MINING INDUSTRY
OF ZAIRE

Katanga Mukumadi Yamutumba

In the course of the colonial period, it was apparent that there was no condensed legislation dealing with investment, except that the colonial authorities applied a system that favored investment in enterprise. Under this system,the profits of companies were taxed on a progressive scale, taking into account the relationship between profits and capital invested. The larger the capital, the less important the relationship, and thus the less significant the progressive scale. This could only favor the self-financing and maintenance of capital. This system, it seems allowed the influx of capital and its maintenance in the country, especially in the mining sector.*

Following the queries of many enterprises settled in the country, after independence, the Government of Zaire promulgated an investment code in 1965, intended to create a favorable climate for the formulation and maintenance of capital. Union Miniere du Haut-Katanga (UMHK) was among the companies that profited from the advantages of the code. When undergoing Zairization, its directors protested vigorously, invoking the treaty between Belgium and Zaire of February 6, 1965, and the 1965 code. Because of the code's obsolete aspects, the Zaire government thought it good to replace it with new legislation on investment in 1969.

---

*See P. Jasinski, Regime juridique de la libre circulation des capitaux (Paris: LGDJ, 1967), p. 7.

THE 1969 CODE (ORDINANCE LAW 69/032)

The 1969 code was promulgated following the Ordinance Law no. 69/032, June 26, 1969. In passing, we note that the promulgation of this ordinance law occurred in a favorable political and economic context. It was at the end of the tribal wars, at a time of the reestablishment of political stability, guaranteeing the security of persons and goods. The climate of confidence was now reinforced by the publication of the new constitution in 1967, which also guaranteed property rights, except in cases where the general interest overrode such rights.

To these encouraging measures, we can add the monetary reform of June 1967, which permitted an equilibrium between public finance and external payments, and the stabilization of the national economy. It also coincided with the rise in copper prices. Thus, the 1969 code was born in a favorable situation, due above all to a healthy national economy.

The code has three objectives: (1) to promote investments tending to modernize and extend those existing enterprises that would contribute to the economic development of the country; (2) to stimulate the mobilization of national finances and orient them toward productive investments; and (3) to attract investments of foreign capital to Zaire by according them special guarantees of transfer.

The code institutes two structures: a general structure and a contractual structure. Admission to the general structure is reserved for enterprises that contribute measurably to six of the areas outlined by Article 6: (1) the amount of value added locally; (2) the amount of employment created; (3) the total of investments and nature of financing; (4) the importance of the effects of the project upon the other sectors of the economy; (5) the effect on the balance of payments; (6) the program for promotion of nationals to leading and specialized posts; (7) localization of the investment; and (8) conformity of the project to the political-economic orientation of the government.

The second structure is that of a contract concluded between the government and the investor. To make use of this structure, the project must first fulfill the conditions of admissibility to the general structure, and second, present a majority interest and the effects of the project on the development of the country. For these, the government accords them greater advantages than those of the general regime. These advantages extend to acquiring a long-term favorable financing regime, which guarantees the stability of all or part of taxation (Article 190).

The code grants several fiscal advantages to beneficiary companies (see Table 16.1). Article 22 of the decree law of August 30, 1965, stipulated that "conflicts born of the interpretation and applica-

TABLE 16.1

List of Fiscal Advantages Provided by Investment Code

Name of Tax	Matter Exempted	Tax (percent)	Length (years)	Article of Code
For all agreed enterprises				
Right of entry	Equipment, machines	b	d	16
Business tax	Same	c	—	16
Professional tax on profits	Profits[a]	4-60	5	12
Tax on property, built-up or otherwise	—	e	—	15
Proportional duty	Act of constitution of SPRL and acts increasing capital	1	—	11
Fixed duty	Act of constitution of companies other than SPRL	—	—	11
For agreed new enterprises				
Exceptional tax on salary of expatriate personnel	Gross salaries	5	5[f]	13
For agreed existing companies				
Shares tax	Dividends of new shares[g]	20	5	14

[a] Exemption for existing enterprises applies only to the part involved in modernization or extension.
[b] Charge varies after the nature of the material or product.
[c] Variable charge: 10 percent in the interior, 6.7 percent on imports, and 7.5 percent on exports.
[d] No precise duration.
[e] Tax varies with classification of locality: 1st class, 30K; 2d class, 15K; and 3d class, 10K.
[f] Length must not exceed five years from the agreement to the production of enterprise.
[g] Issued for extension or modernization.

tion of the contract are within jurisdiction" of Zairois courts. The 1969 code modifies this statement by saying that the arbitration procedure will be under the code of civil procedure. However, it is important to note that the new law permits recourse to arbitration under the convention for the regulation of disputes relative to the investments between states and foreigners that Zaire ratified on August 16, 1969. Moreover, the code underlines the fact that the arbitrated sentence is executed under the law of the Republic of Zaire. The conventions already signed and those being prepared with foreign governments recognize this principle.

As the general regulations make a distinction based on the national origin of the investor, Article 21 reserves to foreign investors the guarantee of transfer of their participation for the value acquired, the annual transfer of the returns from their contribution. This benefit is not guaranteed for resident foreigners: The hope of the government is to provoke the influx of capital from the exterior.

THE PRIMARY RESULTS OF THE APPLICATION OF THE CODE AND THE MINING SECTOR

Tables 16.2 and 16.3 represent all the briefs examined by the commission. They show that since the publication of the code, the commission on investments has had to treat nine mining briefs for a value of 52,486,286 Zaires (1 Z equals $2), which is 24.4 percent of the whole of the projects treated, the total value being 215,013,910 Z.

Note that Gecamines alone has 78 percent. This is explained by the fact that certain important mining investments, like that of Sodimiza, are not in the table. This company was contracted before the code came into force. However, the contract is in the spirit of the code and the ordinance law, and Sodimiza enjoys all the other advantages foreseen by the code or by their valid legislation.

When one takes Sodimiza into account, the part of the mining industry in the whole of investment in course is larger than the 24.4 percent mentioned in Table 16.3. In the long term, with the new research being undertaken, we think that this part will always be most important in the group of projected investments.

Under the code, Gecamines received the following fiscal advantages: (1) exoneration from the export tax, the payment of the tax on business turnover, as well as the taxes on profit and all other duties or charges whatever, relative to production of over 360,000 tons of copper and 11,000 tons of cobalt; and (2) exoneration from import tax on equipment for realization of programs of expansion. Also the decree stipulates that profits that are exempted from taxes and inputs cannot be utilized until the expiration of the period of exemption, ex-

TABLE 16.2

List of Briefs of Mining Enterprises Presented to Commission on Investments Before December 31, 1972

Name	Object	Locality	Assets (in Zaires)	Date	Percent
Gecamines	Extension of capacity for extraction and production of copper	Shaba	40,900,000	2/7/70	78.0
Somitef	Concession of mining research and exploitation	Shaba	375,000	12/10/72	0.8
Ahibraki	Casseritic treatment and new exploitations	Kbili	118,660	—	0.2
Metalkat	Extension of factory for recovery of zinc	Kolwezi	600,000	13/10/72	1.1
St. Zaïroise	Mining research and exploitation	Kivu	1,400,000	5/5/71	2.7
Belgi-Kamines (Zaire)	Extension of activities and casseritic extraction	Kampene (Kivu)	105,000	—	0.1
Seremza	Mining research	Haut-Zaire Kivu-Kasai North Shaba	200,000	Refused	0.4
EMZ	Extension, exploitation of ore and casserites	Kalenge Wulibile (Shaba)	702,000	20/3/72	1.3
Patel-Zaire	Copper smelting	Bas-Zaire	8,085,625	Study	15.4
Total	—	—	52,586,225	—	100.0

Source: List of briefs of the Management of Investments.

299

TABLE 16.3

Investment Projects Presented to Commission until
December 31, 1971

	Number	Assets (in Zaires)	Percent
1. Mining industry	9	52,486,285	24.4
2. Foods industries and agricultural activities	39	28,570,716	13.3
3. Textile and leather industries	14	20,250,598	9.4
4. Chemical industries	14	16,936,884	7.9
5. Nonmetallic mineral industries and construction industry	25	38,737,213	18.0
6. Metal fabrication and electric industries	22	13,587,706	6.3
7. Transport	8	28,984,679	13.5
8. Wood and wood-derived industries	12	6,210,315	2.9
9. Miscellaneous	17	9,249,514	4.3
Total	154	215,013,910	100.0

Source: Table derived from the statistical lists of the Management of Investments.

cept to finance the investments foreseen in Gecamines program of development or for the development of loans contracted for the realization of this program.

CONCLUSIONS

Everybody agrees that the Zairois code is very liberal. The mining enterprises profit from many concessions. For these enterprises, which do not need protection to be profitable, to be given generalized exemptions is a luxury. Zaire must make a balance between the concessions accorded and the contribution of these investments to national production and the formation of capital. Vigilance must be the basic rule of our political economy—for, despite our political desire to end foreign monopolies, the current situation obliges us to deal with international capital. For Zaire, the maintenance of order, stability, and security constitutes the essential guarantee for the attraction of foreign capital.

SOME PRESENT PROBLEMS
OF THE ZAIROIS
MANUFACTURING INDUSTRY
Luabeya Kabeya

This chapter attempts to evaluate the major features of current problems of the manufacturing industry of Zaire, as they appear seven months after the measures of November 30, 1973. The industrialization of Zaire was fundamentally the work of Belgian colonization. This occurred in two distinct phases. The first covers the period between 1920 and 1945, broken by the Great Depression of 1930. During this period the country began to produce cement (1929), soap (1922), sugar (1929), cigarettes (1929), beer (1926), and cotton cloth (1929).[1] The manufacture of cigarettes (1929) disappeared in the Great Depression and did not reappear until after World War II. The second phase started after World War II and ended in 1957. It saw the establishment of the current manufacturing industry of Zaire, including margarine (1950), cigarettes (1948), and metal containers (1948).[2]

This industrial development is centered around two poles. The first, the industrial center of the Shaba, started with the mining and metals industry. The reasons for development of this center include distance from coasts; the low grade of exploited minerals, which barred economic transport over long distances; the high volume of mining employment, creating demand for consumer industries; the need for regular and economic provisioning of the mining industry; industry attracted by osmosis. It constitutes the present industrial base of the country. It is principally characterized by an industrial triangle: (1) industries of basic products (copper, cobalt, zinc, various metals, cement, sulfuric acid, chlorate of soda, glycerine, explosives); (2) textiles industries (mostly cotton); and (3) consumer industries (bakeries, flour mills, breweries). The second center, at Kinshasa, is characterized by a strong consumer industry (food products and drinks, metal fabrication, transportation of plastics, footwear industry, provisions and equipment industry, cement, heavy metal fabrication, boats, woodworking tools, buckets, and water tanks, textiles industries, forestry, and paper industries).

By June 30, 1960, the country had fundamentally two categories of industry: a primary industry, producing basic industrial products from local raw materials. This includes the nonferrous metals and related metals industry; the chemicals industry tied to the mining and metals industry; the nonmetallic minerals industry (bottles, cement, other construction materials); the textiles industry; the forestry industry; and the fabrication of sugarcane and oils. The other is a secondary industry, the transformation of basic industrial or agricultural products, mostly imported. These include metal fabrication, soap factories, breweries, margarine factories, plastic transformation, bakeries, paper and container factories, and confection industries.

The nonferrous metals industry, and the allied metals and chemical industries tied to it, provide none or almost none of their base products for the transformation industry in the second category, which produces mostly for the internal market. The exception is copper, for which there exists a small unit of transformation; Latreca makes cables and electric wire from Zairois copper. Aside from this, there is no direct relation between these principal basic industrial products of the country and Zairois transformation industry.

In the colonial economy, mining products, basic metal products, and some agricultural products (tea, coffee, cotton, oils) were destined solely for export although very small amounts were used for the internal export market.* Currency from the sale of these products serves, among other things, for buying raw materials needed to run the country's transformation industry. Thus, the development of the transformation industry of Zaire depends among other things on the country's capacity to import (which, itself, is a function of the amount exported, its structure, and respective prices on the world market) and the division of such currency between various sectors needing it in the country.

In 1954, however, it became apparent not only that the rate of growth of the manufacturing industry was greater than that of the export sector (14.3 percent against 7 percent a year between 1950 and 1957), but also that the pursuit of industrialization required, among other things, the use of increasing amounts of currency resources for the import of raw materials, to the detriment of other needy sec-

---

*For example, Latreca buys copper from UMHK at the world price, including assurances and transport, when Latreca has its seat of operations in Lubumbashi, like UMHK. Again, Zairois palm oil is exported, and some Zairois consumers import it back from Belgium.

tors of the country.[3] This dilemma became obvious with the drastic
reduction in the country's capacity to import following the first years
of independence. The scientific circles of the country, first, and the
government, thereafter, rapidly took note.[4]

The result was a development strategy centered on the installation in Zaire of provisions and machine industries, whatever their
economic feasibility (the financial feasibility, at least, would be assured, thanks to the customs protection and the readjustment of
prices). Although this is not the place to discuss this strategy, it did
at least have the merit of making an effective start.

In effect, the effort to industrialize in the course of the postindependence period consisted mainly of establishing provisions industries, with the exception of the mining and energy industries (hydroelectric dam at Inga). There are Sozir (petroleum refinery in Moanda,
established in 1963 and entering production in 1967), Goodyear (entered
production in 1972), Ciza (the state cement factory in Kimpese, inaugurated in May 1974), the iron and steel industry of Maluku and the
Kisangani textiles industry (to be inaugurated November 24, 1974, if
all proceeds as planned), Midema (flour mill at Matadi, built in 1969,
entering production in 1973), and other units of modest dimensions.

This, in sum, is the direct contribution of independent Zaire to
the manufacturing industry of the country, a contribution that will
greatly expand when the electricity from Inga becomes available and
the numerous projects just undertaken are completed. The indirect
contribution consists in the establishment of instruments capable of
stimulating the industrialization of the country, which will be possible
when other conditions are realized. At the present time, there are
three such instruments. There is, first, the investment code, promulgated in 1969 and legally revised in 1974. Whatever its impact on the
country's industrialization, the most important lesson to be taken from
it is that an investment code, whatever its degree of liberality, does
not in itself constitute a general development policy for the country, or
unite all the favorable factors that attract investors, whether national
or foreign.

In the second place, there is Sofide (development finance company, established in 1970 with the help of the World Bank) to specialize in financing investment projects. According to the report of the
Bank of Zaire, Sofide has approved, with a view to financing, 42 projects since its creation, for a sum total of 12.7 million Zaires (1 Zaire
equals $2). After deducting for repayment, the intervention of Sofide
comprising both loans and participations adds up to 10.3 million Z by
June 30, 1973, of which 8.7 percent is in the form of participation.
At this date, the volume of credits accorded and already used has
risen to 6.6 million Z. Table 17.1 shows the division of economic
activity assisted by Sofide, by sector.

TABLE 17.1

Percent of Sofide Participation, by Sector, July 30, 1973

Sector	Participation
Manufacturing industry	64.2
Construction and public works	26.2
Transport	5.4
Agriculture	1.3
Extractive industries	1.0
Commerce	1.0
Miscellaneous	0.9

Source: Bank of Zaire, Rapport Annuel 1972-73, p. 234.

The third instrument is the IGP (Institute of Investment Administration) created in 1969 and tied to the presidency of the republic. This institution is charged with managing all state investments, direct and indirect, in commercial enterprises established in Zaire and outside. When the importance of participation by the Zairois state in the country's mining, industrial, and agricultural activity is recalled, participation that sometimes reaches the level of that in the period of the Independent State of the Congo, * the importance of the IGP and the role it may eventually play in the orientation of the national economy can be understood.

A common characteristic emerges from the group of Zairois manufacturing industries, with the exception of metals industries. Fundamentally, all these enterprises are turned toward the interior. Besides the export of cement, which seems to be quite regular, there are no regular exports from this sector. Neither the tax structure nor the enterprises themselves are organized to facilitate competition on the external market. An effort is now being made through the creation of the National Center of External Commerce to study all questions relative to the promotion of Zairois exports, particularly industrial manufactured products. The importance of the Zairois manufacturing sector is indicated by its contribution to employment, to the gross domestic product (GDP), and to the treasury.

*Editor's Note: The Colonial Government was extensively involved in a wide range of economic activities in the Congo.

On the question of employment, the growth in employment has taken place in the modern sector at the expense of the subsistence economy. The continued expansion of employment depends on the modern sector, primarily through diversification of its output. At the present time, the manufacturing sector is the largest contributor to employment, after the agricultural sector, as indicated by Table 17.2. Manufacturing industry's contribution to the GDP is third among the productive sectors, after the mining and metals sector and the commercial agriculture sector, as Table 17.3 shows.

In relative terms, the contribution of the manufacturing sector to the GDP is 8.04 percent, 7.98 percent, and 8.55 percent, in 1970, 1971, and 1972, respectively. Manufacturing industry's contribution to the Treasury in 1969 is shown in Table 17.4.

Considered all in all, the manufacturing industry of Zaire has developed admirably in recent years. Between 1970 and 1972, its rate of growth averaged 7.8 percent yearly. According to the Bank of Zaire, the index of its production passed 9.3 percent in 1972 as compared with 1971. According to the same source, this encouraging evolution is explained by the entrance into production of many investments made from 1970 to 1971, the improvement in local supply, the indirect protection established by a temporary tax on imports, and the rise in external prices following the adjustment in parities.

What are the nature of the problems of this industry? The Bank of Zaire, in its Annual Report for 1972-73, points out that industrial producers face a problem because the cost of raw materials and certain other charges are constantly increasing. This problem seems real, especially if one takes into account the recent increase in internal prices of fuel, which on average exceeds 50 percent. Those enterprises that consume a lot of fuel are in serious difficulties as long as the price of their products remains fixed. In this regard, it seems that Ciza has suffered losses on the order of 50 percent.

TABLE 17.2

Percent of Permanent Employees in Zaire's Economy, by Sector

Sector	1969	1970
Agriculture	27.3	28.8
Mining and metals industry	15.3	14.7
Manufacturing industries	27.5	26.7
Construction and public works, water and gas	5.0	6.0
Tertiary sector	25.0	24.0

Source: Bank of Zaire, Rapport Annuel, 1972-73, pp. 63-64.

TABLE 17.3

Contribution of Different Sectors to GDP
(in millions of Zaires at 1970 values)

Sector	Value Added		
	1970	1971	1972
1. Commercial agriculture	79.1	85.3	88.6
2. Mining extraction and metal-lurgy	211.5	219.1	230.1
3. Manufacturing industries	77.5	82.0	90.1
4. Construction and public works	30.0	34.3	30.7
5. Water, electricity	8.7	9.5	9.6
Total	406.8	430.2	449.1
6. Transport and telecommunications	75.9	83.8	75.8
7. Commerce	116.0	126.9	132.3
8. Services (including state services)	222.7	236.2	245.6
9. Estimated bank production	-4.8	-4.9	-4.8
Services sector	409.8	442.0	448.9
Total goods and services	816.6	872.2	898.0
Import duties and taxes	51.4	56.7	58.8
GDP commercialized at market price	868.0	928.9	956.8
10. Noncommercial agriculture	79.8	80.6	81.4
11. Noncommercial construction	15.0	17.2	15.3
GDP	1,962.8	1,026.7	1,053.5

Source: Bank du Zaire, Rapport Annuel, 1972-73, p. 6.

Tied to this problem of price is the problem of those enterprises that want the rate of return on their investment to rise to more than 20 percent. For these last enterprises, the reason for their complaint is not a growth in costs that they are unable to cover but rather an inability to attain a rate of return equal to what they expected. In a situation where the amortization of capital is very rapid (three to five years), it is possible to see what will happen to the general price level if all these complaints are to be satisfied.

A more fundamental problem seems to us to be the environment surrounding the development of economic activities in Zaire. Despite

TABLE 17.4

Government Revenue of Zaire by Sector, as Percent of
Total Revenue, 1969

Sector	Tax (percent)
Commercial agriculture	4.30
Mining and metals extraction	42.80
Manufacturing industry	24.00
Energy	0.80
Banks and other financial establishments	3.70
Immovable businesses	0.08
Transport	17.05
Other services	0.68
Commerce	6.60
Total revenue	100.00

Source: Rwanyindo Ruzirabuoba, Le Role de l'Impot sur le Revenue dans le Mobilisation des Ressources et dans le Developpement Economique. Cas de la Rep. du Zaire (Kinshasa: University of Zaire Press, 1974), p. 253.

the measures of November 30, 1973, it seems that the framework in which economic activity must evolve in Zaire is still not clearly defined, at least for everyone. And, in the absence of a clearly defined development plan, it is difficult to organize investment efforts. Tied to this environmental problem is the problem of communications infrastructure in the country, which certainly does not fulfill, in its present state, the needs for rapid development of the country. The second fundamental problem of the manufacturing industry of Zaire is that it is very diversified, and its impetus depends on external forces. Moreover, the monopolistic character of existing units does not permit harmonious development. The consideration of such problems seems to us to require more fundamental development of the manufacturing industry of Zaire.

NOTES

1. B. Luabeya Kabeya, "La Structure des Importations et Son Impact sur le Developpement Economique: Cas du Congo," Memoire de Licence, 1966, pp. 109-28.

2. Ibid., pp. 129-52.

3. See J. L. Lacroix, Industrialisation au Congo: La Transformation des Structures Economiques (Paris: Mouton, 1966). See also Luabeya Kabeya, op. cit., p. 51.

4. The work, in particular, of J. L. Lacroix, is at the base of this strategy.

18

THE EFFECT OF
THE MINING INDUSTRY ON
AGRICULTURAL DEVELOPMENT
Lumpungu Kamanda

The evolution of Zairois agriculture in the course of these last years is a constant worrying preoccupation for those responsible for the economic policies of the country. During the agricultural campaign in 1967-68 (proclaimed as the year of agriculture) agricultural production expanded considerably. Unfortunately, the rate of growth has not been maintained. In 1969, production fell from 1968, and, from the end of 1970 to the present, the agricultural recession has been accentuated and prolonged, leading to certain very difficult problems in provisioning the urban centers. What are the causes of this disquieting situation? What is the role the mining industry can play in encouraging the development of agriculture, first on the regional scale and then on the national scale? We will try to answer the second question.

IMPACT OF THE MINING INDUSTRY ON THE
OTHER INDUSTRIES AND ON AGRICULTURE

The Effect of the Mining Industry on Other
Industries Established in the Shaba

Economists, used to the models of industrial relations for diverse branches of economic activity, always try to establish the influences of the growth or recession of the sector on others. Such impact is easy to make out when it relates to an import group (the mining industry is certainly one) and above all when it relates to linkage effects, both forward and backward. But, on other industries, the influence of the mining industry is not so easy to see.[1]

The Role of the Mining Industry in Agriculture
in the Shaba and in Other Regions
of the Republic of Zaire

We will try to see the weight of the mining industry on the start
and the adaptation of agriculture to the imperatives of growth and development. Some figures taken from an agricultural inquiry[2] recently
can perhaps give us the elements for analysis and perhaps an answer.
Taking 1958 as a base year, the areas cultivated by region grew by
1970 by 402.2 percent in the region of Bas–Zaire; 88 percent in the
region of Kasai–Occidental; 76 percent in the region of Kivu; 5 percent in the region of Bandundu; 20.75 percent in the region of Kasai-
Oriental; and 13.63 percent in the region of Shaba. Other areas, by
contrast, fell 17.92 percent in the region of Haut–Zaire and 27.5 percent in the region of the Equator.

Productivity remained weak and stationary for the whole country, except for some capitalist plantations. Unhappily, the difficulties of the last two years accelerated the degeneration of Zairois agriculture. In general, one finds the agricultural economy called
"primitive" with deteriorating techniques and production increases
due only to the extension of cultivation to new territory. The agriculture of the Shaba progressed only by 1 percent a year. This is an insignificant increase, far inferior to the demographic increase. For
the Shaba, as for the other regions of the republic, general figures
hide large disparities on the level of subregions and zones. The agricultural zones of the Shaba are situated on the margin of the copper
zones presently being exploited. In the southeast are the zones of
Sandoa–Kaniama, and in the north the zone of Kabongo and the subregion of Tanganyika. The divorce between the mining industry and its
agricultural hinterland is total.* It is possible to find an explanation
of this state of affairs in the demographic distribution in the interior
of Shaba and in the complementarity of agricultural regions.

Demographic Distribution in the Shaba

The two large agricultural subregions of Shaba are also subregions with high rural density relative to the regional average. The
larger part of the mining region has long been a zone where rural
density was only three inhabitants per square kilometer. The realization of underground riches could take place only through labor re-

_____

*We used the term "agriculture" in the original sense, exclud-
ing cattle rearing, for which the Shaba occupies third place (following
Haut–Zaire and Kivu).

cruited from other parts of the country. It thus appears that since the mines are not surrounded by a swarming rural population, their problems have not been sympathized with by those responsible for mining exploitation. One has long sought the solution in regional complementarity.

The Complementarity Between Industrial and Agricultural Regions

The development of industry was based on the creation of cities whose food needs could not be satisfied by local production alone.* The liaisons of Shaba with the other regions became imperial. As the neighboring regions did not experience the phenomenon of rapid urbanization, the complementarity between agriculture and industry did not seem a necessity at the heart of the Shaba region. Progressively, however, following the pressure of events, this conception has had to change. In 1959, the Shaba supplied half its own maize needs and received the rest from Kasai (8,570 tons), Kwilu (3,600 tons), and Zambia (5,270 tons). But in 1967, when the import of maize by the Shaba reached 68,000 tons (of which about 2,000 tons came from the subregion of Tanganyika), local commercial production reached $\pm$ 25,000 tons.

The shipments from Kwilu and Kasai practically ceased during the second half of 1960. This is the problem that confronted the Shaba Flour Mills and with them the companies charged with safeguarding the health of city workers. Agriculture/mining-industry relations seemed to start afresh. Continuing imports led to the loss of currency, which the country needed, and also, since the priority was never established, the import of essential food products was submitted to negotiations that often lengthened the period of difficulty.

The mining industry in nonferrous metals only influences agriculture in an indirect way, through demand for food products by urban centers and through diversification of the products of the chemicals industry indispensable for the treatment of metals. For the first point, the action of industries established in the Shaba is part of the general framework of actions in the local rural environment or on those of further regions whose contact makes possible the regular provisioning of the urban centers.

The credit facilities accorded at other times by the flour mills to the traders of Shaba, especially those in the agricultural zones,

---

*It should be noted, however, that the production of vegetables is quite developed in the Shaba, but their importance over all is not great.

were part of this framework. One can only regret that such procedures have been supplanted without being replaced by other, more
efficient ones. It is probable that if research on the possibilities of
local provisioning was undertaken with the same ardor as importation,
a big step would have already been taken on the road to coordination
of the two key sectors of the national economy. On the national and
regional scale, one is more and more persuaded that the solution of
Shaba's food problem must be found, not in importing but rather in
the expansion of local and national agriculture. Taking into account
the importance of the mining industry in this part of the country, can
one hope that the agricultural programs begun in this part of the country will know complete success without the participation of the mining
companies?

It seems impossible to reply in the affirmative, not only because
the manufacturing industry must have large markets to expand but
above all because the workers, like the other inhabitants of the Shaba,
risk suffering in the future from the difficulties of importing and an
unbalanced diet, which would affect the workers' productivity.

Finally, there is the chemicals industry, whose role in agricultural development needs no demonstration. Some doubt, again, the
receptivity of the African peasants in general and Zairois peasants in
particular to innovation. Numerous examples show, in addition, that
one cannot reduce agriculture to a docile instrument in the hands of
technicians and bureaucrats.

Without agricultural openings, the chemicals industry is con-
demned to disappear when the mines are exhausted. And Shaba thus
risks playing the role, not of the Ruhr, but of the Alsace-Lorraine of
Zaire. If the Shaba wants to play, as in the past, the role of balancing
metropolis and pole of growth, it needs to establish a certain symbio-
sis between its industry and its agricultural hinterland, the only way
to assure a durable economic development.

CONCLUSIONS

It is very difficult to draw definitive conclusions. One must,
however, remember that (1) the start of the Zairois economy through
the mining industry is an incontestable historical fact; (2) the contribution of the mining industry to the regional agricultural development
has not been of the same weight as in the national economy; (3) the participation of the mining industry in agricultural projects in the Shaba
is necessary, but changes in this participation remain to be detailed in
a debate open to those interested.

NOTES

1. "Economic Situation," Zaire, Department of the National Economy, 1971.

2. Agricultural Census by the UN Food and Agriculture Organization (FAO) in Zaire; results obtainable from the section on Agricultural Economics of the Faculty of Agronomy, Zaire.

V

THE EFFECT OF 51 PERCENT GOVERNMENT OWNERSHIP OF MINES AND MANUFACTURING ON ZAMBIAN DEVELOPMENT

19

THE LEGAL FRAMEWORK OF COPPER PRODUCTION IN ZAMBIA

C. M. Ushewokunze

The past 20 years have seen the political transformation of Zambia from colonial status to independent statehood. Its colonial economy was dominated by the British South Africa (BSA) Company and other foreign mining companies. These enterprises essentially dictated economic policies and activity, by their hold over the mineral resources of Zambia. During the last 10 years the state has made and continues to make considerable efforts to assert sovereignty over its natural resources, to secure that benefits thereof are enjoyed by its own people.

The process of change in economic control was launched in 1968 and followed in 1969 by specific reforms of the mining industry. This reformation demanded a restructuring of the legal system within which economic activity is undertaken, with a view to facilitating Zambian state control of the exploitation of its mineral resources, and the utilization of the proceeds thereof. In this chapter, the object is first to identify the substantive aims of government policy relating to the mining industry, and, second, to study the main provisions of the mining law and attempt to determine their effect in relation to government policy. Where possible, an attempt will be made to suggest alternative approaches to solving the essentially legal problem of securing state control.

In the exploitation of natural resources, and in particular of copper, the law grants rights, imposes duties, and confers powers to facilitate production and the use of profits thereof. To a considerable extent, the business of economic development is dependent upon the policy objectives and the nature of the law and legal framework within which it is undertaken. Thus, for example, whether the exploitation of Zambian copper develops its economy, in the sense of promoting expansion, increase, and diversification of productive industries, or, conversely, underdevelops the economy by preventing, stagnating, or destroying the growth of such industries, largely rests upon the rights, duties, and powers the law confers on the participants to facilitate their operations. Our contention here is that the mining legal framework as presently structured does not adequately enable the state to protect the interests of its national economy in development. The problem facing Zambia and indeed other developing countries dependent on mineral resources for their economic development is not whether but how to structure a legal framework of production that will effectively restore to the state on behalf of its people the right freely to dispose their natural resources in accordance with their national interest.

The Economic Reforms initiated by President Kenneth Kaunda in 1968 and expounded and extended in 1969 were in his view intended "to reorganise the Zambian economy so as to increase the capacity of Zambians to control their own economic destiny." In the mining sector, the reforms were to enhance the nation's economic independence. The independence sought consists in the right and power of the Zambian state to own all minerals in its soil; to determine by whom and on what terms any prospecting and mining for minerals is undertaken; and ultimately to decide how much of the proceeds of sale the investor of capital can retain for his own use and what goes to benefit the industry and the nation.

In Zambia, this goal was to be achieved by immediate complete control of mining rights and progressive control of the mining industry. It is against this background that any legal measures enacted or agreed upon to give effect to the policy must be judged. The function of law, then, is to lay a foundation for progressive control. In this respect the legal substratum must be so structured as to give the state power, which can be exercised flexibly to meet the exigencies of a developing economic situation. Hence it ought to provide a general power framework for the achievement of state policy goals, while at the same time safeguarding the interests of foreign investors to the extent necessary to achieve the state's goals.

LEGAL PRESCRIPTION OF ECONOMIC DEPENDENCE: PRE-1969 MINING LAW

In the preindependence era, all rights to search and mine for minerals in the territory of Zambia were acquired by and vested in the BSA Company incorporated in London in 1889. This "title" was based on agreements concluded with African chiefs or individuals in Northeastern Zambia between 1890 and 1891; and concessions secured from Lewanika the Litunga of the Barotse, in 1900 and 1909. From

1895 to 1923, the company operated both as a commercial company and as the government of the territory. It relinquished its governmental role in 1923 and continued as a commercial company trading in mineral rights until 1964. As owner of mineral rights, it determined by whom and on what terms and conditions any mineral prospecting or mining was to be carried out. In return, it was paid a royalty on the value of minerals produced.

The company was absolute owner of mineral rights, except to the extent it had relinquished its rights absolutely. It maintained an interest in the share capital of mining companies and imposed minimum expenditure obligations. In this way, it retained powerful instruments for controlling and influencing mining development policy and mineral production. Further, it retained the power to make special grants, which were outside the terms of the mining legislation, and this proved to be crucial in the development of copper mining in Zambia. Thus, in order to encourage large companies, with know-how, finance, and organization, to undertake through prospecting, and to develop any discoveries to production, the company used the power to issue exclusive grants of prospecting concessions. The areas ranged from 18,000 to 5,000 square miles. These concessions were conceived as a form of security to justify an extended program of work. And they covered all the present Copperbelt Mines, from which the company derived most of its revenue. The end of the 1920s saw the emergence and dominance of Rhodesia Selection Trust Company Ltd., now Roan Selection Trust (RST) and Anglo-American Corporation of South Africa, over mining, proper, in Zambia.

Strictly, the protectorate colonial government had no right or title to the ownership of mineral rights, or to any of the royalty and rents recovered by the company. Its only power was that of taxation of company income, until 1950, when the company conceded 20 percent of post tax royalty. The company had a controlling influence, both in the Imperial British Government and in the territorial government, which ensured that no change could be made in mining law or taxation without its concurrence. Thus for over 64 years, it acquired close to 135 million pounds before tax and 75 million pounds after it in royalty payments. By 1964, its income was almost 7 million pounds a year after tax. Few of the profits were ploughed back into the country's economy, and there was no legal power by which government could compel the company to do so, except increasing the rate of taxation. The latter in any case remained very low, largely because of the controlling influence of the company in territorial and British government circles.

In the circumstances, the independence government of Zambia expropriated the rights of the company for a compensation of 4 million pounds contributed equally by itself and the British government. The basic argument was that the company had no valid title over the country's mineral resouces. Hence, in the period 1964-69, all mineral rights previously held by the BSA Company were vested in the president on behalf of the republic. Legally the Zambian government inherited and operated within a mining legal framework built by BSA to serve its own interests. They had not expropriated that company's share holdings in the various mining companies. And they could only make prospecting grants in areas open to prospecting. That excluded areas covered by valid and operative grants previously made by the company. Large areas of the country were held in perpetuity under such grants, in which neither prospecting nor mining was taking place. Government had no legal power to compel such development nor could they compulsorily acquire the land and terminate the concessions. Further, the rights of the concession companies were safeguarded in the independence constitution. This could only be amended by a referendum in which the state had to secure 51 percent support of all voters on the voters roll.

The state was now entitled to royalty and rents and could make special grants and grant prospecting and mining licenses on their own terms. Government revenue increased considerably. In addition to ordinary company income tax, which was increased to 45 percent, and to a royalty on the value of minerals produced, the state imposed a special copper export tax. This new tax was intended to take account of increased world prices of copper. It was applied to when the London Metal Exchange (LME) price per ton of exported copper exceeded K600 (K1 equals $1).

Despite increased revenue, Zambia did not control the mining industry as such. It had no legal power to determine the volume and rate of production or investment by the mines. Nor could they dictate the level of dividend distributions and force the reinvestment of surpluses for further development in the industry or indeed other sectors of the economy. In addition there were no means by which the state could influence management policies to secure the training and employment of Zambians in technical and managerial jobs. The industry was basically owned and operated by foreign companies, which imported investment capital, and technical and managerial manpower skills, to utilize the huge pool of cheap labor and export the profits.

1969 MINING POLICY AND LEGAL REFORMS

On June 17, 1969, the necessary majority voted in a national referendum to amend the independence constitution, giving the government power to exercise greater control over the mining companies themselves, through a series of new laws. The enactment of the Mines and Minerals Act, 1969, terminated all concessions and special grants, without compensation, except that existing mine companies were given a first option to retain up to 12.5 percent of their concession or grant area. Largely, they released areas in which they were not carrying out mining operations. Areas retained were held on temporary licenses valid for six months, to allow for compliance with the requirements of the act. The act itself created institutions, defined rights and duties, and laid procedures for the acquisition, holding, and termination of mineral and mining rights. Second, the state negotiated a takeover of 51 percent equity shareholding in existing mines. A series of agreements were thus concluded between itself, Zimco, and the newly created companies, Roan Consolidated Mines (RCM) and Nchanga Consolidated Copper Mines (NCCM), on the one hand, and the mining companies, Roan Section Trust, Ltd., and Zambia Anglo-American Corporation, on the other.

Third, the Mineral Tax Act changed the basis of taxation for the new mining companies from a royalty charged upon the value of minerals produced, in general, despite profitability of the mine, and an export tax, to a mineral tax of 51 percent of profits and a company income tax on the balance of the same. All capital expenditure was a 100 percent deduction from profits for tax purposes. No attempt was made to classify the type of capital expenditure, or its utility.

It may be noted that by controlling the ownership of all mineral rights in Zambia and reducing the rights of every other holder to a specific term of years, the state placed itself in a position to assert its sovereignty over its vital natural resources. Furthermore, the acquisition of 51 percent equity shareholding suggests an intention to determine the level of mining activity in all its different aspects and to place the national interest in development as the paramount consideration in exploration and mineral production and development policies. In general corporate law, 51 percent shareholding at least entitles the holder to appoint a majority of the board of directors, which is the policy-making organ. And these possess a controlling influence on the appointment of the managing director and the policies he pursues.

GENERAL STATUTORY FRAMEWORK

The five basic officers or institutions under the Mines and Minerals Act are the minister, chief mining engineer, chief inspector of mines, director of geological survey, and the Mining Affairs Tribunal. The minister is head of the Ministry of Mines and Industry. Under the act, he decides whether to grant or reject application for prospecting licenses, mining licenses, and amendments thereto. Below the minister is the chief mining engineer, who is appointed to "supervise and regulate the proper and effectual implementation of the provisions of the Act." All applications for mining rights to the minister are made through him. He decides in consultation with the director of geological survey, whether to grant exploration licenses and whether to accept amendments to all types of licenses. Although formerly responsible for registration of mining rights granted or transferred, he has delegated the function to the director of survey. The engineer determines the form and content of all licenses, considers and decides on amendments to development programs. Lastly he settles all mining disputes subject to appeal in the High Court.

The chief inspector of mines has general responsibility for safety of prospecting, exploration, and mining operations. He is head of a force of mine inspectors with wide powers of entry and inspection. The director of geological survey and survey officers possess similar powers of entry, search, and inspection of all prospecting and exploration activity. These are intended to enable them to undertake systematic regional mapping of the country and prospect on behalf of the republic. The director advises the minister on geological matters and assists members of the public seeking information on such matters. He maintains laboratory, library, and record facilities. The wide powers of entry, search, and inspection provide the state with ready means for securing information upon which to base its policies.

Above the licensing institutions is the Mining Affairs Tribunal. It is a judicial body composed of a judge and four mining experts. Its jurisdiction is to hear and determine appeals from decisions of either the Minister or the Engineer. It is not yet operational, however. (The tribunal is potentially a powerful institution in mining development. It may be called upon to decide on both technical matters and matters that may be a strong element of the national economic interest. The fact that it can give binding directions to a minister, politically charged to look after that interest, may be objectionable. This potential conflict between judicial and technical propriety, on the one hand, and the national interest may have contributed to delay in setting it up. Nevertheless the framers of the act appear to have intended to assure likely foreign investors that the state will abide by its own law.)

MINING RIGHTS UNDER THE ACT

The categories of minerals for which mining rights may be acquired include mineral oils, diamonds and gold, industrial minerals, building minerals, and all other minerals. The latter class subsumes

copper. The searching for and mining of copper must be licensed.
At each stage the applicant for a particular right must produce a program of intended operations and must possess adequate financial resources, technical competence, and experience in order to carry out
effective operations. To induce a license-holder to develop the land,
the act requires him to make minimum yearly direct expenditures per
square mile: prospecting license: K25; exploration license, first
year: K2,000, second year: K4,000, third year: K6,000; renewal
period: K10,000. Penalties of up to K5,000 may be imposed for failure to conform to the approved program of development. Ultimately,
a holder is liable to forfeiture and termination of rights. The state,
in law, has considerable power to ensure that the mining industry and
its development operate in the interest of the national economy. The
cumulative effect of the provisions of the 1969 act has been to release
large areas of the country for prospecting and mining by companies
other than existing ones. Consequently, there has been general activation of interest in prospecting and mining in Zambia among foreign
mining companies.

To effect 51 percent state ownership of the preexisting RST
group and the Zambia Anglo-American Company (Zamanglo), two new
companies were created: RCM and NCCM to hold the mining, smelting,
and refining assets of the RST and Zamanglo groups. The state,
through Zimco, acquired 51 percent shareholding in each of the new
companies, which was credited as fully paid, and in return issued
Zimco Bonds and Loan Stock respectively. The state interest was
initially vested in the Mining Development Corporation but is now directly vested in the government through the minister of finance.

The minority interest in NCCM is held by Zambia Copper Investments (ZCI), a wholly owned subsidiary of Zamanglo. Both ZCI and
Zamanglo are registered in Bermuda. In RCM, the 49 percent is split
so that RST (International), a wholly owned subsidiary of American
Metal Climax (Amax), holds 20.4 percent; ZCI holds 12.25 percent;
and the general public (U.S. small investors) holds 16.35 percent.
Both RST and Amax are incorporated in the United States.

The master agreement between RST and Zamanglo, on the one
hand, and the government and Zimco, on the other, sets out the basic
terms and principles for the acquisition of the 51 percent interest.
All other legal instruments were subsidiary to this agreement, including the following: indenture and trust deed creating the bonds and
loan stock, and the terms thereof; management and consultancy contracts between RCM and RST, NCCM, and Zamanglo; memoranda and
articles of association of RCM and NCCM; sales and marketing contract RCM/RST; NCCM/Zamanglo; arbitration agreement; memoran-
dum of understanding, and schemes of arrangement under the Companies Act; acts of Parliament—for example, Miners Acquisition Acts,
1970 to 1973.

The ordinary share capital of each of the mining companies is divided between "A" government and "B" minority shareholders, as different classes. Each class is represented on the board of directors by "A" and "B" directors, respectively.

The takeover regime, in contrast to the 1969 act, is contractual. The ultimate terms are largely determined by the initial bargaining strengths of the parties. The key question is whether the agreements were so structured as to provide a power framework for the achievement of the state's policy objective of control of the mining industry. Payment for the 51 percent share of ownership acquired by the state was in principle based on the book value as it appeared in the audited accounts of each group on December 31, 1969. For Zamanglo it was $178.8 million; but problems over accounting principles resulted in an agreed valuation for RST of $117.8 million. RST was issued Zimco bonds redeemable over 8 years, and Zamanglo Loan Stock repayable over 12 years, both at 6 percent interest. All repayments were to be in U.S. dollars and in equal half-yearly installments.

The mining companies welcomed the decision of the state to participate in mining, because it guaranteed maximization and enjoyment of profits. In having conceded majority shareholding, their primary interest was to provide effective security for repayment. In this regard, they deployed several legal mechanisms to take care of potential exercise of state sovereign political and legislative power, together with possible fluctuation of copper prices. Government, on the other hand, appears to have meekly accepted its being placed in sovereign bondage as inevitable.

Compensation was to be paid out of the government's share of mining profits. Repayment in U.S. dollars was intended as a safeguard against Kwacha devaluation. The state was bound to pay a guaranteed minimum annual sum of U.S. $19 million and $15 million to Zamanglo and RST respectively, even if the price of copper fell to zero, whether or not the mines were actually producing salable copper (or any copper at all, as for example when Mufulira mine sustained a cave-in in 1970 and production was temporarily suspended). In such an event, the Zambian taxpayer would have been bound to pay, or else the state would have had to borrow and pay.

The payment was to be accelerated if dividends received by the state exceeded a certain sum—that is, $28.6 million—after the first year. Two-thirds of excess above this figure had to be applied in redemption. But if the converse situation arose—if there was a drastic fall in dividends—no provision was made to decelerate payments by suspension of or extension of period of repayments. Evidently this was highly inequitable. The state was not even in a position to save for repayments in lean years. To that extent the risk of a fall in prices and natural disasters was considerably increased.

Furthermore, certain events such as the following constituted default and rendered the bonds and loan stock immediately repayable: failure to pay for 30 days; sale by Zimco of RCM or NCCM of all or a substantial part of the assets of the latter two companies; bankruptcy, winding up, amalgamation, compromise, or arrangement of Zimco, Mindeco, RCM, or NCCM—prejudicial to bond or stock holders; unlawful cancellation or abrogation by RCM or NCCM of management consultancy, sales, and marketing contracts; unlawful abrogation or breach by government or Zimco of obligation under the master agreement; material failure by RCM or NCCM to abide by any of the provisions of their articles of association intended to protect B shareholders; failure by RCM or NCCM to pay net of profits dividend to the extent required by the articles of association; unlawful cancellation or abrogation by government, Zimco, RCM, or NCCM of obligations to arbitrate disputes.

Significantly, immediate redemption was here made a general sanction for defaults, relating to substantially separate and distinct agreements signed by different parties. It applies to defaults by government, Zimco as holder of the state shares, RCM, or NCCM. And, further, it applies, both in respect of financial and nonfinancial breaches. In each case, the government was bound to pay and redeem the bonds and loan stock, despite the separate personalities and existence of the RCM and NCCM. The liability of government was both as guarantor of the bonds and loan stock and the government was thus liable only if the principal debtor failed; and as owner of the holding company, Zimco, which was equity controller of the mining companies.

The common element in the breaches, nevertheless, appears to be the fact that, each affected either the total amount payable on redemption—for example, where government reimposed tax and exchange control laws contrary to undertakings in the master agreement; or the ability of government to pay if there was an unlawful cancellation of management or sales and marketing contract. In this case the effect seems to have been considered as likely to reduce productive efficiency.

In addition to these safeguards, the companies secured that the state sovereign legislative power to enact new and change existing laws was strictly limited in several respects. Thus under the master agreement, no law was to alter or affect that agreement or laws giving effect to it, except to the extent contemplated by it. The agreement contemplated only laws giving effect to it. The same provision applied to the memoranda and articles of association of NCCM and RCM and to tax and exchange control laws permitting unrestricted remittance of dividends, interest, and capital to shareholders resident outside Zambia.

We have indicated that the primary interest of Zamanglo and RST was to ensure full and punctual repayment of the bonds and loan stock. But it seems to emerge from the safeguard arrangements that they subjected to the same security items that were not essential to secure payment—for example, failure by RCM or NCCM to pay dividends out of profits to the extent required by the articles of association and the freezing of tax and exchange control laws with respect to remission of dividends and management fees.

The freedom from tax meant essentially that the companies were liable only to mineral tax—that is, 51 percent on profits, and company income tax of 45 percent on the balance. These taxes could not be increased until the bonds were redeemed. Furthermore, dividends declared and fees paid could not be charged a withholding tax of 20 percent until redemption of the bonds. The exchange control rule that only 50 percent of dividends could be remitted did not apply to these companies. Inevitably, large sums of revenue were lost to the state.

The ultimate insurance for repayment was government's unconditional guarantee of the bonds and stock. This was coupled with an arbitration agreement under the World Bank Convention by which the International Center for the Settlement of Disputes, an agency of the bank, is the tribunal for arbitration of disputes between Zambia and RST or Zamanglo, relating to the takeover agreements. Any award by an arbitration tribunal constituted under the auspices of the center is enforceable through the machinery of the World Bank, the International Monetary Fund, and other financial markets, on which Zambia is considerably dependent. Consequently, the essentially private agreements were tied to the complex of Zambia's international financial relations. The credit-worthiness of Zambia was directly put at stake, as the ultimate security. The terms of the agreement seem to indicate that state negotiators were negotiating from weakness. Possibly the fact that the state asked to be invited to participate was exploited to full advantage by these multinational companies. They possessed a large reservoir of expertise. It would appear as if they simply presented the state with a draft, which it substantially accepted with little amendment.

As part of the package, Zambia permitted the shipping out of the country of nonmining assets of RST and Zamanglo. These included K17.9 million for RST and K27.5 million for Zamanglo—all liquid cash, together with their pension funds, previously accumulated in Zambia largely because of normal exchange control restrictions. The only conditions stipulated were that remission was to be by monthly installments over two years for RST and two and a half years for Zamanglo. Second, each company was to hold available for investment in Zambia K15 million and K12 million respectively over five years.

On the whole this was money being taken out of the Zambian economy and essentially defeating the aim of exchange control. The result was to increase the international liquidity of these companies, facilitating investments elsewhere. It deprived Zambia of much needed investment finance, which perhaps might have been utilized more cheaply than overseas borrowing.

On August 31, 1973, the president declared that the takeover contracts were working against the national interest, and therefore:

> In accordance with the mandate given to me by the nation, I have decided that with immediate effect:
>
> 1. Outstanding bonds shall be redeemed.
>
> 2. Steps should be taken to ensure that RCM and NCCM revert to the old system of providing for themselves with all the management and technical services which are now being provided by minority shareholders.
>
> 3. A new copper marketing company wholly owned by the Government should be established here in Zambia.
>
> 4. The minister responsible for mines shall be the chairman of both RCM and NCCM.
>
> 5. The Government will appoint the managing directors of both RCM and NCCM.
>
> 6. Mindeco shall cease to be the holding company for RCM and NCCM.
>
> 7. The minister responsible for finance will hold share in RCM and NCCM for and on behalf of the government.
>
> 8. The rest of the mining operations which are not connected with NCCM and RCM will continue to be administered by Mindeco as is the case at present, and the status of Mindeco therefore will be equal to that of NCCM and RCM.
>
> 9. Normal taxation provisions and exchange control regulations will apply to RCM and NCCM, and the minister responsible for finance and the Governor of the Bank are instructed to take appropriate measures.

Following this declaration, both the bonds and loan stock were redeemed in September 1973, in part using funds borrowed from the Eurodollar market. (These securities were freely marketable on the international financial markets in New York, London, and Paris. Further, there seems to have been no prohibition on the state purchas-

ing them on the open market, perhaps at a discount. Consequently
the cost of redemption might to some extent have been reduced, by effectively redeeming the securities in this way. However, the state,
in fact, announced its intention to redeem, and redeemed at par.)

Only these agreements or terms of agreements that were specifically tied to redemption, however, were effectively revoked by redemption of the bonds and loan stocks. Several terms were specifically tied to bond redemption. The most important were restrictions
on state power to terminate, alter, or affect the operation of the master agreement. Legislation giving effect to it could not be repealed
by law. For example, the Mines Acquisition (Special Provisions) Act,
1970, exempted from exchange control payments and remittance outside Zambia of dividends declared by RCM and NCCM, interest and
principal on the bonds and stock, and fees under management, contract, together with transfers of B minority share to nonresidents.
This act could not be repealed. Second, the state could not by legislation increase the aggregate direct tax to which RCM and NCCM were
liable. And the dividends paid to nonresident shareholders were to
be free of any withholding or any other tax.

By redemption, government has regained the power to increase
rates of taxation on the mining companies and their shareholders and
greatly increased its bargaining power. Lately normal taxation and
exchange control restrictions have been reimposed by the Mines Acquisition (special provisions) Amendment Act 1973.

On the other hand, provisions in the constitutions of RCM and
NCCM do not seem to be dependent upon redemption. Under the master agreement, neither the memoranda nor the articles of association,
it seems, could be altered by law. If this is correct, they continue
to be binding. The articles of association, regulating internal administration of a company, provide specific methods for their alteration.
The intention was to make amendments difficult so as to protect minority shareholders. Thus both A shareholders (government) and B shareholders must support the measure with a three-quarters majority each.
The form may be a written consent of the shareholders, or a special
resolution passed at separate meetings of each class. Further amendments also require support of a majority of the A and B directors voting separately. Thus in each case the minority must give consent.
Essentially this is the minority's right of veto. As long as the master
agreement continues to be operative, the state apparently has no legal
power to alter by law the constitution of the companies, except perhaps with the concurrence of the minority.

The articles of association continue to require a separate vote
of the A and B directors on certain issues, and to require that such
separate vote be held, if the B directors request it, on matters that
in their opinion affect the interests of their shareholders. Such mat-

ters include winding up or amalgamation or reorganization of RCM and NCCM, disposal of their assets, capital investment, prospecting and exploration expenditure, and so on. These are all issues vital both to the operation of the companies and to the national interest. The veto arrangements were essentially contradictory to state control over the mines and mineral production.

DIVIDENDS AND UTILIZATION OF PROFITS

In general corporate law, shareholders have no legal and enforceable right to a dividend until it is declared. And the board has a discretion as to the amount of dividend. The articles of RCM and NCCM impose a mandatory duty on the board to declare a dividend after making certain listed appropriations. The amount of the appropriations depends on market conditions and the company's short-term liquidity. Appropriations listed include preference shares and their redemption fund, capital expenditure, expenditure on prospecting and exploration, and reserves. The level of appropriations was subject to the veto power, placing the minority in a position to dictate that level. Whatever profits are declared, there is no legal duty to use the mineral profits to develop productive industries in other sectors of the Zambian economy. The state could only use its share of dividends for that purpose. Thus as a self-imposed necessity, the interests of the minority and majority were legally bound to be identical so far as the running of the companies was concerned. Both parties in practice had as a policy the maximization of production, profits, and dividends based on the operation of both companies on a commercial basis. However, it is submitted that the state has a wider interest than merely maximization of profits in disregard of social and national economic development policies (see Table 19.1).

Having thus safeguarded their interests, the minority, as stated by President Kaunda, were extremely scrupulous in that "they have taken out of Zambia every ngwee due to them." For investment purposes, it is understood they borrowed on the Eurocurrency market. To that extent they increased the mining companies' and the state debt burden.

MANAGEMENT AND CONSULTANCY CONTRACT
AND THE SALES AND MARKETING CONTRACT

Remuneration for management services consists of four elements: 0.75 percent of gross sales proceeds; 2 percent consolidated profit before income tax but after mineral tax; engineering fee: all

TABLE 19.1

Performance of Companies
(all figures in million kwacha)

	RCM				NCCM			
	1970[a]	1971	1972	1973	1970	1971	1972	1973
Total sales revenue	178.0	218.8	191	237.2	—	449	348	363
Profit before taxation	108.7	84.9	53.7	75.7	—	204	100	100
Profit after taxation	38.8	48.8	43	47.2	—	97	68	77
Profit after tax plus extra item	—	—	—	—	—	97	68	83
Dividends declared	15	22	20.5	—	—	51	36	36
Capital expenditure	12.7	28.4	336	—	—	43	42	59
Total loans borrowed	10.3	13.2	31.5	—	—	—	—	—
Loans for year	6.9	5.3	7.7	—	—	25.2	47	64.7[b]
Mineral tax paid	—	1.6	2.4	—	—	20	21.8	17.7
	—	4.22	9.8	—	—	19.84	17.7	—
Percent dividend distributions	39	45	49	—	—	53	53	43

[a]Six-month period.
[b]An additional K56.9 was obtained as contractor finance. Also, NCCM in 1973 borrowed K12.1m by issue
of bills of exchange against the value of shipped copper.

Note: K1 = $1.56.

Source: Annual Reports and RCM and NCCM figures.

330

costs incurred in or toward the performance of the services; and recruitment fee of 15 percent gross emoluments of a recruit in the first
years of duty, if in service not less than six months. The estimated
management and consultancy fees of RST between 1970 and 1973 (calculated from Table 19.1) were as follows:

RCM
(million kwacha)

	1970	1971	1972	1973
0.75 percent of sales proceeds	1.335	1.461	1.434	1.779
2 percent consolidated profits before income tax, after mineral tax	.045	.832	.526	.742
Total	2.380	2.473	1.960	2.521

Management and consultancy and sales and marketing contracts
were to run for a minimum of 10 years, and thereafter could be terminated on two years' notice, or after the redemption of the bonds and
stock, whichever is the later. Under them, the former owners provide the mining companies with all managerial, financial, commercial,
technical, and other services in order to maintain the business affairs
and operations of (RCM and NCCM) in a manner "directed towards the
optimisation of production and profit." The services included planning of operations of production and of capital expenditure; engineering
and metallurogical services; expatriate staff recruitment; and purchasing outside Zambia. In marketing, for example, RST was the exclusive agent of RCM in every country in respect of sales and marketing
and could perform the work through existing sales facilities. In general, the contracts ensure that there would be no government interference in the management of the companies. The right to nominate
managing directors for RCM and NCCM, who must be appointed by
the board, is vested in the management companies. That right could
only be terminated on the complete redemption of the bonds and/or
the due termination of management and consultancy contract, whichever is the later.

The phrase "whichever is the later" makes it clear that early
redemption did not give the state title unilaterally either to terminate
the management contracts and sales and marketing contract or appoint
managing directors. They remain to this day valid and enforceable,
despite the president's statement of August 31, 1973.

After redemption, unlawful cancellation imposes a duty on the
state to pay full, prompt, and adequate compensation under the Con-

vention for the Settlement of Disputes. In addition, damages may be claimed for breach of contract. It is perhaps because of the foregoing that the state has appointed a managing director designate, an office unknown to the management contracts. At the same time the managing directors, under the previous arrangements, continue to hold office. The announced intention of the state to secure that RCM and NCCM provide their own management services, and the intended creation of a 100 percent state-owned Metal Marketing Corporation can, in strict law, only be given effect after the termination of existing contracts.

The cheapest method of termination we suggest is to secure an agreed termination and agreed compensation for loss of rights. The ultimate solution depends on a process of bargaining between the parties.

CONCLUSION

The explanation given for the terms of these contracts and the veto power is that the state had a vital interest in optimizing production and profit. Productive efficiency had to be ensured to avoid certain political and economic repercussions. This policy position seems to have been rendered inevitable by the severe shortage of Zambian managerial and technical manpower at the time.

But these agreements did not lay an adequate foundation for progressive control of the mining industry as was originally intended. Terms were heavily weighted against the state and did not provide legal obligations to train and employ Zambian manpower during the life of the contracts. Nor was a duty imposed on the mining companies in order to use some of the mineral profits to develop productive industries in other sectors of the economy. Management's legal complete discretion to determine the level of dividend declaration, marketing, and purchasing policy on behalf of RCM and NCCM, is not, we submit, conducive to the establishment of appropriate industries whose goods could replace the volume of imports. The minority are entitled to 100 percent deduction from gross profits of all capital expenditure, in mining operations. This factor may result in the state's losing considerable revenue over inflated valuations or over capital expenditure of doubtful utility to the industry. At present, negotiations are proceeding for the termination of or variation of terms of the contracts. The state ought to use effectively its power to tax both profits of the companies and dividends to shareholders to secure a favorable compromise.

A COMMENT: GOVERNMENT NOW MANAGES
ZAMBIA'S MINES, BY E. ALEXANDER

This note discusses three interrelated subjects: (1) changes
that have taken place since the Lusaka conference to transfer the management of the two companies, NCCM and RCM, to Zambian hands;
(2) the system of pricing and paying for copper generally used outside
the United States; and (3) the role of foreign capital in Zambia.

Zambia's New Mining Structure

The outstanding Zimco Loan Stock, 1982, by which the Zambian
government paid for its share of the Anglo-American controlled mines
was redeemed in September 1973 by payment of a sum of $148.9 mil-
lion to Zambia Copper Investments (ZCI), a Bermuda-based company
formed in Bermuda as part of the 1970 reorganization to hold the foreign interests in the mines. Zimco issued the loan stock to ZCI and
ZCI in turn issued back-to-back loan stock to the former owners.
The holders of ZCI loan stock were the public, 50 percent, and Zambian Anglo-American (in which Charter Consolidated, Anglo-American's London-based associate, has a 23 percent interest), 50 percent.
The $148.9 million was therefore split up among these holders in the
indicated proportions.
    The outstanding Zimco bonds, 1978, issued in payment for 51
percent of the mines administered by Roan Selection Trust (now
wholly owned by American Metal Climax) were likewise redeemed in
October 1973, at a total cost of $114.1 million.
    The original Zimco liabilities, which carried an interest rate
of 6 percent, were refinanced by two Eurodollar loans raised by the
Zambian government of $50 million and $100 million at floating market rates of interest. The balance of the total repayment of $263 mil-
lion was financed by drawing upon Zambia's foreign exchange reserves.
    Legislation was introduced in September 1973 to make RCM
and NCCM subject to normal annual foreign dividend limitations under
which the amount remitted may not exceed 50 percent of that part of
the profit after taxation attributable to nonresident shareholders or
30 percent of the equity capital attributable to these shareholders,
whichever is the less. Dividends are also subject to 20 percent withholding tax, and foreign dividend remittances are restricted to once a
year. NCCM dividends attributable to the 49 percent foreign share-

---

This section was written after the Lusaka Conference as a comment on the chapter by Chris M. Ushewokunze.

holders have been close to these limits in the period from 1970, so the only real change has been the withholding tax and the provision that dividends can only be paid once a year.

Further legislation repealed the tax allowances and the system that existed prior to the 1970 agreements is now being applied "temporarily" pending a decision on new legislation. Capital expenditure on new mines continues to be allowed 100 percent in the year in which it is incurred. In general, for capital expenditure on old mines there is an initial allowance of 20 percent and 20 percent per annum on the reducing balance is allowable thereafter. That is, 40 percent is allowable in the first year. The provision for taxation for the year is K16 million higher than would have been required if the 100 percent allowances had remained in force.

In October 1973, a metal marketing corporation, wholly owned by the Zambian government, was formed with the aim of handling all metal sales for the Zambian mining industry, thus replacing the sales organizations of the minority shareholders who handle the copper sales. As Ushewokunze noted, early redemption did not give the Zambian government the right either to unilaterally terminate the management contracts, sales, and marketing contract or to appoint managing directors. The cheapest method of termination would be to secure an agreed termination with the ultimate solution depending on a process of bargaining between the parties.

As far as Anglo-American Corporation is concerned, this ultimate solution has been almost fully achieved. NCCM will now manage itself and will pay Anglo-American Corporation a sum of K33 million over three years to compensate for loss of profits that it would have made from the sales, marketing, and consultancy agreements to the end of 1979. NCCM will regulate its own affairs under a government-appointed managing director and market its own metal through the State Metal Marketing Corporation of Zambia Ltd. (Memaco). As it wishes, it will draw on staff, particularly technical staff, whom Anglo-American Corporation will make available on a best-endeavors basis. At the time of writing, the parties are negotiating detailed documents to give effect to the understandings reached. Thereafter, it will remain to obtain ZCI shareholder approval for the new arrangements and new articles of association for NCCM. The position as far as American Metal Climax is concerned remains to be settled. However, it is probable that the Zambian government will achieve a similar arrangement to that with NCCM.*

*Editor's Note: The government's proposal for compensation, not at the time accepted by RCM, was reportedly K22 million (Times of Zambia, August 16, 1974).

Implications of New Structure

There is not likely, however, to be any sharp change in management policy. The international price of copper is still the purse from
which all expenditure has to be met, and its fluctuations impose a dis-
cipline on costs. It will be important to remember in the days of
state control that it is efficient mining that generates the surpluses.
The Zambian mines, or the "cash register" as they have been named,
require an increasing amount of capital and expertise for them to
maintain current production levels, without providing for any increase.
The 49 percent foreign shareholdings however remain in NCCM
and RCM. The Zambian government has gained control but not an increased share of the so-called surpluses; these, however, are not
large. In NCCM's case, of profit before taxation of K278 million in
the financial year ended March 1974, taxation amounted to K175 million, K46 million was retained in the business, and K67 million was
paid in dividends, of which GRZ's share was K29 million and the 49
percent shareholders' share K27 million.
The total available for distribution—that is, profit before tax,
in the 1970-74 period was K688 million, of which K94 million or 12.8
percent was due to the minority shareholders. Over the years this
represented a return on equity invested (share capital plus foreign
equity share of reserves) of 12 percent. Essentially the Zambian government has had control of the major part of the revenues* from the
mining industry in Zambia. It has now supplemented that with complete control of the industry's running both day to day and long term.

The Copper Pricing System

It is probably true to say that a certain frustration at the system
of pricing and paying for copper was one of the factors encouraging
the government's action to redeem the Zimco bonds and take control
of the industry. This factual exposition may clarify a few minds.
Throughout 1971, 1972, and into 1973 Zambia suffered from the low
international copper prices. There was therefore understandable im-
patience to see the fruits of the long steep rise in the copper price
starting in January 1973. The producer's revenue however will not
equal the tonnage produced times the LME average and will react to
changes in the LME price with a lag for the following reasons. The
often-quoted LME cash settlement price is for electrolytic refined

---

*Editor's Note: Government also paid compensation from its
share.

copper of specified quality cast as a wirebar. Producers sell a variety of shapes refined to varying degrees as in the case of NCCM in 1973: anode and blister, 10.9 percent; cathode, 15.7 percent; and wirebar, 73.4 percent.

Anode and blister sell at a discount, the amount being related to refining charges and negotiation between buyers and seller. The normal world discount in 1973 was in excess of 50 pounds per ton. Cathodes are settled at a producer discount slightly below the wirebar price.

Average prices for the year are not however applicable. Copper is sold on an annual basis to consumers. Producers sell toward the end of one year their planned production for the next year on the basis of estimated monthly shipments. In the event of a surplus or deficit of metal, resort may be made to the London Metal Exchange. The settlement price for a small proportion of customers is the monthly average LME settlement price minus any discount for the type of copper involved. The bulk of customers however have the right to choose the LME price for their copper. This they can do in general over a two-month period starting with the contractual month of shipment of their copper, subject to a maximum percentage of their quota on one day. Informed consumers will therefore take advantage of price fluctuations to price their copper and will usually "beat the market" on a rising price trend and vice versa on a falling price trend. The level of proceeds per ton of copper will therefore be lower than that of the LME.

Proceeds will also lag behind the LME price on which they are based because copper priced on any one day is generally paid for six weeks to two months later. The reason for this is that customers pay when the copper arrives at European and Japanese ports but price in the month of shipment from African port, the voyage time being six weeks on average. The first payment will be on a provisional invoice based on pricings made when the ship leaves port. The final invoice issued at the end of the pricing period is adjusted, plus or minus, to the provisional settlement amount. The payments that have to be made to consumers in times of falling prices can cause difficulties for companies managing cash and countries managing their foreign exchange reserves.

The net result of these factors is summarized on the graph showing the LME price and NCCM proceeds per ton of copper sold since the beginning of 1973. As this period was one of a rapid rise and fall of the LME price, the LME/proceeds-differentials are exaggerated. In general, the producer's proceeds will reflect the LME price of the previous month. However, there are notable discrepancies, the most obvious being at the high point of prices when consumers were able to anticipate the turning point of the market most successfully. It is clear that customers benefit substantially from pricing rights. They argue that such rights are essential when they face a price that fluctuates violently and that the pricing structure of the fabricating industry is based on this.

These aspects of the pricing system, some unsatisfactory, are faced by most copper producers. Mineroperu and Codelco sell on the same basic terms as Anmercosa Sales and Ametalco, which have been handling Zambian sales. The fact that copper producers cannot dictate their terms in the same way as the oil producers is caused not only by their lack of organization, but, more fundamentally, by the basic weakness of their position.

The issue of appropriate levels of output for each country will have to be tackled seriously in the context of any attempt to stabilize world copper prices. It is clear that overexpansion of production leading to surpluses on the world copper market can only weaken producers' ability to tighten up in pricing and payment terms, which is the first step to wringing the maximum benefit from the existing price-setting system, and second, to raise that price by producer action.

Foreign Capital in Zambia

The 49 percent foreign share-holding will continue in both NCCM and RCM, and this proportional representation indicates the part that foreign investment in Zambia is likely to continue to play. The capital and management expertise does not exist in Zambia to handle the many possibilities open to the country in furthering copper development, nickel, and iron and steel making, to name a few. It will therefore be necessary to continue to play host to foreign capital and expertise.

Much of the material in this volume emphasizes the negative effects of foreign investment on the host country. I think that in many cases the adverse effects have been overstated at the expense of the benefits, and also the implications of the alternative development strategy of developing without foreign capital were not examined. Silitshena makes the former point very clear in his paper on mining and development strategy in Botswana. (See Chapter 23.) The point is clearly that foreign investment must be regarded as a balance of interests. On the one side, the host country requires revenue and economic development; on the other, the foreign investor requires a rate of return commensurate with the risks involved. This fact is very clearly recognized in Zambia, and foreign capital will certainly continue to play a role in partnership with government.

POSTSCRIPT: TERMINATION OF THE MANAGEMENT
AND CONSULTANCY CONTRACT AND SALES AND
MARKETING CONTRACT, TOGETHER WITH THE
SUBSEQUENT AMENDMENT OF THE ARTICLES OF
ASSOCIATION OF NCCM, BY C. USHEWOKUNZE*

After lengthy negotiations, the above contracts relating to sales and marketing by NCCM were finally terminated by agreement on November 15, 1974. That company is now responsible for its own management; in effect, it now recruits its own managers and technical staff. However, in the majority of cases these are on secondment from the Anglo-American Group, but on request. The marketing and sales functions are to be performed by the Metal Marketing Corporation (Memaco), a 100 percent state-owned company.

The articles of association have been altered so as (1) to give power to the majority A directors, instead of the B directors to nominate the managing director, and (3) to modify the veto power of the minority shareholders in general meeting and that of the minority directors in board meetings. In the latter cases a simple majority is sufficient. And, in the former, a special resolution passed by a 75 percent majority of the votes cast is required for any alteration to the capital structure, and provisions of the constitution of NCCM. This is a normal requirement under the Companies Act.

It should be noted, however, that minority directors have, by special agreement, been granted the right to prevent board decisions, which in their opinion are against the interest of NCCM as a whole, by deliberately absenting themselves from a meeting, so as to deny it a quorum. In effect, this is a right of veto in another form, to be used to reinforce an extremely discretionary minority decision as to what is in the interests of the company. It may be suggested that there may arise situations in which the interests of the company may conflict with national economic interests. The political implications of using this power, it may be hoped, will be sufficient deterrent against its exercise.

Coupled with this veto power is the right of the holders of at least 5 percent of the A or B ordinary shares to refer disputes to arbitration at the International Center for the Settlement of Investment Disputes. As noted in the body of the chapter, even in the face of such a small minority, the Government of Zambia has either to accept its demands or submit to arbitration. The international credit-worthi-

---

*Editor's Note: This postscript was written after Ushewokunze had seen the preceding comment by E. Alexander, although it is not a direct rejoinder.

ness of Zambia remains at stake. Significantly no provision is made for appropriations by NCCM toward industrial development in other sectors of the Zambian economy. The main compensation for the termination of the management, marketing, purchasing, and engineering services contracts is a sum of 33 million kwacha payable in six equal quarterly installments of 3 million kwacha followed by six equal quarterly installments of 2.5 million kwacha. Furthermore, NCCM must pay an indemnity to the Anglo-American (AAC) group for any compensation the latter may become liable to pay, in the event of their terminating contracts entered into in the performance of duties under the 1970 agreements. This applies if either NCCM or Memaco do not take over such contracts.

Presumably this additional term is intended to cover purchasing, recruiting, engineering, and sales and marketing saliency agreements made by the AAC group. Although exact information on these contracts is not publicly available, it is arguable that this term considerably increases the amount of compensation. Alternatively, it obliges NCCM or Memaco to maintain contracts that might not be as profitable, or perhaps that might be politically compromising to Zambia—for example, if made with South African sister companies within the AAC family.

NCCM is not to be taxed with respect to the amounts paid as compensation—that is, payments are to be deducted from gross profits in computing taxable profits. It is not clear whether they will be taxed as income of the AAC group. If not, this will be yet a repetition of the errors in the 1970 agreements. In fact, a repetition seems to have occurred insofar as installment payments are concerned.

Payments in this regard are out of profits. But there is no safeguard for NCCM in the event of a drastic fall in copper prices, or perhaps natural disasters in the mines. Hence, the state and ultimately the taxpayer will continue to be potentially liable, if inadequate profits are made as a result of the occurrence of these events.

In conclusion, the agreement to terminate management and related contracts seems to have laid in law a power framework within which to secure control of the mining industry. But its effectiveness will largely depend on the political consciousness and professional efficiency of Zambian managerial and technical manpower. It is to be hoped that this basic legal structure will be built upon administratively and technically as experience is gained and that it will not result in bureaucratic complacency.

20

**COMPARATIVE
FINANCIAL POLICIES:
ZAIRE AND ZAMBIA**
Raj Sharma

Unlike many developing countries, the rich copper base of
Zaire and Zambia generates sizable revenues to their governments.
With the capacity expansion programs now under way in both countries and more effective state control of the mining sector, contribution of copper to government budgetary resources may be expected to
increase significantly in the future.

It is well known that the price of copper is subject to wide and
erratic fluctuations. The heavy dependence on tax revenue of Zaire
and Zambia on international trade makes them vulnerable to changes
in world market prices. It becomes very difficult for them to plan effectively the level of their expenditure and hence the investment programs. In a situation like this, it is extremely important to set aside
during boom periods sufficient reserve funds as a cushion against
slumps in copper prices. This will greatly facilitate an even path of
development.

Copper, like other minerals, is a wasting and hence exhaustible
asset. Presently, both Zaire and Zambia are heavily dependent on
copper for their export receipts, budgetary revenue, and indeed, a
substantial proportion of their gross domestic product. The path of
sustained economic development in their case lies in using the rich
investable surplus generated by copper to diversify rapidly their economies by developing other sectors, which would then become the basis of self-sustained growth on a continuing basis.

This chapter seeks to examine the contribution made by copper
to the budgetary resources of Zaire and Zambia and to analyze the
utilization of this surplus for the development of their economies.

Table 20.1 brings together data on the financial operation of
Zaire and Zambia. The heavy dependence of Zaire's budgetary receipts on mineral exports, especially copper, is borne out by the rela-

tive importance of tax payments by Gecamines. As can be seen from the above table, these payments rose sharply between 1968 and 1970 to reach 54.2 percent of total tax receipts in the latter years. With the fall in copper prices, they declined by about 37 percent (at an annual rate) in 1971, when during the first half of the year they accounted for only 36 percent of total tax revenue.

A study of Table 20.1 further brings out the expenditure pattern of Zaire. It is clearly evident that the current expenditure has registered sharp upward movements. After having increased by 79 percent in 1968, it rose by 25 percent in 1969, by 24 percent in 1970, and was estimated to have increased by 11 percent in 1971. It is understood that during the period under review all categories of current expenditures have increased substantially. The more significant increases were registered in wages and salaries, maintenance and materials, expenditure by embassies abroad and transfer to provinces, mainly larger payments of wages and salaries.

The capital expenditures also increased with the increase amounting to about 100 percent in 1969, but only 19 percent in 1970. They are, however, estimated to show only a very small increase to Z63 million in 1971. It is understood that since 1969, about half of these expenditures have been undertaken by the president's office, with the Ministries of Public Works, Energy (mainly for the Inga Dam and power lines), and Transport accounting for the largest single shares. The share of the agricultural sector, on the other hand, comprised a very small portion of the total capital expenditure.

As can be seen from Table 20.1, copper revenue on an average has contributed about 50 percent to budgetary resources in Zambia. The share declined in 1971 and 1972, when copper price slumped on the world market. This, however, contains only part of the story. The sharp decline in revenue is also attributable to the loss in copper production as a consequence of Mufulira disaster as well as from the very generous (100 percent amortization in the same year) capital investment allowance provided for under the master agreement.

A study of data assembled in Table 20.1 reveals that in the case of Zambia, recurrent expenditure rose even more sharply than that for Zaire. According to one estimate, net recurrent expenditure (net of public debt amortization) in Zambia increased by 168 percent during 1965/66 and 1971. The classification of recurrent expenditure of Zambia (Table 20.2) is very revealing. The recurrent expenditure has risen nearly fourfold during 1964/65 and 1974. It is important to note that personal emoluments account for a large proportion of the budget (over 25 percent in 1974) while expenditure on subsidies, grants, and other payments is equally large for a population of about 4.5 million. The constitutional and statutory expenditure has also registered sharp increases during the period under review.

TABLE 20.1

Summary Data on Government Financial Operations, Zaire and Zambia, 1968–73

Zaire
(Zaires million)

	1968	1969	1970	1971*
1. Recurrent revenue	186.2	269.5	315.3	289.5
2. Revenue from Gecamines	95.6	132.4	159.4	100.6
3. 2 as percent of 1	51.3	49.1	50.5	34.7
4. Recurrent expenditure	165.7	206.6	257.6	294.0
5. Recurrent surplus or deficit (–)	20.4	62.9	57.7	(–)4.5
6. Capital expenditure	25.7	52.7	62.4	63.5

Zambia
(kwacha million)

	1968	1969	1970	1971	1972	1973*
1. Recurrent revenue	306.1	401.2	432.4	309.0	315.2	384.8
2. Revenue from copper	176.1	235.2	251.1	114.1	55.7	107.6
3. 2 as percent of 1	57.5	58.6	58.1	37.0	17.7	30.7
4. Recurrent expenditure	225.7	233.2	275.0	350.3	363.1	394.3
5. Recurrent surplus or deficit (–)	80.4	168.0	157.4	(–)41.3	(–)47.9	(–)9.5
6. Capital expenditure (excluding TanZam Railway)	193.3	156.3	138.8	127.4	124.4	109.9

*Estimate.

Sources: Bank of Zaire, Annual Reports; Republic of Zambia, Estimates of Revenue and Expenditure; Bank of Zambia, Annual Report; Republic of Zambia, Monthly Digest of Statistics.

TABLE 20.2

Zambia: Recurrent Budget, Comparative Figures
(kwacha)

	1964/65 Actual	1969 Actual	1973 Budget	1974 Budget
Personal emoluments	33,563,424	61,519,672	98,585,670	104,890,019
Recurrent departmental charges	26,549,616	65,016,635	75,161,848	87,076,093
Grants	15,547,824	21,697,766	15,625,450	14,909,540
Subsidies		20,714,242	42,950,400	33,708,450
Other payments		239,111	2,554,031	1,920,651
Special expenditure	4,286,850	7,104,916	1,674,348	2,295,590
Pensions	8,718,914	4,375,338	4,500,200	4,723,200
Emergency expenditure	—	—	—	20,320,000
Contingencies fund	—	—	—	5,000,000
Subtotal	88,666,628	180,667,680	241,051,947	274,843,543
Constitutional and statutory expenditure	22,947,706	54,830,929	115,571,150	161,121,453
Total	111,614,334	235,498,609	356,623,097	435,964,996

Source: Republic of Zambia, Estimates of Revenue and Expenditure, 1974

343

As against the very considerable increase in current expenditure, capital expenditure in Zambia during the period 1969-73 has consistently registered a downward movement. It is understood that a large proportion of the capital expenditure since independence is due to the transportation, communication, and power category, which reflects a major effort by Zambia to reduce dependence on Rhodesia. This sector continued to take more than half of all capital expenditure. The share of the agriculture sector has also registered significant increase, but the bulk of investable funds in this sector has gone in the form of loans (or equity participation) to various statutory boards.

The conclusion that emerges from a study of the comparative financial policies of Zaire and Zambia is that in both countries, a large proportion of the surplus generated has been utilized for meeting current expenditure for the benefit of the present generation. The capital expenditure, on the other hand, has increased but only marginally in the case of Zaire, while in Zambia it in fact declined during 1969-73. This means that both the countries need a reappraisal of their financial policies and need to adapt these to the fulfillment of the main objective set out above, namely to use the surplus generated by copper in capital investment programs with a view to diversifying their economies.

This brief analysis underlines the need to study in depth the comparative financial policies of all the copper-producing countries in the developing world. The basic idea should be to study the size of the total investable surplus generated by copper and its distribution as between the multinational companies and the governments. This would call for a study of the structure of taxation, the evolution of concession agreements, and the state participation and control of the mining sector. It is equally important to analyze the manner in which the multinational companies as well as the governments utilize the surplus accruing to them.

21

THE MULTINATIONAL CORPORATIONS AND A THIRD WORLD HOST GOVERNMENT IN A MIXED ENTERPRISE: THE QUEST FOR CONTROL IN THE COPPER-MINING INDUSTRY OF ZAMBIA

George K. Simwinga

Basically, the Mulungushi Economic Reforms, otherwise known as Zambia's Economic Revolution had their roots in the exploitative nature of the expatriate businesses operating in Zambia. Taking advantage of the "artificial" profits arising from the supply shortages following from the Rhodesian Unilateral Declaration of Independence (UDI), they repatriated large amounts of these windfall gains, while at the same time they borrowed locally to finance their business activities.

The Mulungushi Economic reforms included, among other things, the limitation of local borrowing to expatriate-controlled companies, exchange controls on remittances of foreign-controlled businesses, government acquisition of a 51 percent shareholding in 27 companies, and also the restriction of expatriate retail trade to specified areas. More important, for the purposes of this chapter, government expressed dissatisfaction at the lack of mining development since independence. As President Kenneth Kaunda declared, "I want to say to the mining companies that I am very disappointed at the virtual lack of mining development since independence. . . . Instead of reinvesting they have been distributing over 80 percent of their profits every year as dividends."[1]

Sixteen months after the Mulungushi Reforms, the president in a dramatic speech announced at Matero: "I have decided that I shall ask the owners of the mines to invite the Government to join their mining enterprises. I am asking the owners of the mines to give 51 percent of their shares to the state."[2]

Professor Norman Girvan in his study of the bauxite industry in six Caribbean countries concluded that

The companies, through the medium of these agreements and of legal provisions, have been able to secure the active collaboration of the governments of the
countries in institutionalizing the company-country
relations which are overwhelmingly to the benefit of
the farmer. This, in turn, is part of a wider strategy
by which the companies attempt—and in a large measure succeed—in adapting the political environment
both at home and abroad to the needs of their survival,
growth and profitability.[3]

He identified four main concessions that foreign companies seek to
acquire from host governments, namely the resource accessibility,
the payment stream, operational conditions, and status (legal) of
agreement. He defines resource accessibility as the concessions
sought by companies that pertain to the companies' specific rights to
the natural resources of a country. The operational condition refers
to the employment policies, use of profits, and foreign exchange. The
payment stream covers the company's liability to the host country in
the form of income taxes, royalties and rates, and so on.

It is interesting that nearly all these concessions are found in
the agreements signed between the Zambian government and the mining companies. It is the intention of this chapter to highlight the factors that, in the view of this writer, have added to the already overtowering position of the management insofar as policy decisions are
concerned in the Zambian copper-mining industry. Perhaps the greatest impediment to the Zambian government's desire to control the industry, as professed, through the board of directors is the veto power
the B directors have over certain key decisions.

THE VETO POWER

The veto power of B directors as a strong weapon of the managing companies has been a great source of anxiety in the board rooms,
especially when a conflict of interest between the two sides arises,
for example, an issue arose over the use of "formed coke" in processing lead and zinc at Broken Hill Mine. The A directors, mindful of
their obligations to the international community on sanctions against
the Rhodesian regime, proposed that Waelkz Kilns should be constructed to produce locally "coke," which could be used in reprocess-
ing slag. This, it was anticipated, could prolong the life of Broken
Hill Mine by at least eight years, since the reserves were diminishing.
This process would cost approximately K70 to produce a ton of coke.
The cost of importing a ton of coke from the Rhodesian Wankie Coal

Mines cost about K27. The B directors favored leaving the Wankie market open. In the first place, it was argued that it was cheaper to do so, but, at the same time, the AAC directors' own interests in the Wankie Collieries made them even more adamant about continuing to import coke from Rhodesia.

The deadlock appeared in the end to be solved in favor of the construction of the Waelkz Kilns after the B directors were convinced that the deteriorating relations between Zambia and Rhodesia—with countereconomic sanctions on Zambian-bound goods from Rhodesia—left them no choice. The border closure by Rhodesia, then later by Zambia, put the finishing touches to the solution of the dispute. Investigations into the technical feasibility of producing a suitable quality of coke are proceeding even faster following the border closure. [4]

CONTROL OF EXPERTISE

The Zambian government's lack of control over the policy-making bodies of the mining companies is only one aspect of the company's overtowering powers to control the mines. Much more control lies in the hands of the management. Management is used here in the wider sense to refer to those who are actually involved in the day-to-day administration of the mines.

Many writers in public administration have recognized and written extensively on the enormous influence the bureaucracy, for example, exerts on the policy-making in modern government. The bureaucracy derives power from many sources, including its expertise. Max Weber, for example, identified this as a distinctive attribute that gives bureaucracy its enormous influence in modern government. [5]

Weber asserts, "Under normal circumstances, the power position of a fully developed bureaucracy is always overtowering. The 'political master' finds himself in a position of 'dilettante' who stands opposite the expert facing the trained official who stands within the management of administration." [6] Francis E. Rourke, attributes a number of causes for bureaucracy's acquisition of this expertise. Among them he identifies the division of labor within a large organization that allows groups of people to acquire specialized expertise even if they may not themselves have unusual technical qualifications. He also points to the concentrated attention to specific problems they deal with daily as responsible for giving public agencies an invaluable kind of practical knowledge. This knowledge the agencies acquire puts them in an especially advantageous position to influence policy when the facts they gather cannot be subject to independent verification or disproof. [7]

In the same way, the "management" in the copper-mining industry of Zambia can be said to derive its source of power to control from the following bases: (1) its monopoly of the expertise, (2) the control of information, and (3) its absolute privilege to hire and fire.

Copper mining is a highly complex and technical undertaking. It requires, among other things, a high degree of specialized technical skills in a wide range of fields. In Zambia, the great volume of copper mined each year requires sinking shafts in order to bring the copper-bearing ore to the surface. This entails great weight being attached to the safety factor as hundreds of people work underground. Additionally, the ever increasing cost of mining requires that maximum productivity be maintained if the industry is to continue operating profitably. These factors together call for a high degree of expertise and hence give the people with such kinds of rare qualities—that is, technical proficiency in specialized fields such as mining, electrical and mechanical engineers, metallurgists, and so on—tremendous power in the running of the industry.

The mining industries manpower allocation, as of December 1971, and five-year forecasts are highly skewed as between the expatriate and local labor force. While the Zambians are concentrated in areas dealing with unskilled and semiskilled jobs, the areas requiring expertise remain a complete monopoly of the expatriate labor force. Thus, positions from mine captain up to those of mine superintendent, which constitute management are an exclusive expatriate paradise (see Table 21.1).

As Michael Burawoy points out, "Whereas the ultimate responsibility for decision making rests formally in the hands of the head office in Lusaka, power may well reside at the mines where production takes place."[8] He goes on to argue that "because the capitalist organization formulates its goals in terms of profit making, and because profit is made through production, so those who control production are crucial actors. The more indispensable and irreplaceable they are, the more power they have."[9]

Expertise and its monopoly in the Zambian copper industry has since the 1920s played a significant part in shaping the policies in the industry. From about the beginning of World War II, the expatriate labor force—through their labor unions—relied heavily on their expertise to safeguard and improve their lot. For three decades, the expatriates succeeded in preventing any form of African advancement they considered as inimical to the interests of the white miner. In spite of the advantages that might have resulted in the use of cheap local labor in place of some lower-level expatriate labor force, the companies and the colonial government could not afford to go against the wishes of the "expertise monopolists." "The profits of the companies were not high enough to stand a protracted stoppage on this

TABLE 21.1

Local and Expatriate Employment in Zambian Open and Underground Mining Establishment, and Forecasts, 1971–76

	Establishment 1971		December 1972		December 1973		December 1974		December 1975		December 1976	
	Expatriate	Local	Expatriate	Local	Expatriate	Local	Expatriate	Local	Expatriate	Local	Expatriate	Local
Mine superintendent	9	0	9	0	9	0	9	0	9	0	9	0
Assistant mine superintendent	4	—	5	—	5	—	5	—	5	—	5	8
Underground manager	25	—	26	—	26	—	26	—	26	—	26	—
Assistant underground manager	35	—	35	—	41	—	41	—	41	—	41	—
Mine captain (technician/specialist/trainee)	62	—	68	—	68	—	67	1	67	2	65	4
Mine captain (RBS)	86	—	97	4	90	14	84	20	74	36	60	44
Mine captain (other)	29	17	26	25	22	30	17	35	12	40	6	46
Mining engineer (senior/assistant/junior)	46	9	38	5	35	7	32	11	29	14	26	17
Shift boss (technician/superintendent/technician)	102	16	112	31	103	39	95	48	89	54	79	65
Shift boss (RBS)	92	174	77	222	56	253	35	275	23	286	16	293
Shift boss (other)	1	202	—	214	—	210	—	213	—	215	—	216
Engineering superintendent of research (engineer/head of engineering)	9	—	10	—	10	—	10	—	10	—	10	—
Assistant engineering superintendent	7	—	8	—	8	—	8	—	8	—	8	—
Divisional engineer	16	—	16	—	16	—	16	—	16	—	16	—
Assistant divisional engineer	28	—	30	—	30	—	30	—	30	—	30	—
Sectional engineer	137	—	153	—	153	—	151	1	151	3	150	3
Engineer (senior/assistant/junior)	127	6	122	11	117	18	103	27	103	30	97	37
Sectional engineer (maintenance planning)	3	—	2	—	2	—	2	—	2	—	2	—
Maintenance planning officer	50	—	54	—	55	—	53	—	53	1	52	2
Clerk of works	38	—	43	—	43	1	42	1	42	2	41	2
Engineering foreman	297	1	300	2	300	3	300	4	300	7	297	10
Assistant engineering foreman	715	7	754	17	752	36	707	50	707	70	671	100
Technician artisan	690	102	639	171	490	316	267	404	267	470	214	523
Apprentice	—	294	—	183	—	65	—	—	—	—	—	—
Craft trainee	—	182	—	215	—	249	—	284	—	284	—	284
Winding engine driver	2	139	—	157	—	161	—	161	—	161	—	163
Trainee winding engine driver	—	25	—	5	—	5	—	5	—	5	—	5
Training officer winding engine driver	7	—	6	—	6	—	5	—	3	3	—	2
Foreman (other) training	8	22	8	22	5	23	3	25	3	25	3	25

Notes: (1) there are no assistant mine superintendents for open-pit mining; (2) mine captain (other) is missing in open-pit mining; (3) senior soils engineer and soils engineer have no equivalents underground.

Source: Compiled by author based on data obtained from companies.

issue, they were not in a position to risk a European exodus, nor were there anywhere enough Africans available for even a limited amount of general advancement."[10]

Today, three years after the partial nationalization of the mines, expertise still plays a vital role in influencing the course of action, particularly in the mining industry. As pointed out earlier, the safety criterion is just as important as—if not more important than—productivity. This gives an extra weight to the status of those individuals who have this expertise. As Charles Harvey points out, "Safety is a factor in most industries (although not in commerce), but it is a central feature of mining and it would be a brave and probably foolish government official who tried too hard to hasten localization against expert advice in such conditions."[11] In sum, therefore, expertise plays a crucial role in complex and highly technical organization, the monopoly of which gives tremendous advantage to those who wield it over those who lack it.

CONTROL OF INFORMATION

Management's source of power to control the copper-mining industry derives also from its monopoly of the information upon which the decisions are based. The management's grip over the industry is strengthened through (1) the government's distance from the information center and (2) lack of specialized bodies to analyze critically the data presented. Having the monopoly of the accessibility to this information, the management is in a strong position to use the information to its own advantage.[12]

The government's distance from the information center stems from the fact that the terms of the management and consultancy agreements give the managing companies (RST and AAC) the mandate to prepare company reports, financial statements, production and marketing financial matters, and so on. Although the government has a right to demand these documents, data presentation is subject to various interpretations, especially in the absence of careful analysis. This then leads us to the question of the lack of government's sufficiently specialized bodies to examine critically the data available to it from the operating companies upon which policy deliberations revolve.

The existing arrangement in the mining industry is that experts in different fields from every division are required to prepare reports about the project they want to undertake.[13] After recommendations are made and approval is gained from the operating company, the report is then submitted to the industry's Technical Liaison Committee comprised of three people: a Mindeco representative and one representative each from two companies, RCM and NCCM, respectively. After the Technical Liaison Committee's deliberation, the project is passed on to the board of directors for discussion and approval. Bear in mind that there is need for approval of the two groups of directors, voting separately, if the project involves the areas prescribed in the contract. The main point here is that the government representatives are not only outsiders in the sense of proximity to the production area, but do not even have a sufficiently elaborate body of specialized expertise to deal with all technical aspects of the mining industry to argue convincingly against the information provided by the "men on the spot." In other words, the management's absolute control of the system of information (upon which policy decisions are usually based) is a powerful weapon to their advantage, particularly in view of the limited nature of the government's capacity to question such information with technical authority.

THE POWER TO HIRE AND FIRE

As pointed out earlier, the management contract gave exclusive rights to RST and AAC to run the mines, including the exclusive right to hire and fire employees who work for the mines. The power to hire, in my view, is yet another weapon available to management to consolidate its power. The management could, for example, maintain the status quo—that is, monopolizing expertise by a careful selection of entrants and by reducing the pace of advancement for a particular category of workers. By restricting the privilege of gaining more experience and expertise to the persons who share the "values" of management, while at the same time keeping out the "potential dissenters," management could easily consolidate its power. Since the partial nationalization of the copper industry, the lack of clauses in the contract concerning the training and localization in the industry meant that each company continued to train its own personnel.

In May 1972, a Manpower Planning, Training, and Zambianization Committee was formed to control manpower planning, training, and Zambianization in the industry on a centralized basis. The committee's main objectives were (1) to achieve the optimum rate of Zambianization; (2) to rationalize (reorganize) training activities; (3) to develop an industry manpower plan; (4) to implement a common labor statistics system; and (5) to improve productivity by the more efficient use of manpower.[14]

In response to the first objective, training programs were mounted at technical institutions and the University of Zambia to prepare Zambians for careers in the mining industry. On completion of their studies, candidates would be distributed throughout the industry's divisions according to need.[15] Although the manpower planning and training function has been centralized, the power to hire is still the responsibility of individual divisions. They implicitly retain the power of advancement of their employees, although the progress of graduates sent to them by the centralized body is monitored jointly. The success and advancement of these graduates when sent to divisions still remains to be seen, as the program is only one year old. But if the past is anything to be guided by, the majority of Zambians who joined the mines at about the time of independence with the basic formal qualifications as engineers have not progressed beyond the rank of mine captain. Although no quantifiable evidence was available to this researcher in support of the alleged slow rate of progression to indigenous employees of the mines, all the Zambians interviewed expressed dissatisfaction at the pace of their progress.[16] In making any promotions—be they local or expatriate—the management has always defended its actions as purely based on merit, given the high level of know-how that is required for the efficient running of the industry. It has vehemently denied that promotions to some senior positions in the mines are mere "window dressing" aimed at appeasing the critical Zambian public. If indeed merit is the yardstick used, it becomes hard to understand why the more recent appointments of three Zambians by RCM to the posts of assistant general managers came about. Two of these appointees were, in fact, outsiders to the mining industry, although the third had been a personnel superintendent within the mines.[17] The post of assistant general manager is the second most senior below the manager of a division; if this job could be handled by an outsider (as implied by virtue of the appointments), what could prevent a man who has not only got the necessary formal technical education but has had behind him some working experience within the mines from being promoted?

The second problem insofar as the hiring and firing privilege of management is concerned is that of retention of the local personnel. The turnover of the trained Zambian personnel from the industry has been high. One of the reasons often quoted by management for the high turnover of the Zambian staff is that the mines, being a relatively large organization and requiring the employment of several hundreds of people with professional qualifications, cannot afford to pay as high salaries as do businesses in other sectors. This argument, in my view, needs qualification. According to the Times of Zambia,

> As of January 1, 1973, the overall requirement for graduates in the industry [mining] was 860 and only 35 were Zambians. In the technologist field to date there are only 65 Zambians against a total staff strength of 854. Out of the total requirement of

> 1,887 artisans, only 192 are Zambians. In fact,
> one expert has staked his reputation and forecast
> that Zambianization of the mines would be complete
> by the year 2085.[18]

If the management can afford large salaries for the large numbers of expatriate staff, what difficulty would there be in raising the salaries of their few Zambian counterparts to a level closer to theirs? Observers often ask why there is no determined effort to retain Zambians with the industry.

The point being made insofar as the hiring function of the management is concerned is that while the industry continues to hire Zambian personnel, no efforts are being made to retain the Zambian personnel with expertise. It can be construed, therefore, that this could be a deliberate effort to maintain the status quo, whereby the expertise should continue to be monopolized by the expatriate labor force, which is largely in sympathy with the objectives of the two mining companies. This, in the final analysis, strengthens the hand of management in its efforts to influence the ultimate policy decisions with which to govern the copper-mining industry in Zambia to the advantage of the minority shareholders.

CONTROL OF BUYING AND SELLING POLICIES

The companies' power to dominate policy concerning localization is only one aspect of the story. In other areas such as profits and expansion of production, they occupy a dominant position. Both the management and consultancy contract and the sales and marketing contract were awarded to Amax and AAC. For providing these services, each company received 0.75 percent of the gross sales as remuneration on each of the contracts. Although quite substantial portions of these services are provided locally, large sums of money are involved, and the calculations are based on gross sales rather than profits. This contract affects the profitability structure, thus reducing both the company's taxable income and the size of the profits due to the Zambian government.

Another important clause in the contract that gives tremendous advantage to the mining companies is the concession that the two companies are free to remit in full all the net profits without exchange restrictions. This has resulted in the companies' remittance in full of all the net profits, either for distribution as dividends or for reinvestment in other countries. An official of the Mining Development Company in an interview said they were borrowing money from abroad in order to finance mining development projects,[19] and yet this was a period when the price of copper was very high so that they could normally expect reasonable profits with which to finance, at least partially, mining developments.

Second, the sales and marketing contracts also give Amax and AAC a mandate to sell copper and buy mining equipment to and from whomever they please. Therefore, the traditional factor markets of South Africa, the United States, and Great Britain are favored regardless of the existence of competitive markets elsewhere. These are areas in which Amax and AAC have sister companies.[20]

What has been astonishing to inquiring minds in Zambia has been the sharp rise (20 percent) in the cost of mining in Zambia between 1970 and 1972. Suspicion has since been activated by preliminary findings of two government economists working in the imports and exports of Zambia. In this project, they have been trying to compare the records of Zambian imports and exports to and from selected countries. These were compared with the records of those countries' exports and imports to and from Zambia during a given time period. Allowing for variables such as transportation costs and change in the value of currency, they discovered that wide discrepancies existed between these records from the South African and American countries. In contrast to this, countries like Japan and Germany, which had traded with Zambia as much as the United States and South Africa had, showed little discrepancy in their records. Although no definite conclusions can be made at this point in the absence of more detailed data, such information points out the possibility of transfer pricing as a mechanism by which the companies have tried to evade the country's taxes, as well as reduce the profit base upon which the government accounts for the 51 percent shareholding.

PRODUCTION POLICIES

The minority shareholders' power extends to production policies. As pointed out earlier, all appropriations for capital exploration or prospecting expenditures require the approval of both the A and B directors, voting separately.[21] This means that even if the Zambian government wants the companies to expand production by either opening up new mines or extending the existing ones, they may be restrained by the minority shareholders. The Zambian government can do nothing unless it abrogates the agreements of the contract, and the consequences of such an act have been clearly spelled out to the government. Insofar as extension or development of the mines is concerned, the Zambian government depends on the cooperation of their minority shareholding partners.

CONCLUSIONS

We have examined the consequences of the institutional arrangements that were arrived at following President Kaunda's Matero pronouncements. The Zambian government's objective was to put the activities of the foreign mining companies under the control of the state. It has been argued in this chapter that the contractual agreements arrived at do not put the Zambian government in any stronger position to realize its original objective.

A few conclusions can be drawn from this study. The desire to shift control from the foreign-owned multinational corporations does not lie in the superimposition of majority vote in the policy-making bodies. The problem really runs deeper than that.

In the first place, care ought to be taken in the drafting of legal documents that would eventually bind the activities of the policy-making bodies. Insofar as the Zambian government's anticipated objective of controlling the copper industry was concerned, certain clauses such as the B directors' veto power over crucial policy areas have been highly restrictive of government action to realize its objective. Real control of the industry lies in management because this is the area that shapes and directs the ultimate policy decisions. Understandably, the Zambian government could not at the time of the takeover handle the management of the mines. To place the management function with the very people whose activities proved inimical to the interests of the state and then expect them to be remote-controlled by a show of hands at the board of directors meetings looks like a pipe dream.

Since management is the crucial aspect in this matrix of control, a careful selection of individuals to staff the management is crucial. Individuals who share the values of the recruiting organization— in this case the Zambian government—should be selected. A management whose values are closely tied to the foreign corporations cannot be expected to serve the Zambian government's interests. If Zambianization of the copper-mining industry is being proposed for this purpose, then it is a worthwhile undertaking. On the other hand, this would go a long way toward explaining the foreign mining companies' reluctance to accelerate Zambianization.

The Government of Zambia must begin moves to take over the management of the mines as well as the sales and marketing. The present arrangement does not in any way bring the government nearer to its original objective. My reasons for appealing for controlling the "management" as a second step in effective control of the mines is not shrouded in blind nationalistic demagoguery. It is rather based on my understanding of relationships that exist between the foreign institutional corporations and the host government and the conflict of interest between the two, all of which finally interact in the environment of the "management."

The management, it has been argued, has resources that put it in a stronger bargaining position vis-a-vis the majority shareholding Zambian government. Management is a segment or subsidiary of a multinational corporation. As such, it becomes part and parcel of a comprehensive plan of the mother body in its search for maximizing profits in the different geographical areas throughout the world and in all the areas in which they have diversified their economic activities. To give a hypothetical example, a multinational corporation might not only have copper mines in Zambia, the United States, and Peru, and so on, but also might have copper refineries in Britain and South Africa, coal mines in Rhodesia, copper-wire fabricating plants in Kenya and the West Indies, steel mills in Pennsylvania, and so on. All these connections would have significant effects insofar as the multinational corporation's global operations are concerned. Macro-level plans for the entire global organizational complex, taking place either in New York or London, integrate the activities of each and every subsidiary company into the entire system. Thus it might, for instance, be decided that a copper-fabricating plant be located in one country (X), while the other country (Y) would continue to provide the copper metal in raw form. Other subsidiary companies of this cross-country organization would continue to provide the raw materials necessary to produce copper in country Y.

This action is rational insofar as the interests of the multinational corporation are concerned. To maximize returns on their capital, they have to minimize costs by making full use of the existing equipment, especially if such plants have already achieved economies of scale. From their point of view, political boundaries mean little insofar as the entire operations of the company are concerned.

This perception of the problem undoubtedly brings them into conflict with the host countries' national interests. The location of a copper-fabricating plant outside Zambia, for example, and the preference of Rhodesian to Zambian coal not only loses Zambia a number of jobs but also constitutes a loss in foreign exchange. But to relocate these economic activities would involve costs—at least initially—that may not be welcome to the companies. Yet, with the "management" as representative of the interests of the multinational corporation while at the same time enjoying a tremendous position of influence over the majority shareholders, namely the Zambian government, management will undoubtedly foster the interests of the mining companies at the expense of the Zambian government.

Thus, the Zambian government at present faces not only the multinational corporation as a rival but also the management, which poses an even greater threat to the interests of the government by fostering the foreign companies' interests under the pretext of expertise. A complete removal of this "cog" in the wheel will not only remove this extra tool available to the multinational corporation but also bring the Zambian government closer to exerting appreciable influence on the running of the country.

In my view, the takeover of the management function of the mines will, to a large extent, reduce the power base of the corporation in the Zambian mining industry. Of course, this is not to suggest that the problems of external manipulation will be over. The multinational corporation will continue to exert pressure on Zambia as it does in all the small trading countries, through manipulation in the international market. But, for Zambia, one problem solved would mean one obstacle less. It would entail control of industry, at least at home, leaving obstacles only on the international scene.

NOTES

1. Kenneth Kaunda, Zambia Economic Revolution (Lusaka: ZIS, April 19, 1968).

2. Kenneth Kaunda, Zambia Towards Economic Independence (Lusaka: ZIS, August 11, 1969).

3. Norman Girvan, "Making the Rules of the Game," in Social and Economic Studies (Jamaica: Institute of Social and Economic Research, University of West Indies), pp. 378-420.

4. Mindeco, informant who preferred to remain anonymous.

5. C. W. Mills and H. H. Garth, "From Marx to Weber," Essays in Sociology (New York: Oxford University Press, 1949), p. 214.

6. Ibid., p. 232.

7. Francis Rourke, Bureaucracy, Politics and Public Policy (Boston: Little, Brown, 1969), p. 42.

8. Michael Burawoy, The Colour of Class in the Copper Mines (Manchester: Manchester University Press, 1972), p. 91.

9. Ibid.

10. Bromwich, cited in Burawoy, op. cit., pp. 13-14.

11. Mark Bostock, Economic Independence and Zambian Copper (New York: Praeger Publishers, 1972), p. 8.

12. Compare Rourke, op. cit.

13. Mindeco official in an interview, June 1972.

14. Circular to the division managers from the managing directors of RCM and NCCM, respectively, May 15, 1972.

15. Interview with T. G. Mann, manager of Manpower Planning, Training, and Zambianization, June 1972.

16. Seventeen employees who held either diplomas or university degrees in various technical fields at Nchange, Rokana, and Luanshya divisions expressed disenchantment at their rate of progress. They felt they were discriminated against.

17. Times of Zambia, July 13, 1973, p. 5.

18. Ibid.

19. Interview with top-ranking Mindeco official, June 1973.

20. Compare K. Nkrumah, Neo-Colonialism: The Last Stage of Imperialism (New York: International Publishers, 1966).

21. Mark Bostock and C. M. Harvey, Economic Independence and Zambian Copper (New York: Praeger Publishers, 1972), pp. 219-55.

22

THE NEED FOR A LONG-TERM INDUSTRIAL STRATEGY IN ZAMBIA
Ann Seidman

THE PROBLEM: ZAMBIA'S DISTORTED MANUFACTURING SECTOR

The importance of manufacturing growth is widely agreed upon in Zambia, as elsewhere.[1] In the postindependence era, Zambia's manufacturing sector actually did grow at a rate exceeding that suggested by the UN experts as critical for attainment of the goals of the "Development Decade" of the 1960s.[2] But the rapid expansion of manufacturing industry did not contribute significantly to the spread of increased productivity in all sectors of the Zambian economy. This chapter seeks to explain why this is so and, on the basis of that explanation, to suggest an alternative industrial strategy that could contribute more effectively to the attainment of a more balanced, integrated national economy capable of spreading productive employment opportunities and increasing the levels of living of the population in all sectors.

The Inherited Manufacturing Sector

At the time of independence, Zambia inherited an archetypal dual economy. Its limited export enclave was devoted almost entirely to the export of crude copper, which constituted 90 to 95 percent of its exports. It imported almost all the manufactured goods consumed. Outside of the limited enclave built around the copper belt and about a thousand line-of-rail settler estates, the rural areas stagnated.[3] The drain of tens of thousands of young men, forced to seek wage employment by colonial taxes and regulations hindering other forms of African participation in the so-called modern sector, disrupted and undermined the existing system of production in these regions.[4]

Zambia's manufacturing sector at independence contributed less than 7 percent of the total gross domestic product (GDP),[5] about half as much as was typical of other countries with the same per capita income.[6] It was dominated by the beverages and tobacco industry, which produced almost a third of total manufacturing value added. (Value added here refers to the gross value of the output of domestic manufacturers, minus the value of imported inputs.[7] The beverages and cigarettes industry produced about a third of the value added in Ghana, where total manufacturing value was about the same proportion of GDP as in Zambia, and about 15 percent of value added in Kenya, where manufacturing produced about 13 percent of the GDP.)[8]

Postindependence Expansion

Several factors stimulated a rapid expansion of Zambia's manufacturing industry after independence:[9] the Rhodesian Unilateral Declaration of Independence (UDI) in 1965; the oil pipeline from Dar-es-Salaam and a major expansion of the road haulage industry; the doubling of government expenditures financed by rising revenues due to high copper prices and new tax policies in the 1960s; and the expansion of consumer demand, due to increased wages and salaries. The estimated contribution of manufacturing to GDP in money terms had almost quadrupled by 1972. (The actual increase in real terms was probably considerably less, for official data indicate that prices of manufactured goods had risen about 40 percent between 1966 and December 1972).

Closer examination, however, reveals several disturbing features of the manufacturing sector.[10] First, it became increasingly dependent on imported parts and materials. Local value added actually declined from half to about a third of the gross output of manufacturing industry (see Table 22.1). Second, the composition of manufacturing remained much as it had been before independence, except that the beverage and tobacco share of total value added by manufacturing increased from 27 to 41 percent. Third, the number of establishments reported by the Census of Industrial Production dropped by a third from 1966 to 1969, while the number of workers per establishment almost doubled. The average amount of fixed capital invested per employee also doubled in the same period. This suggests that immediately after independence a large number of would-be entrepreneurs tried to establish simple manufacturing shops, but that many of them quickly went out of business. The average value added per establishment multiplied over three times, from about K90,000 to over K300,000. In the competitive battle spurred by the postindependence boom, the larger firms with more capital and larger output—

TABLE 22.1

Value Added in Manufacturing in Zambia and Percentage of Total
Manufacturing Value Added by Industry
(in current prices)

Manufacturing Industry	1965[a]		1972[b]	
	(millions of kwacha)	(percent) of total)	(millions of kwacha)	(percent of total)
Food manufacturing	6.6	13.7	23.6	14.3
Beverages and to-bacco	13.0	27.0	67.9	40.7
Textiles and wearing apparel	3.9	8.1	12.6	7.6
Woods and wood products, including furniture	2.4	5.0	3.3	2.0
Paper, paper products, publishing, and printing	2.1	4.4	6.5	3.9
Rubber products	0.8	1.6	5.4	3.2
Chemicals, chemical petroleum, and plastic products	2.8	5.8	7.6	4.6
Nonmetallic mineral products	6.1	12.7	10.6	6.4
Basic metal products	5.8	12.0	2.3	1.3
Fabricated metal products, machinery, and equipment	4.4	9.1	24.4	14.8
Other manufacturing	0.1	0.2	0.3	0.2
Total	48.0	100.0	164.5	100.0

[a]There was some change in the system of national accounts in 1970, so 1965 data are not entirely comparable.
[b]Provisional.

Note: K1 = $1.56.

Source: Calculated from Monthly Digest of Statistics, July 1973, Table 54, p. 54.

mostly foreign-owned—appeared more able to survive. Finally, Zambia's industry, became, if anything, more concentrated in the export enclave. Of a total of 532 industrial establishments still in existence in 1969, only eight, or 1.5 percent, were established in other provinces. The rest were in the three line-of-rail provinces. Manufacturing employment in the line-of-rail provinces increased marginally from 96.7 percent of total manufacturing employment in 1966 to 97.3 percent in 1969.

In 1972, after the world copper price had fallen to a postindependence low, stringent exchange controls and import licensing were introduced in 1972 to reduce resulting balance-of-payments deficits. The closure of the Rhodesian border in 1973 further reduced imports. At the same time, the wholesale price index of domestically used goods rose about 26 points from 1968 to 1972, and another 4.2 points in the first five months after the border closure. In these circumstances, the total value added by manufacturing continued to rise by about 10 percent a year from 1970 to 1972. Although about a fourth of this apparent increase was attributable to rising prices, the continued expansion of the manufacturing sector was one of the brighter features of the economy in this period. Partly due to the stagnation of the GDP, as well as to its own growth, manufacturing value added increased to about 13.4 percent of the total GDP. It somewhat offset the dramatic decline of mining's contribution, caused primarily by the fall of the world copper price, from over 40 percent of GDP in 1969 to about 24 percent in 1972. Yet examination of the limited data available for 1972 suggests that the pattern of industrial growth changed little. The manufacturing sector had contributed very little to restructuring Zambia's inherited dual economy.

THE EXPLANATION: DEPENDENCE ON
EXISTING "MARKET FORCES"

It is crucial to explain this situation so that new policies may be devised to change the role of manufacturing in the future. The explanation appears to lie in fundamental aspects of the planning and implementation process itself.

The Inadequacies of an "Import-
Substitution" Policy

Both the First and Second National Development Plans explicitly adopted a policy of expanding social and economic infrastructure and encouraging the establishing of import-substitution industries. Even in theory,[11] however, an import-substitution approach appears more likely to aggravate the dualistic features of an economy like that of Zambia than to restructure it. In Zambia, as in other Third World countries, the lopsided structure of the economy is reflected in the distortion of the existing market for imports. Thus, the composition of consumer imports is almost entirely shaped by the demands for luxury items of the 10 percent of the population that accumulates over half, and perhaps as much as three-fourths, of Zambia's national income in the form of high salaries, profits, interest, and rent. The remaining 90 percent of the inhabitants either earn little more than bare subsistence wages[12] or produce a major share of their needs on semisubsistence farms. They buy a distinctly limited range of goods, most of which are already produced in Zambia.[13]

To build industries to produce goods previously imported primarily to supply the demands of the "elite" in Zambia is to perpetuate an economy geared to their needs. It is almost inevitable that the technologies of such industries will be relatively capital-intensive and reliant on imports of parts and materials. The production of refrigerators, automobiles, air-conditioners, radios and television sets, and so on is technologically complex. It requires a level of industrial development that cannot be achieved in Zambia for years to come. "Manufacturing industries" producing such items are likely to consist primarily of last-stage assembly and processing of imported parts and materials. This inevitably reduces the potential spread effects.

The argument that industry should fulfill "consumers' demand" may hinder expansion of local production in two ways: First, the import of items considered by high-income groups to be "superior" may directly reduce the market available for those domestically produced. (This danger is aggravated to the extent that foreign managers of importing firms seek to maintain the Zambian market for their overseas affiliates' produce.)[14] Second, the import of several different "brands" or "makes" of a given item reduces the possibility that the demand for a single standardized product will expand sufficiently to permit the achievement of economies of scale in local production of standardized materials and parts.

That government efforts to encourage construction of import-substitution projects in the context of Zambia's distorted market will tend to aggravate the inherited lopsided pattern is illustrated by two examples chosen from the list of projects incorporated in the Second National Development Plan:[15] the passenger car assembly plant in Livingstone and the expansion of the Kafue Textiles Plant to produce rayon and cotton-polyester materials.

The car assembly plant has already been built by the Italian firm Fiat. The government provided 70 percent of the capital. Its output of 5,000 passenger cars, a thousand less than the total number of pas-

senger vehicles newly registered in 1971, aims to meet the expanding domestic demand of the high-income groups for private automobiles.
The few hundred workers employed in the Fiat plant do little more
than assemble a kit imported from Italy. Essentially, when in full
operation, the plant's main achievement is to guarantee that Fiat,
by providing 30 percent of the investment, will have a near monopoly
of the limited Zambian market.

     This is not to argue that, if production of private cars ought to
be high on the priority list of Zambian manufactures, Fiat is the
wrong car to introduce. Its small size and relatively low cost make
it preferable to many other possible alternative car models. In fact,
it can only be hoped that, as the Livingstone plant achieves full production, the government will prohibit the import of competing models
so that it can operate at full capacity and that an increasing share of
the parts may be locally produced. (The worst horror story of the
import-substitution automobile "manufacturing" industry is probably
that of Chile, where, by 1967, 19 firms had been established to assemble parts and materials to "produce" 16,400 vehicles—although the
economies of scale of integrated production dictate a minimum output of at least 100,000 vehicles.[16] Obviously, none of the firms could
operate at capacity, the possibilities of backward linkages were
thwarted, and the impact of the "industry" in restructuring the economy to reduce dependence on copper exports was nil.)

     But several questions would appear to require consideration:
First, should Zambia, at this stage in its development, use scarce
government investment funds to produce cars for the few people with
high incomes? Or should it now give priority to the production of
simple tools and equipment to increase rural productivity and levels
of living: for example, animal-drawn ploughs, maize-grinding machines, equipment for drilling bore holes to provide all-year-round
water supplies, and so on. Or, alternatively, should the government
invest in more rapid expansion of lorries and buses to provide cheap
transport for low-income rural-dwellers and the goods they buy and
sell?

     Second, the extent to which specific plans have been made for
backward linkages with the rest of the economy remains unclear; but
the Zambian market appears too small to make some of the linkages
viable for a considerable length of time to come.

     Third, should Zambia, given its limited existing market, em-
bark on this project alone? Would it not have made more sense to
combine efforts, say, with neighboring Tanzania (which also established a Fiat factory) to build one project for both countries, which
could take advantage of much larger economies of scale? A larger
regional output would facilitate the establishment of associated plants
to produce the necessary parts and equipment, reducing imports from
the Italian factories and increasing local employment and value added.

The second example, the doubling of Kafue Textile Plant capacity from 12 million meters of cotton fabrics to 25 million meters per year of cotton polyester and rayon fabrics presents a different set of issues. The most critical question is whether Zambia should at this stage in its development expand the production of rayon and cotton polyester fabrics at all. This will, of necessity, require the import of new machinery and equipment, for the plant's present equipment is primarily designed to weave and spin cotton fabrics. In addition, it will require the continuing import of rayon and polyester materials for processing. Zambia itself will not, for a long time, have the technological capacity to produce these materials, which require advanced chemicals industries with large economies of scale. Add to this the high cost of skilled manpower required to manage the project and operate the new machinery, as well as the profits of the participating foreign partner, and the balance of payments burden is likely to be considerable. The bulk of the consumers of the resulting relatively expensive textiles is likely to be the tiny high-income group, which tends to insist on "quality" regardless of cost. Furthermore, the import of parts and materials for these "higher-quality" materials will limit the possible spread of effects that could be generated by production of more cloth woven from cotton, already being grown by Zambian small farmers.[17]

Implementation of an approach geared to expanding cotton production and processing in Zambia would require the government to prohibit the import of competitive materials. This might sharply reduce "consumer choice," but since only 10 percent or less of the population really have any choice as to materials purchased, it does not seem to be too great a sacrifice to make.

These examples illustrate how adoption of the catch-all phrase "import substitution," given Zambia's distorted market, fosters the establishment of manufacturing projects producing goods primarily to meet the demands of a narrow high-income group associated with the export enclave. They do not produce items that might help to increase the productivity or better the standards of living of the 90 percent of the population living at bare subsistence levels. They cannot contribute much employment or local value added, for they rely primarily on the import of sophisticated parts and materials, which cannot, for a long time, be produced in such a narrow local market.

The Reinforcing Impact of Implementation Machinery

The role played by the set of institutions shaping critical investment decisions affecting manufacturing reinforces the negative impact

of the import substitution policy. A peculiar combination of ministries, manned by civil servants, and autonomous parastatals facilitates the operation of existing "market forces" to maintain the lopsided, externally dependent export enclave, rather than contributing
to decisions designed to alter it.

The term "market forces" does not have the same meaning in
Zambia as in traditional Western theory, which assumes the existence
of many competing private firms seeking to maximize their profits.
In a country like Zambia, the private firms that seek to maximize
their profits by establishing manufacturing projects are far from competitive in this orthodox sense. On the one hand, the local market is
so narrow that one or two large firms, taking advantage of even minimum economies of scale, can supply most of the demand in any given
field. Many of the larger firms with the capital and know-how to invest in manufacturing are associated with multinational corporations*
whose profit-maximizing decisions are based on consideration of their
worldwide assets and marketing networks rather than on local needs.

The Ministerial Setup

In Zambia, as in other former British colonies, "ministries"
have replaced the "departments" of the colonial governments, but
their functions have remained much the same. Each ministry deals
essentially with one or another aspect of administration and infra-
structure-creation: transport and power, public works, trade and
commerce, education, health, and so on. This approach is built on
the assumption that the local economy lacks capital for investment in
productive sectors; hence, its primary task is to create a "hospitable
investment climate" to attract foreign investment, particularly for the
manufacturing sector.

In reality, however, Zambia, with the highest per capita income
in sub-Saharan independent Africa, produces investable surpluses totaling some K500 million to K800 million every year.[18] This is more
than double the total 1972 assets of Indeco, the government's primary
instrument for participating in the manufacturing sector. But the necessary institutional changes have not been made to ensure that a significant share of these surpluses are invested to reduce dependence on

----

*A study of the 80 most successful businessmen in Lusaka, Zam-
bia's largest single market area, showed that most of them were engaged in retail trade and real estate. Almost none were engaged in
manufacturing activities, other than simple repair shops, because of
lack of technical background as well as the larger amounts of capital
necessary to establish modern manufacturing industries.

copper exports and increase employment in other sectors of the economy. A major portion of the investable surpluses remaining in the private sector after taxes has been shipped out of the country, even after the Economic Reforms of 1968 and 1969, largely in the form of profits, interest, dividends, compensation for government acquisition of shares in industries, and salaries for expatriates. Altogether, these totaled almost K200 million in 1971* and were provisionally reported to total K167 million in 1972. The "invisible" losses during these two years alone exceeded Indeco's total 1972 assets by about 40 percent. They equaled about a third of Zambia's investable surpluses, far more than foreign interests have invested in the country as a whole, including the mines, in any one year since independence.

Traditional attitudes and institutional arrangements have tended to hinder the investment of the increased tax revenues obtained by the government in directly productive activities. Although since independence, the government has succeeded in capturing a significant portion of the investable surpluses produced in the country in the form of tax revenues, it has invested little in manufacturing. In the peak postindependence year of 1970, the government actually obtained tax revenues totaling some K432 million, over a third of the GDP, or about K108 per capita. In the 1970s, with the fall in the world copper price and the change in base of taxes on the copper mines, government's total revenue dropped to a little over K300 million, about a fourth of the GDP. Even in 1971-72, Zambia's per capita tax revenue was K75, considerably more than triple that of her East African neighbors.[19] Yet the Second National Development Plan projected only about 8 percent of total public-sector investment in manufacturing. Private investment was expected to exceed public investment in this sector by about 30 percent, constituting about 21 percent of all private investment.[20] (About 52 percent of private sector investment was expected to be in mines.)

Zambia's government today, it is true, spends far more on social, as well as economic, infrastructure than did the former colonial government. This reflects in part the government's response to the postindependence political demands of Zambia citizens. It also fits in with the perception and training of civil servants as to their role. But the failure to expand the productive sectors outside of the mines leaves the national economy precariously dependent on world copper prices and sales. This danger was sharply illustrated in 1971-72, when the fall in world copper prices led to a one-third drop in government revenues and a cutback in both current and capital expenditures for social and economic infrastructure.

---

*Another K47 million was subsequently reported lost on account of "net errors and omissions."

Ministerial intervention in the area of manufacturing remains restricted, primarily, to indirect measures, like licensing, quotas, tax holidays, and specific infrastructural projects perceived as incentives for stimulating desirable growth patterns or inhibiting those contrary to enunciated goals. Here, the division of responsibility among the ministries creates possibilities of conflicts and contradictory policies. A perusal of the list of activities of various ministries at the back of the Second National Development Plan[21] illustrates this: The Ministry of Development Planning and National Guidance is responsible for overall planning, although no machinery is provided to enforce any decisions it might take. The Ministry of Trade and Industry is responsible for Indeco, the government's holding company for dealing with industry and for the establishment of an import-export agency, which may be expected to influence the import of machinery, equipment, and supplies for the manufacturing sector, as well as finished goods that may compete with its output. The Ministry of Power, Transport, and Works is responsible for providing important infrastructural facilities that may crucially determine whether and what kinds of manufacturing project may be established in various parts of the country. The Ministry of Rural Development is expected to establish intensive development zones, which presumably will include rural industries, as well as agricultural plans to stimulate production and marketing of essential agricultural raw materials. (The list is not really very different from the "departmental shopping lists" that characterized colonial plans.)[22]

But these ministerial activities appear to be inadequately coordinated and are sometimes even in conflict with each other. At the planning level, for example, the Second National Development Plan reports[23] that Rucom, an Indeco subsidiary, is to establish eight units to machine-form and burn 160 million bricks. The same document asserts[24] that Indeco's Steelbuild Division has on hand studies for two large-scale mechanized factories to produce 160 million building bricks.

Conflicting policies and measures may also arise at the implementation stage. The Ministry of Finance is responsible for expanding tax revenues to meet the government's growing budget requirements In recent years, it has been seeking ways of increasing the revenue from the mines, which produce over half the investable surpluses available in the economy. The Ministry of Mines, on the other hand, has been seeking to reduce taxes on mines to induce foreign investors to expand their investments and increase copper exports. Without entering here into the complex question of who is right,[25] it is self-evident that the two ministries appear to be working in opposite directions.

There is a wealth of literature to suggest that, everywhere in the world, bureaucracies are designed to tend the machine, not to change it.[26] In Zambia, as in other former colonies, the inherited system was designed to provide infrastructure, leaving the productive sectors, particularly manufacturing, to take care of themselves. This apparently remains the main function of the ministries today. Simply to append new policies and programs onto the old structure appears unlikely to ensure that they will change their roles.

Indeco: An "Autonomous" Parastatal Holding Company

The parastatals that hold the government's shares in various productive sectors of the economy remain several times removed from government supervision. The overall parastatal holding company, Zimco, consists of several subholding companies. Of these, Indeco has been designed as the government's main instrument for manufacturing industrial development.[27]* Since the Mulungushi Reforms, Indeco's role has been significantly enlarged. Yet it seems to have relatively little impact in altering the pattern of manufacturing growth in the nation. This leads one to ask whether Indeco, as it is organized, is really adequate for its task.

Indeco is a lineal descendant of the Northern Rhodesia Industrial Loans Board, established in 1951. The board was reorganized in 1960 as the Northern Rhodesia Industrial Development Corporation. In its new form, it functioned much like the development corporations established in other British colonies to stimulate private investment in industry.[28] It was again renamed the Industrial Corporation of Zambia, Ltd., and commenced participating with foreign firms in establishing new large-scale industries to help "ensure the efficient and profitable operation of such ventures."[29]

Indeco's role in the economy was greatly enlarged as a result of the economic reforms. In 1968, Indeco's net group assets totaled about K35.5 million. By 1972, these had multiplied over six times to K233 million. (It might be noted that, even before the reforms, Indeco's assets were as large as those of Tanzania's National Develop-

---

*Zimco's other holdings include Mindeco, which until recently held the government's shares in the copper mines, although these are now held directly by the Ministry of Mines, leaving Mindeco responsible only for small mine development; Findeco, responsible for the government's financial holdings; the National Hotels Corporation, which handles most of the government's tourist facilities; and the National Transport Corporation, which handles the government's transport interests.

ment Corporation after the implementation of the Arusha Declarations.[30] By 1973, Indeco's assets were about six times the size of
those of Tanzania's NDC.) Indeco's productive divisions employed
about 45 percent of all manufacturing workers in the latter year.
Total profits of Indeco subsidiaries had multiplied almost 10 times,
from K283,000 on a turnover of K1,863,000 (that is, 15.1 percent)
in 1968 to K27,086,000 on a turnover of K286,002,000 (that is, 10.5
percent) in 1972. Indeco's own share of profits had jumped from
K340,000 to K7,083,000 in the same period. Government's tax revenue for Indeco subsidiaries totaled another K11,714,000. Indeco's
foreign partners received K8,269,000 as their share of after-tax
profits,[31] most of which they remitted to their home countries, along
with the compensation they received for the government's acquisition
of shares.

Indeco has continued to operate essentially as an autonomous
parastatal, primarily to facilitate investments initiated by its foreign
partners.[32] Its board of directors is composed of government ministers and members of the business community. It is expected to make
decisions much like a private firm, primarily concerned with the economic viability of projects it undertakes. Accepting import substitution as a main goal, its directors appear to take the existing market
as given.

The Second National Development Plan explicitly declared that
Indeco's role should be to accumulate profits from projects established within the framework of the existing high-income urban market
for investment in industries designed to contribute more effectively
to spreading productive activities to other sectors of the economy.[33]
The chairman of Indeco's board of directors, A. J. Soko, has recently
reemphasized that Indeco has mounted a "sustained drive towards import substitution."[34]

The organizationof Indeco appears likely to reinforce, rather
than diminish, tendencies to shape decisions about manufacturing ventures in accord with the dictates of prevailing "market forces." It is
subdivided into eight divisions, each of which holds several of the
nearly 50 productive companies in which the government, together
with foreign investors, has shares. Each subsidiary is operated in
accord with a separate agreement with the foreign partners. The foreign partners have minority representation on the boards of subsidiary
companies in accord with their shareholdings, but more important
provide the day-to-day management.

A leading European commercial analysis, Prospects for Business
in Developing Africa, reports Western investors believe that "local
participation will ensure that managerial control lies with the foreign
investor and at the same time satisfy many of the requirements of
'national interest.'"[35]

Indeco has been given autonomy in dealing with foreign investors and financial interests.[36] This creates an environment in which Indeco subsidiary managements may make decisions relating to individual projects without considering the imperatives of any kind of overall strategy directed to restructuring the economy.

It might be noted that elsewhere the grant of discretion without very specific guidelines and the continual spotlight of public information has facilitated corrupt practices.[37] That foreign firms investing in Africa are not averse to taking advantage of opportunities thus created is indicated by Prospects for Business in Developing Africa:[38]

> One long-established British trading firm simply wrote off a large investment in Liberia because bribe demands from officials were going beyond accepted African standards. An American firm in Nigeria has become so used to paying out "dash"—or minor bribes —upon demand, that such bribes are considered a normal business expense on P&L statements. An Italian manufacturer takes the long-term view that . . . any "extraordinary expenses" incurred today will be more than compensated for by profits from tomorrow's expanding market.

Examination of the details of Indeco's operating divisions (see Table 22.2) suggests that, for the most part, they follow the pattern of already established manufacturing industries. Of the eight Indeco divisions, only five are actually engaged in productive activities. Even among these, some subsidiaries are still more involved in the import and distribution of parts and materials than in local production. They tend to concentrate on producing or importing items demanded by higher-income line-of-rail customers or for the mines and associated activities. Only Rucom Holdings, the smallest of the eight divisions in terms of assets, is explicitly designated to invest in projects off the line-of-rail. All Indeco projects, even Rucom's, seem to be considerably more capital-intensive than the average of the nation's manufacturing firms. Indeco Breweries is the largest Indeco division, in terms of both employment and profits returning to Indeco itself.

Three Indeco divisions are not engaged in any form of productive activity. Of the two largest, one, Indeco Trading, clearly has a vital role to play in ensuring increased government control and direction of imports as well as regulating prices. Presumably its goal is to hold down prices of necessities to consumers, as well as to facilitate the spread of specialization and exchange throughout the economy. Unless there are clearly defined criteria delineating its role, however, it is

TABLE 22.2

Indeco Divisional Statistics, 1972

	Brew-eries	Chemi-cals	Indus-trial Hold-ings	Steel-build Hold-ings	Rucom Hold-ings	Indeco Trad-ing	Indeco Real Estate	Other	Total
Turnover*	70,395	77,646	32,323	19,026	27,464	60,989	804	355	286,000
Profit (loss) before tax*	8,015	8,789	1,304	3,658	2,885	2,379	(273)	289	27,046
Profit (loss) after tax*	5,786	5,437	275	1,883	989	1,078	(310)	194	15,332
Net profit (loss) to Indeco*	2,829	2,611	(179)	1,073	268	589	(310)	182	7,063
Net assets*	31,441	94,802	13,873	37,780	12,578	14,288	21,751	(3,320)	223,193
Net employees (including asso-ciated companies not consoli-dated)	4,924	3,620	2,841	2,910	3,557	3,555	283	226	21,917
Profits as percent of assets									
Before tax	25.5	9.2	9.4	9.6	22.9	16.7	—	—	12.1
After tax	18.4	5.7	1.9	5.1	7.8	7.5	—	—	6.1
Value of assets per employee*	6.3	2.6	4.8	18.0	3.5	4.0	7.6	—	10.0

*In thousands of kwacha.

Note: Figures in parentheses indicate loss.

Source: Indeco, Ltd., Annual Report, 1973 (Ndola: Falcon Press, 1973), p. 40.

conceivable that attainment of this goal may on occasion conflict with Indeco's policy of establishing profitable projects. (Since this was written, this division has been separated from Indeco.)

The other of the two largest nonproductive Indeco divisions, Indeco Real Estate, permits use of Indeco funds to subsidize housing for higher-income Zimco employees. Its total assets are almost double those of Rucom Holdings. It continually operates at a loss. The inclusion of this division in Indeco hardly seems justifiable, given the necessity of redirecting every kwacha of Zambia's investable surpluses to investment in projects designed to increase productive employment opportunities in all sectors of the economy.

In short, Indeco's policy of import substitution has contributed to perpetuating, rather than altering, the undesirable postindependence trend of manufacturing growth. Indeco's organization as a system of autonomous parastatal holding companies inserted in layers between the government and the foreign managers of individual subsidiaries is inadequate to introduce socially oriented criteria for the expansion of old and establishment of new projects. Analysis of the available information relating to Indeco's operating subsidiaries appears to substantiate this conclusion.

Other Institutions Influencing Manufacturing Development

An explanation of the causes of the distorted manufacturing growth in Zambia would not be complete without looking at the impact of those institutions that control import trade and the banks. Import and internal wholesale institutions help determine the kinds of goods imported to support and develop the manufacturing sector, as well as those that may compete with local output. The banks play a major role in determining the availability of credit for the day-to-day operations of specific manufacturing projects. While it is not possible here to examine these institutions in depth, it is necessary to touch on those aspects that have fostered the distorted growth of Zambia's manufacturing sector.

Import and Internal Wholesale Trade

Zambia's inherited trading institutions functioned primarily to sell imported manufactured goods, produced mainly in Rhodesia, South Africa, and England, to the high-income consumers along the line of rail. After independence, Zambia's government sought to break its trade links with Southern Africa for political reasons and as a precondition of developing its own manufacturing industry. It was also essential to increase the import of capital machinery and equipment for the Zambian manufacturing sector, while reducing the import of luxury consumer goods.

The government took major steps to reorganize its import and internal wholesale trading mechanism. The urgency of breaking trade relations with the south was increased by Rhodesia's UDI and Zambia's efforts to implement the UN boycott as rapidly as possible. The construction of the Tanzama pipeline, the Tazara Railroad, and the Great North Road in cooperation with Tanzania were all explicitly designed to provide alternative routes to the sea through a friendly independent neighbor. The government also introduced import licensing. As part of the 1968-69 reforms, it took over a majority of the shares of the major importing companies, Consumer Buying Corporation (CBC, a subsidiary of the British firm, Booker McConnell), ZOK (a South African firm), Mwaiseni, and the National Drug Corporation. It established the National Import and Export Corporation on the foundation created by the Zambia National Distribution Corporation, which was initially set up to handle the import of goods from China.*

The government's shares of ownership in the big importing houses was lodged in the Indeco Trading Division. Initially, the foreign partners provided the management of the firms in which they still hold the minority of shares. Indeco Trading Division's turnover of about K61 million in 1972 equaled about two-thirds of the value of all consumer goods imported into Zambia but less than 20 percent of the country's total imports. The rest of the capital goods, machinery, parts, and materials imported were handled by separate parastatals, government agencies, and private firms. Overall state control was only exercised through foreign exchange controls and licensing.

After the 1973 border closure, all importing through Indeco Trading was turned over to the Export-Import Corporation under the management of the British firm, Bookers, the parent company of CBC. The government purchased ZOK outright. All imported goods were distributed by the Export-Import Corporation through the various other Indeco trading subsidiaries and private retailers in accord with their projected requirements based on past experience.

---

*To enable the People's Republic of China to acquire the necessary Zambian currency to finance the local costs of building the Tazara Railway, the Zambian government agreed to facilitate the sale of Chinese goods in the country. From Zambia's point of view, this merely meant the import of Chinese goods instead of those from other sources, like South Africa. At the same time, this enabled China to grant Zambia a long-term loan for the railroad, which included the local costs of labor and material.

These measures were particularly effective in reducing trade with Rhodesia. While overall imports multiplied 2.5 times from K156 million in 1964 to K403 million in 1972, Rhodesia's reported share dropped from over a third at the time of independence to less than 3 percent in 1972. (Most of the imports from Rhodesia in 1972 were in the form of hydroelectric power from the Kariba project, which had deliberately been constructed, before independence, on the Rhodesian side of the border.) By March, a few months after the 1973 border closure, and with the expanded output of the Kafue hydroelectric project, reported imports from Rhodesia dropped to 1.5 percent of Zambia's total imports.

The government's measures were less effective in reducirg South African trade. The absolute value of Zambia's imports from South Africa more than doubled from 1964 to 1968, exceeding the overall rate of expansion of imports into the economy. South Africa's share of total imports declined after the Economic Reforms, from 23 percent to 15 percent in 1972. In absolute terms it remained about double what it had been in 1964. The 1973 border closure reduced South Africa's share of Zambia's total imports to slightly over 9 percent by March of that year.

Government appears to have been considerably less successful in reducing the import of consumer goods compared to capital equipment and machinery.[39] The import of goods for fixed capital formation in the manufacturing sector expanded at about the same rate as imports, so that its share of total imports from 1966 to 1970 remained between 5 and 7 percent, falling to 5.3 percent in the latter year. Imports of materials and parts as inputs for manufacturing sector kept pace with the overall expansion of imports, an aspect of the growing dependence of the manufacturing sector on imports of intermediate goods, but as a share of total imports, they increased only marginally, from 14 to 16 percent.

Imports of consumer goods, on the other hand, declined only marginally from about 28 percent in 1966 to about 26 percent of the total much larger import bill of 1970. Foodstuffs actually increased from 3.6 to 7.7 percent of the total.

The tightening of exchange control and import licensing restrictions in 1971–72, to reduce the balance-of-payments deficits as the world copper price fell, also aimed to change the composition of imports more significantly. The main shift in the composition of imports over the next two years was an increase in the share of machinery and transport equipment from 38.5 to 42.9 percent, and an absolute decline in the value of imported mineral fuels and oils, as well as a fall in their share of total imports from 11.7 to 7.5 percent.[40] Presumably this latter was primarily accounted for by the Tanzama pipeline.

At the same time, items continued to be imported that could have been produced by local firms. For example, Indeco's Kafue Textiles suffered financial losses because of the import of cotton piece goods.[41]

The available evidence suggests that the importing agencies have not yet given sufficient attention to altering the composition of imports to support the kinds of manufacturing industries that might spread productive employment opportunities throughout the economy. This may reflect the lack of an explicit industrial strategy in the context of which specific projects could be designated for support by all government agencies. Or it may reflect the fact that, throughout the period under consideration, the managements of the importing agencies, typically provided by the foreign partners, were more concerned with purchasing materials and supplies from their own associated concerns in other parts of the world. Whatever the reason, unless imports are explicitly designed to strengthen, rather than compete with, domestic manufacturing, it seems hardly likely that a more appropriate manufacturing sector will emerge.

The Role of the Banks

In Zambia, as in other former colonies, the banking system was established historically to facilitate the expansion of the foreign-dominated export-import trade and associated activities.[42] After independence, the continued operation of the major commercial banks along traditional lines almost inevitably fostered the lopsided expansion of the manufacturing sector shaped by the existing "market forces."

Three big foreign-owned banks, Barclays, Standard, and National Grindlays, remained the primary source of credit for manufacturing projects in Zambia.* Standard and Barclays, together, have some 51 branches located primarily along the line-of-rail, with a few in some of the larger towns in the other provinces. These banks have been locally incorporated in Zambia since 1971, but they are still directed by managers and boards of directors in the context of the parent banks' overall interests. The banks in which the Zambian government participates are far smaller. The wholly state-owned National Commercial Bank, established after the Economic Reforms to "promote the average Zambian entrepreneur by offering him lending policies not readily available in the past,"[43] has opened five branches in

---

*It was announced that the government would acquire 51 percent of the shares of the banks as part of the Economic Reforms, but apparently the negotiations for implementation broke down, and they remain entirely foreign privately owned.

the main line-of-rail towns. The Commercial Bank of Zambia, with some state participation, also has branches in the main line-of-rail towns.

There has been no general shortage of credit for manufacturing since independence. The expansion of credit granted by the commercial banks to manufacturing has actually exceeded that sector's growth, multiplying more than six times over from 1964 to 1971, and continuing to expand in the 1970s. But the criteria for loans made are dependent on the bank managers' assessments of the profitability of individual projects and the security offered, in accord with commercial banking practice everywhere, rather than their potential contribution to the spread of productive employment opportunities throughout the economy. [44] In Zambia, as elsewhere, these credit policies reinforce the tendencies of manufacturing firms to conform to the existing market.

The government's attempts to influence the banks' credit policies have been restricted to indirect, negative sanctions exercised through the Central Bank, [45] except in the case of the two relatively small banks in which it participates directly. In 1969, and again in 1972, the Central Bank attempted to reduce the amount of credit provided by the banks as part of an effort to reduce inflationary pressures. In the latter year, it also tried to reduce the balance of payments deficits resulting from continued expansion of imports in the face of declining export earnings. These efforts to reduce the quantity of credit tended primarily to reduce employment and affect smaller business adversely. [46] Larger firms, especially those with foreign links, have better credit relations with the big commercial banks as well as overseas sources of funds. This may have been a factor contributing to the trend toward the reduction in the number of smaller, more labor-intensive manufacturing firms, which has characterized Zambian industry in recent years.

The Central Bank's efforts to reduce imports through exchange controls in the 1970s were also hampered by the fact that the commercial banks facilitated the efforts of the big firms to obtain overseas credit for imports. As a result, customs data[47] indicate that actual imports exceeded those for which foreign exchange had been licensed by over 20 percent in 1971 and 1972. These unauthorized imports cost the country over K80 million a year in foreign exchange.

In other words, the commercial banks' postindependence policies have tended to foster the shaping of the manufacturing sector by the existing "market forces." The indirect methods of control available to the Central Bank were not designed to alter this effect and in fact may have aggravated the consequences.

The Explanation Summarized

The explanation here suggested for the lopsided growth of man-ufacturing in Zambia since independence is not intended to be exhaus-tive. It aims primarily at presenting available evidence that shows that adoption of an import-substitution policy backed by the inherited sets of institutions in critical areas of the economy foster the shaping of decisions affecting manufacturing expansion by the existing "mar-ket forces."

Zambia's experience tends to substantiate a growing body of theoretical analysis drawn from other Third World countries as to the consequences of an import substitution policy. Efforts to substitute local production to meet the demands of a limited high-income group are likely, in the early stages, to foster production of a few techno-logically less sophisticated items—beer, foodstuffs, and textiles—for which the internal market is relatively broader, in part by put-ting many of the smaller local firms out of business. Continued im-port substitution is likely to become increasingly characterized by the establishment of large foreign firms for last-stage processing and assembly of imported parts and materials, producing a range of con-sumer durables that are technologically complex and require econo-mies of scale that will not be realized in a small market like Zambia's for many years. In other words, government efforts to stimulate the expansion of import-substitution manufacturing projects may be ex-pected to perpetuate and even aggravate the externally dependent dual-istic economy. The sets of institutions in the critical areas of the Zambian economy that shape the manufacturing sector appear to func-tion in accord with assumptions and criteria that inevitably contribute to this outcome.

In short, the explanation for the failure of Zambia's manufactur-ing industry to contribute significantly to the spread of productive em-ployment opportunities into all sectors of the economy appears related to the fact that government efforts—both in terms of policy and control exercised over implementation machinery—have tended to leave the decisions affecting manufacturing to be shaped by the inherited dis-torted market pattern dominated by foreign private firms. (This ex-planation might well be deepened by a thorough examination of the evidence as to the extent to which the emergent class of Zambians, who have since independence assumed control of the civil service and the parastatals, and may in fact be pursuing policies of the type out-lined above in its own perceived self-interest. There is not the space nor has adequate research yet been completed, to include this explan-ation here.)

TOWARD A LONG-TERM INDUSTRIAL STRATEGY

If the explanations indicated above for the causes of the distorted growth of Zambia's manufacturing industry are valid, it follows that a new strategy must be designed embracing two fundamental aspects. First, the proposals should be carefully worked out in the context of a long-term perspective plan for the establishment of specific manufacturing projects that will contribute explicitly to restructuring Zambia's economy. Second, the strategy must be concerned with the specific institutional changes required to facilitate implementation of this long-term perspective plan.

A 20-Year Perspective Plan for Manufacturing Industry

To alter the pattern of manufacturing in Zambia, it is necessary to state explicitly the new set of criteria that must guide the formulation of plans to be implemented over a longer time horizon of, say, 20 years. The above explanation of the past performance of the manufacturing sector in Zambia suggests that the following criteria should form the core of a long-term perspective plan:[48]

1. The outputs of projects established should contribute explicitly to expanding productive employment opportunities and gradual improvements in the levels of life of the 90 percent of the population who now vegetate at subsistence levels.

2. Employment creation should be specifically considered in determining whether the technology used in specific projects should be labor- or capital-intensive. Given the fairly wide range of possible technologies available for specific industries, the choice of a particular technology should tend toward the one using more labor and less capital wherever this is consistent with the other criteria here enumerated.

3. The location of each project should be considered in terms of the possibility of establishing poles of growth and essential linkages in each province and district of the national economy. Obviously, some projects must be located near sources of raw materials, while others may require to be located near existing markets and external economies to be viable. Over the longer time perspective, projects should be located to shift the pattern of markets and external economies to the advantage of the less developed rural areas.

A 20-year perspective plan should be conceived as establishing the appropriate kinds of manufacturing industries in a series of interlinked stages. These stages may be incorporated into successive five-year plans, plans that are in effect sufficiently long to permit establishment of essential projects required to implement the major goals of each stage. Each stage should be designed to overcome specific constraints and open specific possibilities upon which the next stage may be built. The objective of the stages, taken as a whole, would be to reduce dependence on the export of copper and to achieve a more national integrated economy capable of providing increased productive employment in all provinces.

For example, the first stage of the perspective plan in Zambia could be directed to producing simple, essential farm inputs and simple consumer necessities to raise the levels of living in rural areas as well as that of lower-income groups in the cities. Animal-drawn ploughs, wheelbarrows, axes, and shovels would help semisubsistence farmers to increase their output for sale. These tools should be relatively inexpensive and easy to use, so that small farmers could afford to buy them and could operate them efficiently. Attention should be directed, too, to providing adequate seed, fertilizers, and irrigation facilities.

Simultaneously, small processing plants should be established in rural areas to buy the farmers' increased output and process it for sale. These might include simple fruit and vegetable processing projects, perhaps with reusable glass containers produced by the Kapiri-Mposhi glass factory; small-scale rice mills in the several potential rice-growing areas; small grain-milling machines; and meat-processing plants. Wherever possible, projects established in local areas should be fairly labor-intensive and meshed with crop-growing and harvesting seasons so as to increase rural employment opportunities and incomes. These projects would provide the market needed for expanded agricultural output in each province, while substituting local production for items now imported. By increasing farmer incomes, they would simultaneously broaden internal markets for the growing output of the expanding manufacturing sector.

Not all manufacturing projects established in this first stage should be small. A large vertically integrated factory like the Kafue Textiles plant can provide an expanding market for domestically produced and locally ginned cotton, provided competitive imports are prohibited and appropriate measures are taken to stimulate local cotton production. Even for such large plants, however, significantly different degrees of capital- versus labor-intensive technologies exist.[49] Wherever possible, it would seem preferable to use the less capital-intensive technology to conserve scarce capital while increasing employment impact.

Once a larger integrated project has been established to produce a basic material like cloth, smaller industries can be encouraged in every province to produce wearing apparel to meet expanding local demand. The same holds true for a range of industries. A careful study of particular projects should determine which might best be larger, functioning as poles of growth that may set off chains of growth throughout entire provinces. Once these are identified, the linkages must be planned to ensure that potential spread effects do in fact take place.

In the first stage, projects should also be built to produce essential transport equipment and machinery to facilitate the spread of internal markets and reduction costs as domestic production and incomes increase. Again, some of these could be fairly small projects located in regions, as, for example, those producing scotch carts and wheelbarrows. Larger projects could produce lorry and bus bodies to expand transport facilities for passengers and goods over longer distances. The West African "mammy lorry" might serve as a prototype: The bodies are locally made of wood, while the wheel base and engines are imported. If these latter could be standardized through the import of a strictly limited number of types, it should be possible in a later stage of the long-term plan to produce all the parts and equipment and eventually even the engines, domestically.

Repair shops and garages could be established in the rural areas, as Rucom Industries has begun to do, to ensure that transport and farm equipment is kept in running order. Manufacturing projects should be established in each province to produce construction materials to build schools, hospitals, small factories, and improve local housing. The local production and use of timber, bricks, and tiles should be encouraged. Improved tools and equipment should be provided to small local establishments to produce low-cost wooden furniture.

The first stage of the plan, in sum, would be directed to linking new manufacturing establishments to specific improvements in the productivity and levels of living of the low-income population. No new large-scale project would be established on the Copper Belt or along the line of rail until every possibility of setting up smaller units in several regions to produce an equivalent output had been thoroughly canvassed. The need to spread employment opportunities and utilize local raw materials would become primary criteria for the evaluation of any proposed industry.

This is not to say that there would be no room for larger-scale, more capital-intensive projects. It is self-evident that the mines must continue to operate on a large scale using modern, relatively capital-intensive techniques to enable them to compete on the world market. Their role will continue to be to provide foreign exchange for capital equipment for the range of other industries that will be planned throughout the economy. At the same time, local production

of supplies and materials required by the mines should be encouraged
by cutting off imports of competitive products wherever long-run economic feasibility can be established. Some of these projects, too,
may need to be relatively capital intensive. The chemicals plant producing explosives for the mines, as well as the petroleum refinery at
Ndola, are, of necessity, relatively capital intensive. Factories set
up to process copper into finished products, too, are likely to be relatively capital intensive, in order to benefit from economies of scale.
Zamefa, for example, produces wire and cable on a large scale. Expansion of its output to meet domestic needs, as well as for sale in
other countries, should contribute significantly to foreign-exchange
savings as well as earnings. The output of such plants should be systematically examined within the context of the long-term perspective
plan to determine how they can best contribute essential inputs to expanding output in other sectors of the economy, as well as the mines.

In later stages of the perspective plan, greater emphasis should
be placed on somewhat more technologically complex manufacturing
industries. (It goes without saying that such possibilities would be
greatly enhanced by coordinated planning and implementation of industrial development with neighboring countries.) Although there is inadequate space here to discuss these, it should be emphasized that, as
productive employment and incomes rise, the market for a wider
range of manufactured goods may be expected to expand.[50] The parts
and materials for increasingly complex farm and transport machinery
and equipment may be produced in Zambian factories. Expanded hydroelectric facilities should facilitate the use of electric power machinery and equipment in every rural area, thus creating a demand
for more sophisticated power-driven tools.

The specific ingredients of each stage of the perspective plan
should be carefully evaluated in light of the progress made in the preceding stage. Ongoing research and increasing involvement of local
people are essential to discover many new possibilities and utilize
old ones in each province and each village. Rapid expansion of educational facilities is essential to provide technical skills as well as to
foster the spread of manufacturing projects to stimulate national and
local development. Manpower planning must accompany the decisions
to develop the manufacturing sectors in order to expedite the replacement of foreign high-level manpower by Zambian personnel at every
level.

The details of the long-term perspective plan and its stage-by-
stage development cannot be blueprinted here. It requires a careful
study of the available resources of each province and district, necessi-
tating appointment of personnel on a national, provincial, and district
level to bring together existing information and to begin work immediately on further research as the foundation of such a perspective plan.

What can be emphasized, in view of Zambia's postindependence experience as well as the experiences of many other less developed nations, is that without such a perspective plan the pattern of manufacturing industry created is likely to continue to perpetuate and aggravate the external dependence and lopsided growth of Zambia's inherited dual economy.

The Need for Institutional Changes

The explanation here given for the causes of Zambia's distorted manufacturing-sector growth suggests the inadequacy of the entire state machinery for implementing any kind of alternative pattern of manufacturing expansion. A crucial aspect of any long-term industrial strategy, therefore, must be directed to restructuring the institutions and working rules, which have in the past contributed to distorted manufacturing growth. Investment decisions governing the expansion of the manufacturing sector should be increasingly influenced by those who would benefit, particularly the lower-paid wage-earners and the rural-dwellers. This implies that, wherever possible, participation by those who stand to gain should be built into the new working rules and institutional arrangements. [51] Presumably this places a political responsibility on UNIP, the trade unions, and various other organizations of the people. This political aspect is vital as a foundation for formulating and implementing a long-term plan to restructure the economy, but it is not the subject of this chapter, which seeks only to indicate changes in the machinery that seem essential if such a plan is to be put into effect.

First, as a basis for formulating and implementing a long-term perspective plan, primary responsibility must be lodged with a sufficiently centralized agency to ensure that a cohesive, integrated overall plan is worked out and effectively implemented. The cabinet and parliament should approve the long-term perspective plan and each of its successive stages so that it has behind it the force of law. At the same time, a built-in system must be created to achieve ongoing evaluation and two-way channels of communication between planning authorities and all those affected by the plan to correct mistakes and take advantage of new opportunities.

Second, the role and method of operation of the parastatals responsible for manufacturing industries must be integrated into the overall system of plan creation and implementation. Careful reevaluation of the techniques of management and control should be directed to ensuring that day-to-day decisions of parastatals are implemented in accord with the requirements of the long-term perspective plan. In this connection, the ways and means of increasing workers' participation in management decisions should be considered.

Third, the entire set of institutions governing import and internal wholesale trade needs to be reexamined to ensure that it contributes to the long-term strategy proposed. Tariffs, exchange control, and import licensing appear to have been a rather indirect and inadequate means of changing the composition of imports to support the appropriate kinds of domestic manufacturing growth. Leaving critical decisions of what imports to order and from what sources to order them in the hands of the foreign private partners of the trading agencies appears unlikely to facilitate a rapid shift in the import pattern. As quickly as the domestic manpower constraint permits, with a judicious use of imported manpower, these decisions should be more directly controlled within the framework of carefully worked-out criteria in the context of the long-term plan.

The financial system is the final major area of institutional change suggested by the explanation of the causes of Zambia's distorted manufacturing sector. What appears to be required is the creation of a sufficiently centralized authority to formulate and implement an overall financial plan that would effectively direct the investable surpluses already produced in the economy to the implementation of the long-term perspective plan and its successive stages. The financial plan should include the following specific features:[52]

1. Formulation of an overall income policy that would explicitly guide the policies affecting all incomes, taxes, and prices. This policy should govern not only wages and salaries, but profits, interest, and rent as well. Its aims should be to ensure a steady improvement in the real levels of living of the lower income groups in rural and urban areas, while capturing and directing a major share of the inordinately high incomes of less than 10 percent of the population to essential productive activities.

2. Coordination of the collection and utilization of a significant share of tax revenues as well as profits accumulated in the parastatal sector and long-term savings institutions for the purposes of financing planned expansion of the productive sectors.

3. Control of commercial banks to direct domestic credit, as well as any funds that might be borrowed externally on favorable terms, to implementation of specified projects in the context of the long-term perspective plan.

A system of coordinated institutions for formulating and implementing long-term physical and financial plans to restructure the Zambian economy cannot be created over night. Nor will any system designed for this purpose ever be "perfect." What is essential immediately is that priority be given to the creation of such a system, which can start now to formulate and begin to implement the stages of an effective perspective plan.

CONCLUSIONS

An analysis of the available data indicates that Zambia's post-independence manufacturing industrial growth, while apparently exceptionally rapid, has not contributed much to restructuring the Zambian economy. If anything, it has tended to perpetuate and even aggravate the inherited dualism. The explanation for this distortion appears to lie in the adoption of an import substitution policy backed by an implicit assumption that the choice of projects is best left to existing "market forces" dominated by a narrow high-income group and giant multinational corporations; and the failure to make the essential institutional changes required to ensure that a major share of the rather large investable surpluses produced in the country itself are directed toward the implementation of a more appropriate pattern of manufacturing industry.

If the above explanation is valid, it argues strongly for the necessity of formulating a long-term perspective plan explicitly elaborating the stage-by-stage development of manufacturing industries to spread productive employment opportunities into the rural areas, altering the inherited distorted income and market patterns. At the same time, it suggests a fundamental change in the role of various ministries, parastatals, and financial agencies that must assume responsibility for implementing such an approach.

In short, only a long-term industrial strategy, linked with institutional changes required to implement it, can ensure that the growth of the manufacturing sector will contribute more effectively to increasing productive employment opportunities and improving the real living conditions of Zambians in all sectors and provinces.

NOTES

1. Ministry of Development Planning and National Guidance, Second National Development Plan, January 1972-December 1976 (Lusaka, 1971), p. 93; and R. B. Sutcliffe, Industry and Underdevelopment (London: Addision-Wesley Publishing Company, 1971), p. 103.

2. O. Kreye, "The Myth of Development Decades," Die Dritte Welt, Jahrgang 2, no. 3, 1973.

3. W. J. Barber, The Economy of British Central Africa: A Case Study of Economic Development in a Dualistic Society (Stanford, Calif.: Stanford University Press, 1961).

4. A number of recent studies substantiate this point in considerable detail; for example, L. Van Horn, "The Agricultural History of Baretseland," History Seminar no. 14; and C. P. Luchembe, "Rural Stagnation: A Case Study of the Lamba-Lima of Ndola Rural District,"

History Seminar no. 17 (Lusaka: University of Zambia, 1974, mimeographed).

5. All statistics relating to the Zambian economy, unless otherwise cited, are in or calculated from data in Central Statistical Office (Lusaka), Monthly Digest of Statistics 70, 7 (July 1973).

6. H. B. Chenery, "Patterns of Industrial Growth," American Economic Review, September 1960, p. 646.

7. The data re value added by manufacturing in this section are from the Central Statistical Office, Censuses of Industrial Production (Lusaka, 1966 and 1970).

8. Compare A. Seidman, Comparative Development Strategies in East Africa (Nairobi: East African Publishing House, 1972), p. 23.

9. These are discussed in more detail in M. Faber, "The Development of the Manufacturing Sector," in C. Elliot, ed., Constraints on the Economic Development of Zambia (Nairobi: Oxford University Press, 1971), pp. 301-07.

10. The evidence supporting this analysis appears in Central Statistical Office, Census of Production (Lusaka, 1965 and 1970); Central Statistical Office, Monthly Digest of Statistics (July 1973), pp. 54, 20, 19; Central Statistical Office, External Trade Statistics, 1970; Second National Development Plan, op. cit., especially p. 172.

11. Extensive evidence has been gathered in Latin America, where the import substitution process was initiated in the Great Depression of the 1930s; compare UN Economic Commission for Latin America, The Process of Industrial Development in Latin America (New York: United Nations, 1966), p. 29; A. G. Frank, "Economic Dependence, Social Structure and Underdevelopment in Latin America" (typescript), 1969; S. Marario, "Protectionism and Industrialization in Latin America," Economic Bulletin for Latin America, March 1964; N. H. Leff and A. D. Netto, "Import Substitution, Foreign Investment and International Disequilibrium in Brazil," Journal of Development Studies, April 1966; A. O. Hirschman, "The Political Economy of Import Substituting Industrialization in Latin America," Quarterly Journal of Economics, February 1968; C. Furtado, "Industrialization and Inflation," International Economic Papers, no. 12, 1967. This experience has also been repeated in other countries; compare Seidman, op. cit., especially ch. 6; R. Soligo and J. J. Stern, "Tariff Protection, Import Substitution and Investment Efficiency," Pakistan Development Review, Summer 1965; C. F. Diaz-Alejandro, "On the Import Intensity of Import Substitution," Kyklos, 1965; H. G. Johnson, "Tariffs and Economic Development: Some Theoretical Issues," Journal of Development Studies, October 1954; J. H. Power, "Import Substitution as an Industrialization Strategy," Philippines Economic Journal, Spring 1967. The effect of the distorted distribution of income on import substitution has been explicitly

discussed in J. L. Lacroix, "Le Concept d'Import Substitution dans la Theorie du Developpemente Economique," Cahiers Economiques et Sociaux, June 1965, p. 174; and Sutcliffe, op. cit., pp. 267-68.

12. The distribution of income in Zambia is analyzed in A. Seidman, "The 'Have-Have Not' Gap in Zambia—or, What Happens to the Investable Surpluses Produced in Zambia?" (University of Zambia, mimeo., 1973).

13. Central Statistical Office (Lusaka), Consumer Price Index-Low Income Group: New Series, 1969 Base, describes a household budget survey of 2,600 low-income households in the main line-of-rail cities in Zambia. It shows that the families of workers earning the average wage that year could not expect to get through the month without borrowing. The average rise in wages since that period has essentially been canceled out by rising prices.

14. An international business research publication, Prospects for Business in Developing Africa (Geneva: Business International S.A., 1970), emphasizes, in discussing market strategies, that "many companies that want a piece of the long-term action are now establishing African 'beachheads'" (p. 39).

15. Second National Development Plan, op. cit., p. 94.

16. Sutcliffe, op. cit., p. 227.

17. A. A Beveridge, "Converts to Capitalism: The Emergence of African Entrepreneurs in Lusaka, Zambia" (New Haven, Conn.: Ph.D. dissertation, mimeo., 1973).

18. Seidman, "The 'Have-Have Not' Gap in Zambia," op. cit., p. 11.

19. Seidman, Comparative Development Strategies in East Africa, op. cit., p. 263.

20. Second National Development Plan, op. cit., p. 43.

21. Ibid., pp. 197-304.

22. R. H. Green, "Four African Development Plans: Ghana, Kenya, Nigeria and Tanzania," Journal of Modern African Studies 3, 2 (1965).

23. Second National Development Plan, op. cit., p. 96.

24. Ibid., p. 244.

25. The author has attempted to explore that issue in "Alternative Development Strategies in Zambia" (University of Zambia, mimeo., 1973).

26. Compare B. B. Schaeffer, "The Deadlock in Development Administration," in Colin Leys, ed., Politics and Change in Developing Countries (Cambridge: Cambridge University Press, 1969).

27. The information relating to Indeco, unless otherwise cited, is from Indeco, Ltd., Annual Report, 1973 (Ndola: Falcon Press, 1973).

388 NATURAL RESOURCES AND NATIONAL WELFARE

28. For example, see A. Seidman, "Planning for Development: Problems and Possibilities in Sub-Saharan Africa" (New York: Praeger Publishers, forthcoming); also see Y. Kyseimira, "The Public Sector and Development in East Africa," Makerere Institute of Social Research Papers, January 1968; Gold Coast, Part I, Government Proposals in Regard to Future Constitution and Control of Statutory Boards and Corporations in the Gold Coast (Accra: Government Printers, 1956), especially p. 5; Arthur D. Little, Inc., Tanganyika Industrial Development: A Preliminary Study of Bases for the Expansion of Industrial Processing Activities, on behalf of the Ministry of Commerce and Industries of Government of Tanzania under contract with the U.S. Agency for International Development, 1961.

29. Indeco, Annual Report, 1973, op. cit., p. 10.

30. G. Kahama, "The National Development Corporation and the Industrialization Process in Tanzania," Public Lecture, UNIDO seminar, January 26, 1969.

31. Indeco, Annual Report, 1973, op. cit., p. 40.

32. Second National Development Plan, op. cit., p. 194.

33. Ibid., p. 195.

34. Indeco, Annual Report, 1973, op. cit., p. 4.

35. Business International SA, Prospects for Business in Development Africa, op. cit., p. 55.

36. Indeco, Annual Report, 1973, op. cit., p. 6.

37. G. Myrdal, Asian Drama (New York: Random House, 1968), pp. 1126, 1131.

38. Business International, SA, Prospects for Business in Developing Africa, op. cit., p. 14.

39. Data relating to composition of imports, unless otherwise indicated, are from Central Statistical Office, External Trade Statistics, series, 1966-70 (Lusaka, Central Statistical Office).

40. Bank of Zambia, Report and Statement of Accounts for the Year Ended December 31st, 1972 (Lusaka, 1972), p. 39.

41. Indeco, Ltd., Annual Report, 1973, op. cit., p. 32.

42. Compare Seidman, Planning and Development, op. cit., ch. 18; see also N. T. Newlyn and D. C. Rowan, Money and Banking in British Colonial Africa (Oxford: Clarendon Press, 1954).

43. Zambia Daily Mail, October 23, 1973.

44. Compare C. R. M. Harvey, "Financial Constraints on Zambian Development," in Elliot, ed., op. cit., pp. 136 ff.

45. Ibid., pp. 125, 133-35.

46. Compare M. Bostock and C. Harvey, eds., Economic Investment and Zambian Copper: A Case Study of Foreign Investment (New York: Praeger Publishers, 1971).

47. Bank of Zambia, Report, op. cit., p. 36.

48. This hypothesis is well-formulated in B. Van Arkadie, "Development of the State Sector and Economic Independence," in D. P.

Ghai, ed., Economic Independence in Africa (Nairobi, East African Literature Bureau), especially pp. 108-12. Some of the theoretical aspects are further analyzed in J. Saul, "The Political Aspects of Economic Independence" in the same book, pp. 123-50.

49. The possibilities of such an approach are discussed more fully in Seidman, Planning for Development, op. cit., Chs. 6 and 7.

50. Summary data relating to two textile mills with the same output but very different labor-capital ratios is provided by National Development Corporation, Third Annual Report, 1968 (Dar es Salaam: National Printing, 1968), p. 55.

51. Compare R. H. Green and A. Seidman, Unity or Poverty? The Economics of Pan-Africanism (Harmondsworth: Penguin African Library, 1968), passim, for discussion of possibilities; R. H. Green discusses the necessity for further research to realize these possibilities in "Economic Independence and Economic Cooperation," in Ghai, ed., op. cit., pp. 45-87.

52. This argument is more fully developed in Seidman, Planning for Development, op. cit., chs. 3, 4, and is increasingly emphasized by academics (for example, F. Holmquist, "Implementing Rural Development Projects," in G. Hyden et al., Development Administration, p. 228) as well as political leaders (for example, Tanganyika African National Union, T.A.N.U. Guidelines, 1971 [Dar es Salaam: Government Printer, 1971], p. 9). In Zambia, participation is considered a fundamental tenet of humanism (compare K. Kaunda, Times of Zambia, November 13, 1973). The problem is to create the necessary institutional machinery and working rules to implement the idea, ensuring that the broad masses of the working people, the peasantry, and the unemployed are increasingly involved in decision-making to ensure that restructuring in their interests does in fact take place.

VI

THE IMPACT OF NEW COPPER MINES ON TWO UNDERDEVELOPED ECONOMIES

23

MINING AND DEVELOPMENT STRATEGY IN BOTSWANA
Robson Silitshena

Botswana is a large country covering an area of 570,000 square kilometers. It is a landlocked state bounded in the east and south by South Africa, in the west by Namibia, and in the north by Rhodesia and Zambia, the latter through the narrow Caprivi Strip. It has a population of about 650,000, 80 percent of which lives in the eastern part of the country where the average annual rainfall is about 550 millimeters, sufficient for both arable and livestock farming. More than half of the country to the west and southwest consists of the vast semidesert, the Kalahari.

The country achieved independence on September 30, 1966, after 81 years of British protection. With independence, the country inherited "a backlog of deep poverty"[1] because the British had not done much to develop the country, which it was always assumed would be incorporated into South Africa,[2] a dream that South Africa herself cherished right up to 1965. A large part of the country being semidesert and since no major mineral discoveries had been made, it was always assumed that the country had no potential at all, and hence very little, if any "colonial development" had taken place at the time of independence.

The infrastructure and services were poorly developed. The railways were limited to the Rhodesia Railways line from Mafeking to Bulawayo that ran (and still runs) along the eastern part of the country. The only major road was the dusty north-south road that still runs

The author wishes to thank G. Tough, P. Eigen, A. Seager, and W. Pintz and some of his colleagues in the Botswana campus for their useful comments and suggestions on the first draft of this study.

alongside the railway line. Outside the traditional villages there were only two small towns, Francistown and Lobatse, which between them had very few industries. Educational facilities were lacking, and illiteracy (now at about 68 percent) was very high. Hence lack of skilled manpower was and still is one of the most acute problems facing the country.

The mainstay of the economy was cattle farming. Agriculture did not guarantee a steady income, not only because of intermittent droughts but also because farmers still employed "primitive" methods of farming, and many regarded cattle as a form of prestige and not to be parted with easily.

Manufacturing industry was almost nonexistent. Among the inhibiting factors were inadequate infrastructure, small market, lack of skilled manpower, and competition and pressures[3] from the more powerful South African firms, which had free access to the Botswana market, as the country was (and still is) a member of the Southern African Customs Union, which included Lesotho and Swaziland as well. Thus the main source of cash for the majority of people, outside the occasional cattle sales, were the wages of migrant laborers working in South Africa.[4]

SOME ASPECTS OF THE ECONOMY

In 1971/72 the gross domestic product (GDP) was estimated at R87.6 million (R1 equals $1.49). Its composition in percent was as follows:[5]

Sector	Percent
Rural activities (agriculture, hunting, fishing, forestry, and traditional ownership and construction of dwellings)	34
Mining	14
Construction	10.6
Commerce and industry (manufacturing, trade, hotels, and restaurants)	15.4
Private sectors services (transport and communications, financial institutions, real estate, and so on)	12
Government services	14
Total	100.0

Agriculture remains the dominant sector of the economy, accounting for over 87 percent of the population. The per capita income in this sector, between R35 and R40 in 1972,[6] is very low compared to the average income of over R509 in other sectors. There is a very skewed distribution of wealth in the rural areas, and "there is strong evidence to indicate that in recent years the gap between large and small farmers is widening."[7] In the 1971/72 agricultural season 30 percent of the farmers had no cattle at all while the top 10 percent of the farmers owned more than half of the cattle outside the white-held areas.[8] The large farmers are also the main cultivators, being responsible for a very large acreage under crops.

Livestock farming, mainly cattle farming, remains the most important activity. In 1968 livestock products, mainly beef and related products, accounted for 95 percent of total value of exports. However, the 1972/73 figures indicate that Botswana may be moving to a slightly more diversified economy as mineral exports—mainly diamonds—accounted for approximately 42 percent of the total value of exports.

The number of people in wage employment in 1973 was 48,039. Of these only 1.2 percent were employed in manufacturing, which was dominated by the state-owned Botswana Meat Commission abattoir in Lobatse. Central and local government were among the major employers, accounting for about 29 percent of the employed, while about 17.5 percent were employed as domestic servants. Construction, which has increased in importance since independence, accounted for 13 percent of the employed, while mining and quarrying accounted for only 3.4 percent.[9] At any one time, there are about 50,000 Botswana employed in mining and other sectors of the South African economy. Their total earnings are estimated at R8 million, of which only R2 million is remitted to Botswana.[10]

Until 1971/72, up to 46 percent of the government recurrent revenue was derived from the British grants-in-aid. In addition the British grants and loans accounted for a large portion of development expenditure. The reduction and, finally, the complete phasing out of these grants were made possible largely by the huge increases of revenue from customs, excise, and sales duties associated with the massive capital imports for the mining projects, following the revision of the Customs Union agreement in 1969.[11] Thus between 1971/72 and 1972/73, recurrent revenue had increased from about R18 million to about R28 million. Currently this source is contributing about 55 percent of total domestic revenue. The Botswana government, however, does not consider this a stable and dependable source of revenue not only because revenue from this source is expected to drop with the completion of construction of mining projects, but also because, as the vice president put it, "the size of the total customs area pool is largely determined by decision of the South African government over which we have little influence."[12]

One of the major aspects of the economy is this dependence on South Africa. Even the government acknowledges that the economy is a "satellite" of the South African economy.[13] South Africa is the second largest market of Botswana's beef exports; it provides most of the imports and many jobs as noted above and has until recently been the main source of private capital. In addition, Botswana, like Lesotho and Swaziland, uses South African currency.* The current plan aims to reduce this dependence by strengthening the internal economy and establishing diversified external economic links. In this context mineral development is seen as having a vital role to play.[14]

Before the recent mineral discoveries, many observers did not see any chance of economic independence for Botswana.[15] Before focusing attention on the benefits of mining to Botswana, it would be useful to survey the current mineral developments.

RECENT MINERAL DEVELOPMENTS
IN BOTSWANA

Mining is not a new activity in Botswana. The small-scale gold mining (at Tati near Francistown), which ceased in 1964, started in 1869; in addition there has been sporadic small-scale mining of other minerals, chiefly manganese and asbestos in the southern part of the country. However, compared with the current mineral development these activities were of very minor significance indeed.

One of these new developments stems from the discovery by the De Beers group in 1967 of diamond pipes west of Francistown on the edge of the Kalahari. One of these pipes, the Orapa pipe, is 112 hectares, making it the second largest after the Williamson's pipe in Tanzania. The approximate grade of diamonds is 0.89 carats per ton of ground. The proportion of gem to industrial stones is 12:88. The De Beers group has sunk R21.5 million into the diamond mine, which started production in July 1971. The pipe is being mined by open cast methods at an annual rate of 2.8 million tons to produce about 2.4 million carats. Part of the infrastructure (power, water supplies, and township) has been provided by the company. An all-weather road

---

*It was announced in September 1974 that Botswana will issue its own independent currency within two years. Swaziland now issues its own currency, which is freely convertible into land and is backed 100 percent by the land.

and a telephone link from Francistown, at a cost of R2 million, have been provided by the government from a loan raised from the company to be repaid out of dividends from the government's 15 percent shareholding in the company. The De Beers subsidiary in Botswana has been allocated a 4 percent quota of the total sales of the Central Selling Organisation in London.

Perhaps of much greater importance and magnitude than the diamond mine is the copper-nickel project at Selebi-Pikwe, some 88 kilometers southeast of Francistown. These deposits are being exploited by Bamangwato Concessions Limited (BCL), a company 85 percent of whose equity share is owned by the Botswana subsidiary of the Roan Selection Trust (RST) and 15 percent by the Botswana government. RST is itself dominated by American Metal Climax (Amax) and Anglo-American Corporation.

The proven reserves of copper and nickel stand at 40 million tons, with an average ore content of 1.4 percent and 1.1 percent, respectively. The first phase, which has reached a stage of limited production on account of technical problems, is concentrating at the larger Pikwe deposits. The Selebi deposits, 15 kilometers away, will be exploited from 1979. When full production is reached, BCL will mine the deposits by underground and open-pit methods at a rate of about 2 million tons of diluted ore a year.

The company has invested R120 million, of which R33 million has been raised through the sale of its shares, to bring the first phase of the project into production. With the unexpected technical problems and inflation, the final figure is expected to be over R150 million. A large portion of the loan capital has been raised from a consortium of West German banks, while a smaller amount has been raised as suppliers' credit from the Industrial Development Corporation of South Africa.

The mining plant, a concentrator, and a smelter have already been constructed. As part of the financial agreement, the copper-nickel matte will be refined in the Amax refinery in the United States for the next 15 years. This has been forced on the government by Amax, which, anxious that its U.S. refinery not continue working below capacity, made this a condition for providing part of the necessary capital. The Botswana government would have preferred to have the ore processed locally or in Zambia.

The company has also signed another 15-year agreement with Metallgesellschaft AG by which the latter will purchase two-thirds of the annual production of nickel and all the copper produced at the prices ruling in West Germany. Triomf Fertiliser (South Africa) has agreed to purchase 72,500 tons of sulfur from BCL.

The infrastructure for the project, estimated at R59.7 million, has been provided by the government. It consists of a dam and ancillary works on Shashe River and an 80 kilometers long pipeline; a 45-megawatt thermal power station at Selebi-Pikwe burning locally produced coal from Morupule Colliery;* a township and municipal services designed for an initial population of 7,500 people, which was originally expected to rise to 28,000 in 1978 but which is now estimated by some as high as 40,000; a 65-kilometer and 14-kilometer branch railway lines between Serule and Selebi-Pikwe and between Palapye and Morupule respectively, and a gravel road between Serule and Selebi-Pikwe.

The major sources of the government loans for financing this infrastructure have been the International Bank for Reconstruction and Development (IBRD), the Canadian government, and the U.S. government. The IBRD loan, at 7.25 percent interest a year and repayable in 25 years, has been guaranteed by the companies involved in the project. To enable it to meet its debt commitments even in years when company profits are very low, the government has also involved the companies directly in its debt repayments. According to the agreements between the companies and the government (including the newly established Botswana Water Utilities Corporation and the Botswana Power Corporation), the water, power, and railway charges will be fixed in such a way as not only to cover the operating costs but also to meet the government debt commitments. In the case of a shortfall in loan repayments, the company will pay the difference. This formula was thought to be preferable to the one where the government made its own arrangements to pay its debtors out of its mining revenues. In addition the company will make an annual township contribution, which will meet almost in full all the cost of putting up the township, estimated at R6.3 million.

Another recent mineral development is the opening of the small Morupule Colliery (near Palapye south of Francistown), which is being worked by a subsidiary of the Anglo-American Corporation. The mine, which is being worked by underground methods, started production in July 1973. Designed to produce up to 640,000 tons, it will initially produce 200,000 tons of coal, almost all of which will be sold to the Selebi-Pikwe mine and power station. Morupule coal is expected to be used in the Makgadikgadi brine mine as well. The exploitation of these coal resources is another testimony to Botswana's determination to be as self-sufficient and independent of South Africa as possible, for it would have been far cheaper to import South African coal and electricity.[16]

Discussions are currently taking place between the government and the De Beers Group about the possibility of opening a second dia-

---

*Because of technical problems, the power station is not yet burning Botswana coal but rather is using South African coal.

mond mine 50 kilometers southeast of Orapa, and between the government and the RST group about the possibility of exploiting the brine deposits at Makgadikgadi (Makarikari) depression west of Francistown. The mining of both deposits is scheduled to commence before 1978.

MINING IN THE DEVELOPMENT OF BOTSWANA

Some writers on economic development see mineral development as the magic route to fast and sustained economic development for the poor countries. Economic growth through agricultural development is seen as being a "slow and laborious process."[17]

However, as several economists have argued,[18] it is not always the case that mineral development provides a base for further economic development. The inability of mining (or, more accurately, the inability of government policy to control and direct mining) to contribute to rapid economic development is usually attributed to a number of reasons. It is therefore essential to make a realistic assessment of the hopes and optimism expressed by the former director of RST, among others, that the successful bringing into production of the Selebi-Pikwe mine can "change the nature of the economy of Botswana in terms of employment, gross national product, exports, national revenues and secondary consequences of the mining industry."[19]

Mining, because of its capital-intensive nature, will not create many jobs. The Orapa diamond mine employs about 600 people, the coal mine only between 40 and 50, and Selebi-Pikwe, 2,000, of whom 600 are expatriates. Only in Selebi-Pikwe is there any expectation of significant secondary employment being created, and assuming an approximate one-for-one ratio between direct and secondary employment, the number of jobs created for local workers is estimated at 3,500 and is expected to rise to 5,300 in 1980. This is rather unfortunate because one of the major problems facing the country is lack of employment opportunities, which is largely responsible for labor migration. It is estimated that the adult labor force increases by 12,000 each year while only between 5,000 and 6,000 jobs are being created annually.

For several reasons, Selebi-Pikwe is not likely to become a major growth point in the north for quite some time. First, the refining and further processing of the minerals and the setting up of industries—for example, the chemical industries, to service the mining industry, are thought to be remote possibilities on account of the limited local market, but no doubt also because of the ease with which the products of latter industries can be supplied from South Africa. Second, the numbers employed are so small that they do not provide an adequate base for the development of industries producing consumer goods, except for export-oriented industries. High electricity and water charges, because of the absence of economies of scale, may also inhibit secondary industrial growth.

The mining complex and the associated township may in future offer a good market for local agricultural produce. In the case of the former, although the company is bound by an agreement with the government to purchase local products and materials where they compare favorably in price, quality, and delivery dates with imports, there is very little at present that it can purchase locally. Furthermore, it is doubtful whether the company's purchasing practices, involving centralized buying via their Johannesburg office, will facilitate buying even what is available locally. Consequently, Selebi-Pikwe represents a classic case of enclave development, for even ordinary foodstuffs are imported. (It is ironic that farmers at Tuli Block, not far away from Selebi-Pikwe, export to South Africa.)

The diamond mine, because of security reasons, is not opened to the public, and all services (for example, a small bakery and shopping facilities) are being provided by the company. Further, it is situated in a thinly populated part of the country where the climate is not conducive to farming. It seems therefore that, except for the colliery, there will be limited forward and backward linkages resulting from mining, at least in the short run.

The mining infrastructure, particularly transport facilities, will aid other sectors of the economy, particularly agriculture. The roads should provide farmers with access routes to the markets. The transportation system from the brine deposits is being planned to serve other purposes than mining. Francistown is being connected by transmission lines to the Selebi-Pikwe power station.

Revenues to the central government do constitute a major direct contribution of mining. Government revenue through income tax, royalties, and dividends from the Selebi-Pikwe mine is put initially at about 54 percent of the company gross profits. The mine is expected to yield an annual revenue in excess of R4 million when it is in full production. The breakdown for 1977/78 is expected to be as follows: royalties (R750,000), income tax (R2,220,000), and dividends (R1,220,000).[20] Royalties may actually bring in more. Total revenues from the Selebi-Pikwe mine would be much larger if the costs of putting up the infrastructure were not considered as company cost. On the basis of present arrangements now being renegotiated, * govern-

---

*According to the report in the Botswana Daily News of October 17, 1974, "the Botswana Government will shortly be taking 50 percent share in the equity of the Orapa diamond mine." The new agreement

ment revenues (through income tax, profits tax, withholding tax, dividends, and royalties) from diamonds, is about 57 percent of the company gross profits and is expected to increase. If the present forecasts materialize, the government is expected to receive revenues from the diamond industry in excess of R15 million annually.

The mineral revenues are seen as a cornerstone in the country's struggle to achieve its development goals.

From the government's point of view, the country is already beginning to enjoy a certain measure of economic independence from South Africa through the arrangements for the financing of the copper-nickel project, while the diamond mine has strengthened the internal economy. The government hopes to achieve social justice through its key economic strategy, which is to secure rapid and large returns from the capital-intensive investment, mainly mining, and to reinvest these revenues mainly in rural areas. It is only in this way that the benefits of mining can be "distributed equitably throughout the country [and] the glaring disparity in incomes and in the standard of living between those who live in urban and those in the rural areas [can] be reduced."[21]

Among the major aims of government policy[22] with respect to rural development are the following:

1. Self-sufficiency in staple food crops—maize, sorghum, millet, and cowpeas—and horticultural produce, the bulk of which is now imported from South Africa. The newly set-up Grain Marketing Board will regulate the marketing of major crops and pay good prices to producers. It is also hoped that a sufficiently large surplus will be produced so as to supply the export markets as well and thus assure the farmers of even larger incomes. The Ministry of Agriculture says that the results of its Dryland Farming Research Scheme now coming out, if properly implemented, indicate that farmers can be assured of incomes of not less than R400 per year from crop farming alone. To expand the livestock industry, in particular to expand the marketing activities of the Botswana Livestock Development Corporation, a subsidiary of the BMC is to be set up to reach farmers in very remote areas, especially in the northern areas of the country.

2. To curb soil erosion and encourage good range management.

3. To expand a program for providing water supplies to the villages by sinking more bore holes and building small dams and to improve social services (education, medical, and welfare services).

"is expected to raise Botswana's share of Orapa's profits . . . to more than 75 percent."

4. To create new employment opportunities wherever feasible, in particular to promote industries, services, and crafts, so as to reduce the numbers at present without any means of support.

There are very many constraints in agriculture in Botswana. They range from hostile climatic conditions through relatively poor soils and poor infrastructure to lack of agricultural inputs, managerial skills, and capital. Quite a large number of holders do not plant anything or plant late because of difficulty of getting draught animals while a comparable number lacks seeds, and at the beginning of 1973/74 season, there were cases where holders consumed seeds that had been distributed by the government because they had no food left. A large number of farmers cannot afford fertilizers, and as a result of constant cultivation without fertilization, the soil is depleted. The combination of all these factors leads to extremely low productivity.

The funds from mineral revenues will be channeled to the farmers via the National Development Bank (NBD) or to cooperatives through the Cooperative Development Trust. These institutions may have to change their policies radically if they are to be of service to the very poor in the rural areas. There will have to be a lot of risk-taking, as some of the funds will have to be lent on trust[23] to people with nothing to show for collateral. The NBD itself is expected to increase its own staff in the field.

The availability of these extra funds will depend upon a number of factors. First, revenues from other current sources, especially from the Customs Pool, must remain stable; otherwise, the extra revenues from mining will be absorbed into recurrent expenditure. Recurrent expenditure has increased threefold since 1966/67. Second, the country has been accumulating loans for its development projects and partly to offset its balance-of-trade deficit, which increased from about R8.5 million in 1960 to R33 million in 1973/74 (before the full impact of increased fuel prices was felt). If these loans are not managed properly, the country may find itself spending a lot of its mineral revenues in debt servicing.

Third, in this rainfall-deficient country, recurring droughts like the ones that hit the country in the 1960s can easily divert development funds into famine relief and increase rural exodus. It would therefore seem that the "birth" of the mining sector has not obviated the need for diversifying the economy. In recognition of this fact, the Ministry of Finance has set up the Revenue Stabilization Fund to prevent revenue fluctuations.[24]

Fourth, with rapid population growth in the urban areas, especially in Selebi-Pikwe itself, a large portion of government revenues from mining may be spent in urban areas to provide housing, services, and job opportunities. Unless something is done about the present in-

come differentials between urban and rural areas and rural underdevelopment, rural–urban migration cannot be stemmed. Perhaps,
therefore, it would have been more advisable if the government had
adopted a "mixed" strategy of allocating part of the mining revenues
to rural development and part to creating productive investment in urban areas, such as setting up labor-intensive industries. There
would, no doubt, be problems—for example, the small local market—
but this would be a solution more consistent with the present realities. *
The population of the squatter settlement at Selebi-Pikwe, for instance,
is estimated at 20, 000.

Last, the extent to which Botswana will benefit from the present
mineral ventures will depend largely upon the mining policies of the
government. The last part of this chapter is devoted to a brief review of major mining policies in Botswana.

MINING POLICIES IN BOTSWANA

The right of prospecting for, mining, and disposing of minerals
on any land is vested in the state. Until 1967 each tribe held mineral
rights within its own territory. When the RST Group started prospecting in the northeast in 1969 it signed an agreement with the tribal authority of the local tribe, the Bamangwato. Under an act passed in
1967, all tribes surrendered their mineral rights to the state so that
minerals are developed in the interest of the country as a whole.
Only in the freehold areas, which had originally been set aside for
European occupation—the Gaborone, Ghanzi, Lobatse, Tati, and Tuli
blocks—were mineral rights still in private hands. The Mineral Rights
Tax Act of 1972, aimed at taxing all private mineral rights, induced
a number of private mineral rights holders to surrender these rights
to the state. The last to do so were the Tati Company Limited and
Tati Territory Exploration Company, which surrendered these rights
at the end of June 1974. No compensation has been paid to the companies for surrendering these rights. The Tati Territory Exploration
Company has been issued, instead, a short-term prospecting license
subject to government control.

Prospecting for minerals, like any other mining activity, is
carefully controlled by the government under the Mines and Minerals
Act of 1967. Prospecting permits are valid for a maximum period of
three years and longer than that only in special cases by ministerial
decree. A mineral discovery does not confer an automatic mining

---

*This is not to imply that the government has no policy whatever
on industrial development.

right. The state reserves the right to allocate a mining right and to determine the terms and conditions of eventual mining consistent with its mining policies.

The Government of Botswana argues that because it lacks the technical capacity and finance to undertake the exploration and mining of the country's mineral resources on its own, it has to draw on the assistance of bilateral and multilateral agencies and to encourage the participation of private companies. Although the policy of diversifying the companies is beginning to show some signs of success, a large portion of mining and prospecting is still done by two groups, the Anglo, De Beers, and Amax companies. All the signs are that they will dominate the mining industry for a long time to come. In the case of diamonds, the feeling is that since De Beers controls the diamond market, there is very little else that can be done to increase competition. To attract private capital, the government has pledged to "pursue policies designed to maintain a favourable environment for investment and to permit investors to earn a fair return on their capital."[25]

The government's stated aim is to participate in the ownership of the companies. It considers appropriate a grant of shares between 15 percent and 25 percent free of charge with an option of purchasing additional shares.* Current government shareholding in both BCL and De Beers (Botswana) Ltd. is 15 percent. It attaches great importance to participation in the internal decision-making processes of the companies, so as to influence the day-to-day operations of the companies. Two of the directors in both BCL and De Beers (Botswana) Ltd. are government appointed. A clause in the master agreement between the RST Group and the government relating to long-term contracts states specifically that any contract for the sale or processing of minerals for more than one year and for more than R1 million should be submitted to the board of directors for prior approval, and each of the directors representing the government should be given an opportunity to express his views and those of the government and that the final decision should reflect these views. However the government representation being so small, the amount of influence it can exert in decision-making will inevitably be limited. It may therefore find it necessary in the future to increase its shareholding if it is to play a more effective role in decision-making. This is not to deny the fact that the government minority share-holding is backed by the powers of a sovereign state, whose authority the board room decisions cannot flout with impunity.

_____

*Which seems to be what it is trying to do in its current renegotiations with De Beers.

The government seeks long-term relationships with mining companies. One of the key issues in such relationships with the companies is taxation. The government's aim is to design a fiscal package—including taxes, dividends, and royalties—that will assure the government as large a share of company gross profits as possible.[26]

The royalty is the price of the minerals the companies have to pay to the country for exploiting its resources. In the absence of effective competition, a pricing mechanism has to be employed that is related to the profitability of a mining venture. This is not without precedent; for instance, in the United States regulatory agencies determine the price for services, which are not subject to competition, by reference to a fair return on the capital employed by companies rendering these services. In Botswana royalties are fixed at a level that leaves to the company internationally acceptable fair return on their capital. Since the relevant circumstances that determine the profitability of a mine, in particular the price of the mineral produced, cannot be predicted by the government at the time when the initial royalty is set, it is very important for the government to retain sufficient flexibility to adjust the royalty if unforeseen changes in technology or markets result in excessive profits to the company.

In the case of BCL, the government has entered into a long-term tax agreement until 1998. The rate of income tax will normally be 40 percent; the royalty will be 7.5 percent subject to a minimum advance payment of R750,000; in addition, the government will receive dividends from its 15 percent shareholding. However, because this is a long-term tax agreement, a tax "escalator" to deal with the windfall profits arising out of abnormally high metal prices has been provided for. If the operating profits margin (the excess of sales revenue over production cost expressed as a percentage of sales revenue) exceeds 48.5 percent, the income tax will increase by an equivalent percentage up to a maximum tax rate of 65 percent. At that point (when the operating profits margin is 73.5 percent), the aggregate of royalty and tax (excluding dividends) would be 67.625 percent of gross profits, if no capital is being written off.[27] Despite this, the feeling is that by "freezing" the taxes until 1998, the government has tied its hands to a large extent. (With regard to the diamond mine, the tax agreement covered profits tax, income tax, and royalties only and allowed the government to introduce additional taxes; and so in 1972 it introduced the withholding tax on dividends to nonresident shareholders. In addition, the agreement contains a renegotiation clause for abnormal circumstances. On the basis of this built-in flexibility, the government is presently renegotiating the Orapa Agreement.)

Infrastructure and services are provided by the government and controlled by public bodies so that they are operated and developed in the national interest. But the costs involved have to be met by the mineral developers. Transport, in particular, is planned within a context of general regional needs. This is illustrated in the case of the brine deposits. The company wanted to transport its products by road between Francistown and Makgadikgadi, but the government has insisted on the construction of a railway. This railway, besides strengthening the links with Zambia, will be used for the cattle traffic to Francistown (trekking cattle to Francistown leads to their loss of condition in addition to the loss of time) and will certainly make possible the construction of the second abattoir to serve these very remote areas from Lobatse. The railway is also being routed through Matsitama, where copper deposits have been discovered.

In exceptional instances, especially when it has access to "soft loans," the government may prefer to finance some of the infrastructure. This would reduce the capital cost on which the fair rate of return is calculated, which is financially in the interest of the government.

While mining will not create many jobs, the government is committed to examining all the alternative development proposals of the mining companies with the view of creating as many jobs as possible. Within the constraints imposed by high capital costs and lack of domestic market, the government is concerned to see that "the processing of minerals in Botswana is carried to a point that brings the maximum feasible economic benefit to Botswana even if that does not result in the maximum financial return."[28] This domestic processing should, it is argued, stabilize markets for Botswana's minerals, while reducing the transport costs for a landlocked country like Botswana. The siting of the processing plants will take into account the need for balanced regional development. However, until transport facilities are improved considerably, it is difficult to imagine how the last aim can be achieved.

The government is not adopting a "get-rich-quick" attitude. The mineral development will be phased in such a way as to avoid sharp fluctuations in economic activity and employment, especially of the construction industry.

There are other important aspects of the mining policy that have not been discussed because they are peripheral to the main emphasis of this chapter. The two most important ones relate to localization of medium and management positions in the companies and the control of environmental destruction and pollution. With regards to the former, the companies, in cooperation with the government, are expected to embark on training programs of locals, with 1990 as the deadline for complete localization.

CONCLUSIONS

Recent mineral discoveries have raised hopes of faster economic growth and independence for a country whose economy has, since its inception, been an appendage of the South African economy. But the changes will take time. In the meantime, many Botswana who cannot find employment locally will continue to find work outside the country, mainly in South Africa (as long as this option remains open). Botswana is also likely to depend upon South African manufactures for quite some time.

Immediately, the direct impact of mining on other sectors of the economy will be in the infrastructure. Indirectly other sectors, especially agriculture, will benefit through the government strategy of investing mineral revenues in them in order to promote and spread development. Once the initial problem of capturing these surpluses is overcome, the major problem will be how to spend this money more effectively. The government rural development program will need to be designed and carried out in a way that benefits mainly the poor sections of the rural population and not the rich farmers, as other programs so far have tended to do.

The availability of mineral revenues to invest in rural areas will depend partly upon the workings of the formulas that the government has worked out with the mining companies. Botswana has tried to learn from the experiences of other countries. The major problem in negotiating agreements is uncertainty about the future—for example, changes in the market and technology. The Botswana government is therefore keen to insert a renegotiating clause in its agreements with mining companies. On the basis of this renegotiating clause, it has been able to renegotiate the Orapa Tax Agreement in less than five years. The company has been making huge profits and had thus overshot what the government considered a fair rate of return on its capital invested.

It is still too early to judge the Selebi-Pikwe Tax Agreement. It will be interesting, though, to see whether built-in mechanisms such as the tax "escalator" will be sufficient to deal with unexpected situations in the absence of a renegotiating clause.

NOTES

1. Lester B. Pearson, Partners in Development (New York: Praeger Publishers, 1969), p. 25.

2. See for example A. Sillery, Botswana: A Short Political History (London: Methuen, 1974), pp. 120-25 and Ch. 11.

3. One example of this was a soap factory in Lobatse, which "was forced into liquidation by selective price cutting by established

South African manufacturers" in the early 1960s. See P. Landell-Mills, "The 1969 Southern African Customs Union Agreement," Journal of Modern African Studies 9, 2.

4. See, for example, I. Schapera, Migrant Labour and Tribal Life (London: OUP, 1947).

5. Computed from figures given in Republic of Botswana, National Development Plan 1973-78, Part I (Gaborone: Government Printer, 1973), p. 20.

6. Republic of Botswana, Rural Development in Botswana (Gaborone: Government Printer, 1973), p. 1.

7. Ministry of Agriculture, Agricultural Survey 1971/72 (Gaborone: Government Printer, 1972), p. 8.

8. Central Statistics Office, National Accounts of Botswana 1971-72 (Gaborone: Government Printer, 1973), pp. 1-2.

9. Central Statistics Office, Statistical Newsletter no. 2 (Gaborone: Government Printer, 1973), p. 20.

10. One of the most recent works on the subject of migratory labor in South Africa is F. Wilson, Migrant Labour in South Africa (Johannesburg: South African Council of Churches and SPRO-CAS, 1972).

11. See Landell-Mills, op. cit.

12. Budget Speech 1974 delivered at the National Assembly on March 11, 1974.

13. Republic of Botswana, National Development Plan, 1973-78, p. 46.

14. The loans for the infrastructure of the Copper-Nickel Project (see below) have enabled the government to widen its economic links. See also W. Henderson, "Independent Botswana: A Reappraisal of Foreign Policy Options," African Affairs 73, 290 (January 1974).

15. See, for example, I. Griffiths, "Botswana Discovers Its Own Resources," Geographical Magazine, December 1970.

16. International Bank for Reconstruction and Development and UN Development Program, Botswana-Shashe River Feasibility Studies: Final Report (London: Sir Alexander Gibb and Partners, 1968). Letter dated December 9, 1968.

17. A. M. Kamarck, The Economics of African Development (New York: Praeger Publishers, 1967), p. 131.

18. Compare H. Myint, The Economics of the Developing Countries (London: Hutchinson University Library, 1967), 3d ed., 1966-68; Andre Grunder Frank, Capitalism and Underdevelopment in Latin America (Harmondsworth: Penguin Books, 1971); R. I. Rhodes, ed., Imperialism and Underdevelopment (London: Monthly Review Press, 1970).

19. Botswana RST Ltd., Annual Report, 1971, p. 4.

20. Republic of Botswana, National Development Plan, 1973-78, p. 220.

21. A speech by the President, Sir Seretse Khama, reported in Botswana Daily News, June 12, 1973.

22. Republic of Botswana, National Development Plan, 1973–78, Ch. 5; Republic of Botswana, National Policy for Rural Development (Gaborone: Government Printer, 1973); Rural Development in Botswana, op. cit., p. 4.

23. B. C. Muzorewa, "Security for Loans in Rural Development," paper read at a Conference of the Eastern Africa Agricultural Economics Society, Lusaka, May 1974.

24. Republic of Botswana, Budget Speech 1974 (Gaborone: Government Printer, 1974), p. 7.

25. Republic of Botswana, National Development Plan, 1973–78, p. 211.

26. It is pointed out that his is a far much better arrangement than most other existing arrangements. The Liberian case, see R. Clower, et al., Growth Without Development: An Economic Survey of Liberia (Evanston, Ill.: Northwestern University Press, 1966) is often quoted.

27. See also C. Joubert, "Notes on a Botswana Example of the Taxation of Mining," Botswana Notes and Records, vol. 5.

28. Republic of Botswana, National Development Plan, 1973–78, p. 209. One of the effects of this policy may be reduction of revenues that accrue to the government from the sale of unprocessed minerals.

   As developing nations go, Papua New Guinea is fortunate. The
country's 2.5 million people are comfortably spread over 466,000
square kilometers. There is little evidence of hunger or overcrowd-
ing. Per capita annual gross domestic product (including the value of
subsistence production) is nearly U.S. $600.[1] The nation's largest
city, Port Moresby, has a population of only 75,000, and even though
urban areas are growing at a rate of 13 percent annually, it will still
be some time before the country has to face the urban problems that
confront other Third World governments. Ninety percent of Papua
New Guineans still live in the rural sector, and only 55 percent have
any ties at all to the monetary economy. The standard of rural subsistence life, measured in calories per day, in mortality rates, and in
similar terms, is much higher than in most Asian and African countries. And compared to the people of nearby Southeast Asian nations,
those Papua New Guineans who do participate in the cash economy are
not desperately poor; the urban minimum wage, for example, is more
than four times the rate in Singapore or the Philippines.

   And yet many of the problems common to Third World nations
are beginning to appear in Papua New Guinea. Urban unemployment
is increasing. The school system is turning out thousands of students
who will not be able to find jobs suited to their level of training, while
at the same time there is a severe shortage of highly skilled technical
and professional personnel at the top levels. Social unrest and com-
munal violence are much in evidence. Increasingly vocal protests are
being made about alienation of land and the exploitation of forest,
fish, and mineral resources by foreign corporations. In many ways,
the development of the huge Bougainville copper project is the best
example of the kinds of problem and policy choice facing the new Papua
New Guinea government. All the problems just described can be seen

in connection with the copper mine, and any national development policy that attempts to cope with these problems—and with the broader
problems of shaping and encouraging structural change in the economy
—must begin by dealing with the existing reality of the mining industry.

   Papua New Guinea's current leadership sees many of today's
development policy problems to be the result of previous Australian
colonial policies.* As a background to the general problems of development strategy facing the country today, and to the specific problems
of mining policy, a brief review of colonial policies since World War
II and of their results is necessary.

BASIC DEVELOPMENT POLICY

Postwar Economic and Social Policy

   The postwar Labour government in Australia emphasized the
restoration and extension of administrative control and the provision
of social services—especially primary education, health, and communications. In contrast to the prewar regime, under which the territories
of Papua and New Guinea had been expected to be financially self-sup-
porting, the Labour government introduced annual budget grants,
along with a variety of paternalistic measures such as an increase in
the minimum wage, a reduction in hours of work, and the beginnings
of an agricultural extension program. In addition, a major part of
the banking system was brought under (Australian) government control.

   This basic emphasis on government as a social service and administrative mechanism continued through the Liberal-Country Party
government in Australia in the 1950s. Under both the Labour and Liberal-Country administrations, economic development was left largely
to expatriate firms. For example, some $18 million in war damage
compensation was paid to Australian businesses and individuals in
Papua and New Guinea.[2] While Australian firms were encouraged to
locate in Papua and New Guinea to use the territories' resources, there
was a conscious effort to avoid any competition with already established industry in Australia.[3]

---

   *Papua New Guinea is currently a self-governing territory of
Australia. Full independence is due in early 1975. The country includes the former German colony of New Guinea, which came under
Australian control in World War I, and the Territory of Papua, governed by Britain from 1888 to 1905, when it came under Australian administration.

During the 1950s, public sector spending accounted for some
70 percent of gross domestic product, [4] and the bulk of this came from
direct Australian grants and the spending of Australian departments
operating in the territories. In 1959-60, for example, internal revenue accounted for less than one-third of total public spending. [5] The
effect was steadily to increase domestic demand, creating some new
opportunities for growth in the economy but more typically resulting
in rapid increases in the volume of imports.

On the expenditure side, government activity was heavily oriented
toward law and order and social services. In 1959-60, "general administration" accounted for one-third of the total territory budget,
while education, health, and social services accounted for another
35 percent. The commodity-producing sector, including agriculture
and forestry, accounted for only 10 percent of spending, while economic infrastructure (roads and bridges, power, and communications)
was allocated 18 percent of the budget. [6]

Despite the emphasis on social service spending, there were
some important lacks in this period. In education, almost all effort
was devoted to primary education; there were no Papua New Guineans
sent on to universities until 1960, and the shortage of trained local
personnel remains a major problem today. (At the same time a large
Australian civil service in the territories was built up, totaling 3,600
members by 1960.) Papua New Guineans who were in the monetary
sector at all were generally seen as providers of plantation labor
(for which they received some food, clothing, and housing and from
35 cents to one Australian dollar [A$]* a week), as domestic servants
or occasionally as low-level government employees.

In the private sector of the economy, virtually all activity resulted from the operations of Australian firms and individuals. Copra
and other coconut products, produced largely on foreign-owned plantations, accounted for 62 percent of all exports in 1950 and 55 percent
in 1960. [7] Other significant crops, also produced largely on plantations, included cocoa and rubber. Nonagricultural production was
concentrated primarily in forestry and saw-milling and in gold-mining,
which had experienced a boom in the 1930s and was rapidly declining
as a major industry.

Trade and industry were conducted as if Papua New Guinea were
part of Australia. The three major trading companies (Burns Philp,
Carpenters, and Steamships) "discouraged local production of items
that might be in competition with those imported." [8] Australian exporters to Papua New Guinea received export-incentive tax concessions, while most exports from Papua New Guinea to Australia were

---

*A$1 = U.S. $1.31 (November 1974).

subject to normal duty rates. As a result, there was little incentive
to establish manufacturing subsidiaries in the territories.

In agriculture, the dominant thrust of government policy throughout the 1950s was to encourage growth of the Australian-owned plantation sector. Most agricultural extension services were provided to
plantations, not to Papua New Guinean farmers. The government also
provided substantial assistance to plantation development by buying
land from Papua New Guineans for leasing to expatriates, by building
roads and airfields, by conducting research on plantation crops, and,
perhaps most important, by organizing the supply of labor. Financial
incentives to develop plantation land, generally duty-free imports,
and accelerated depreciation for tax purposes were also available.

There were corresponding effects on the development of Papua
New Guinean agriculture. Some labor was diverted from village farming to the plantations. Considerable social disruption occurred in
areas where a large proportion of land was alienated for plantation
use. Extension services and credit that could have been used for supporting local agriculture went instead to service the plantation sector.
Even those portions of government plans that were devoted to Papua
New Guineans are subject to criticism for not using available resources
economically to improve the lives of the maximum number of people.

Capital movements and banking also reflected the colonial government's basic direction. Even in the 1950s, Papua New Guinea was
a net exporter of private capital; most investment was generated from
retained earnings, and the net outflow approached $10 million per
year.[9] Trading banks would not lend to farmers without an individually recorded land title, which meant that they would not lend to
Papua New Guineans, who almost always held land under customary
tenure. Essentially the only credit available to Papua New Guineans
came from the small Native Loans Fund, which made 373 loans averaging A$1,570 between 1955 and 1963, and an ex-servicemen's loan
fund, which made 126 loans averaging A$1,700 to Papua New Guineans
and 148 loans averaging A$46,000 to Australians between 1958 and
1962.[10]

The Emphasis on Growth

The earlier pattern—protection for resident Australian interests,
pacification of Papua New Guineans, and preservation of the economic
status quo—shifted markedly in the 1960s.

Several factors influenced this change. First, the UN Trustee-
ship Council Mission in 1962 created pressure on the Australian colonial government to take positive steps toward social and economic
progress. On the political front, Australia responded by establishing

an elected House of Assembly in 1964, with a majority of Papua New Guinean members. On the economic front, the UN criticism about lack of development was met in the first instance by engaging the World Bank study and then, after receipt of the team's report, by the appointment of an economic planner and the promulgation of a five-year plan in 1968.

Many of the World Bank recommendations were made into official policy with the publication of the five-year plan. The plan proposed rapid expansion of export crops; attraction of foreign investment into large-scale forestry projects, as well as mining; and expansion of light manufacturing industry.

The five-year plan also foreshadowed a changing emphasis in the social services. In place of the earlier attempt to extend primary education widely, there was a crash program at the secondary and tertiary level, complete with establishment of two universities, in an effort to produce at least a small number of trained people to take over positions in the economy (as of mid-1974, Papua New Guinea had about 100 university graduates, the most experienced of whom had been out of the school system for only 10 years).

More generally, the five-year plan was intended to bring Papua New Guinea to "a stage of self-sustaining economic growth, with the ability to finance the greater part of its total budgetary needs from domestic savings. . . . The economy should have reached the stage (by the mid-1970s) that some tapering off in aid is either in operation or in prospect."[11] The plan placed a heavy emphasis on infrastructure development. Papua New Guinea's international borrowing increased rapidly to more than 15 percent of the total budget, primarily in support of major projects like the Highlands Highway, opening up the heavily populated interior of the island, or the major hydroelectric power project, which is only now nearing completion on the Ramu River. As a result of this infrastructure development, Papua New Guinea currently has a fairly well-developed system of internal communications. The broad economic development that was intended to accompany the provision of infrastructure is, however, somewhat less evident.

In conventional terms, the results of the first plan period were impressive. Both GDP and gross national product increased at an average annual rate of 13 percent from 1966 to 1972.[12] By mid-1974, per capita GNP was U.S. $540—fairly well up on the list of developing countries.

These figures lose some of their impressiveness when subjected to more detailed analysis. For example, the subsistence sector, in which some 45 percent of Papua New Guineans live, increased its production at a rate barely equal to the annual population increase. Also inflation was a significant factor. When expressed in constant dollars,

the rate of growth of GNP drops to 5 percent—not a major advance in
a country with a 3 percent annual rate of population growth. And,
largely as a result of the Bougainville copper mine, factor payments
to the rest of the world increased at an annual rate of 47 percent from
1967 to 1972.

Despite the emphasis on financial self-reliance in the original
five-year plan, the actual history of the budget showed more than a
doubling of Australian aid from 1968 to 1973. In 1973, internal revenue still accounted for only 33 percent of total government spending.
The number of Australian civil servants more than doubled to a high
of over 7,000 before a rapid localization program was introduced in
late 1972. One result of this trend was that, despite the expressed
emphasis on building infrastructure as the basis for future economic
development, spending on capital works and maintenance actually was
a smaller proportion of the budget by the end of the 1960s than it had
been a decade earlier.

CURRENT POLICY

The coalition government that took office after the 1972 election
has moved away from the earlier growth-oriented policy toward an approach that places more emphasis on the rural population. The basic
ideology of the government is expressed in its eight aims:

1. A rapid increase in the proportion of the
economy under the control of Papua New Guinean
individuals and groups and in the proportion of personal and property income that goes to Papua New
Guinea.
2. More equal distribution of economic benefits,
including movement toward equalisation of incomes
among people and towards equalisation of services
among different areas of the country.
3. Decentralisation of economic activity, planning, and government spending, with emphasis on
agricultural development, village industry, better
internal trade, and more spending channelled to
local and area bodies.
4. An emphasis on small-scale artisan, service,
and business activity, relying where possible on
typically Papua New Guinean forms of organisation.
5. A more self-reliant economy, less dependent
for its needs on imported goods and services and
better able to meet the needs of its people through
local production.

6. An increasing capacity for meeting government
spending needs from locally raised revenue.

7. A rapid increase in the equal and active participation of women in all forms of economic and social
activity.

8. Government control and involvement in those
sectors of the economy where control is necessary
to achieve the desired kind of development.[13]

As a preparation for the end of the five-year plan period, the
Australian government again requested a World Bank mission to visit
Papua New Guinea at the time the new coalition government came to
power in 1972. This mission, headed by Michael Faber of the University of East Anglia, England, strongly recommended a change of emphasis in development policy along lines coinciding with these aims.[14]

With the coming to power of the new government, supported by
expert advice such as that supplied in the Faber report, the way was
prepared for a change in planning strategy. There was also an effort
to change the structure of planning organization in Papua New Guinea.
A new Central Planning Office was created with a newly appointed director who was directly responsible to a cabinet planning committee
made up of the four political leaders of the coalition government.

INVESTMENT POLICY

The broad outlines of policy toward foreign investment in Papua
New Guinea roughly parallel the stages of overall development policy
outlined above.

Until the mid-1960s, the pattern of foreign investment in Papua
New Guinea was typical of many plantation economies. Three large
Australian-based companies, Carpenters, Steamship Trading Company,
and Burns Philp, owned a large share of the plantation properties and
virtually the entire shipping and commercial sectors. Finance was
controlled by four Australian banks, and there were many smaller
Australian plantation owners and traders. Colonial government policy
was largely restricted to providing basic protection for these interests.
Direct competition with Australian production was discouraged, and
there was no substantial effort at import substitution.

This policy changed sharply in the mid-1960s as part of the
larger push for economic development. Major policy statements were
made by the colonial administration and even, in some cases, ratified
by the House of Assembly. The Investment Capital Guarantee Declaration of 1966 of the House,

> invites and welcomes capital investment for develop-
> mental purposes in the Territory of Papua and New
> Guinea and guarantees to the world that expatriate
> capital investment in the Territory of Papua and New
> Guinea for the establishment of new industries or the
> development of existing industries shall not be sub-
> ject to expropriation, nor to discriminatory taxation
> or other like levies, nor to oppressive trading legis-
> lation, nor to unreasonable limitations on its repatri-
> ation, and solemnly charges future Parliaments of
> the Territory of Papua and New Guinea not to legis-
> late in a manner inconsistent with this Declaration
> unless that proposed legislation has the support of a
> majority of the electors of the Territory expressed
> in a referendum. [15]

Under a newly enacted <u>Pioneer Industries</u> legislation, nearly 100 foreign firms had been granted five-year tax holidays by 1973.

The results of this policy were a rapid increase in manufacturing
industry (from a very small base), with a somewhat less than proportional increase in employment. The value of manufacturing output
rose from A$22.4 million in 1968 to A$40.7 million in 1973. [16] Large-
scale investment in agriculture was attracted, especially in oil, palm,
and tea, despite uncertain world market conditions. And the basis
was laid for Papua New Guinea's emergence as one of the world's significant exporters of copper.

In 1973, as the new government has attempted to put into practice the eight aims, the expressed attitude toward foreign investment
began to harden. In a major policy speech to the Australian Institute
of Company Directors, the Chief Minister, Michael Somare, outlined
some elements of the new approach:

> We are well aware that we are still highly depen-
> dent on imported skills and capital.
> We cannot afford to turn our back on foreign in-
> vestment and loans if we are to meet our commit-
> ments to our people for more schools, more health
> services, and other development.
> We accept that foreign investment will be needed
> to help provide skills and revenue for expanded so-
> cial services but we are resolved to channel that in-
> vestment for the greatest possible benefit to the peo-
> ple of Papua New Guinea. . . .
> There may be certain fields, such as petroleum
> and natural gas, where we will be actively seeking

foreign involvement because only overseas companies
have the capital and organised skills to develop our
resources.

On the other hand, there are likely to be some
areas, like road transport or agriculture, where foreign participation will be actively discouraged and the
field reserved for Papua New Guineans. . . .

We want the benefits to go to the mass of the people. For this reason, there will be increasing emphasis on taxation of resource projects so that the benefits can be redistributed to the people through Government projects. There will also be increasing emphasis on Government ownership of resource ventures,
in partnership with foreign corporations.

I believe we must also ensure that, whenever conditions beyond the control of either the Government
or the resource company result in spectacular windfall profits, the lion's share of these profits will be
kept within Papua New Guinea.[17]

A new coordinating agency, the National Investment and Development Authority, has been established to exercise centralized control
over foreign investors, particularly with respect to enforcement of
the terms and conditions under which companies are allowed to come
into the country and with respect to policing transfer pricing and other
mechanisms that can erode government tax revenue.

The government has also suspended the Pioneer Industries Act
and has taken a noticeably tougher line in negotiations with foreign
companies for new projects. Despite these tougher measures, however, non-Australian business interest in investing in Papua New
Guinea remains high. Japanese interests are currently making a
strong attempt to gain control of a large share of the country's forest
and fish resources and are also beginning to move into mining. U.S.
firms are active in oil and gas exploration and mineral prospecting.
As of November 1974, for example, negotiations are in progress for
development of three major timber areas, for a copper mine with potential output of 100,000 tons a year and for numerous smaller projects. The new tough line on investment does not appear to have reduced the likelihood of substantial capital inflow.

MINING POLICY

The development of mining policy is similar to, and an integral
part of, overall investment policy. Once again, it is useful to separate postwar policy into three stages.

Until the mid-1960s, mining policy and legislation largely reflected the history of the New Guinea gold rush of the 1920s and 1930s. The entire structure of legislation was designed to cope with the problems of sorting out conflicting claims in gold fields, registering individual mining titles and similar issues. Although a few base metal mines had been operated from time to time, the emphasis was on relatively limited goldmining operations.

Taxation laws relating to mining simply repeated the provisions established in Australia, including an allowance for accelerated depreciation and an exclusion from taxation of 20 percent of all income earned from copper (introduced in Australia as a wartime measure to increase domestic production). But, until the Bougainville project, no company reached the point of even evaluating its prospects for commercial operation in Papua New Guinea.

The mining laws were substantially changed in 1966 by addition of new provisions for large projects. These provisions, sponsored by the colonial administration in response to the growing likelihood that prospecting on Bougainville would result in the establishment of a major mining operation, provided for concession agreements to be negotiated over relatively large areas. It was under these provisions that the Bougainville agreement was negotiated in 1967.

More recently, there have been moves to impose stricter controls on foreign mining companies. A set of mining policy guidelines adopted by the government in May 1973 provides for the following: (1) the government should receive at least 50 percent of the profits from mining ventures in the form of taxes; (2) incentives such as accelerated depreciation and the 20 percent income exclusion will no longer be permitted; (3) the government will receive an option to purchase equity in all new mining projects, and any contribution to infrastructure by the government will earn a proportionate equity share; (4) with respect to infrastructure, all facilities will normally be provided by the mining company, and will become the property of the government in time; (5) disputes between the company and government will be resolved within the Papua New Guinea legal system; (6) the government will establish controls to ensure the maximum possible amount of employment, local purchasing, development of Papua New Guinean-owned ancillary businesses and further processing within the country; and (7) all agreements will be subject to regular review to take account of changing circumstances.[18]

These broad guidelines form the background for the recent renegotiation of the Bouganville agreement and for the negotiation of agreements for new mines, like the project currently being proposed by the U.S.-based Kennecott Corporation, which would produce 100,000 tons of metal per year from a deposit near the border between Papua New Guinea and the Indonesian province of Irian Jaya.

THE BOUGAINVILLE AGREEMENT

The Bougainville mine was the Australian administration's greatest success in attracting foreign investment. The A$400 million spent on the mine was far above the total of all other investment up to 1972. In view of criticism in the UN Trusteeship Council and elsewhere that Australia had not provided an adequate basis for the future economic development of its territory, the colonial administration saw the Bougainville project as a highly desirable operation, and the generous terms of the agreement reflect in part the desire of the Australian government negotiators to ensure that the project went ahead.

The major provisions of the agreement were as follows:

Financial: The company received a three-year tax holiday, plus the accelerated depreciation provisions of the Australian mining law. This had the effect of making the total tax holiday period some six to eight years, depending on the level of profits in the years in which the accelerated depreciation would have been taken. In addition, the 20 percent exclusion of copper income from any tax was enshrined for the life of the agreement. Once tax became payable, it would have started at a rate equal to the general corporate tax rate (currently one-third of profits) and then risen over four years to a level of 50 percent. In addition, the government could impose a dividend withholding tax of 15 percent on dividends actually remitted abroad. Any other taxes that might be imposed by the government were to be allowed as credits against the corporate tax liability. The government also received an option to buy 20 percent of the equity in the company at par, payable in cash before the end of the construction period. This option was expressed, and the government borrowed A$26.5 million for its 20 percent share. Royalty was fixed at 1.25 percent of freight on board (FOB) value of copper and gold produced. As a result of these provisions, the total government share of the company's 1973 profit of A$158 million amounted to only A$29 million—the bulk of which was the government's share of dividends on the equity it had paid for.

Infrastructure: In addition to its payment for equity, the Papua New Guinea government was also committed to spend A$42 million for development of a township serving the mine area, communications, education, and medical facilities. As in the case of the equity payment, this money was largely raised by borrowing.

Further Processing: The company's only commitment in this area was to a feasibility study of smelting at an indefinite time in the future. In the event, the mine's total output of 180,000 tons of metal per year has been committed for sale in the form of concentrates on long-term contracts that expire in 1982 or later.

New Mines: The company was granted an absolute right to additional mining leases in the general area of its operations. It is estimated that one deposit in this area may prove to be as large as the deposit currently being mined (that is, reserves of 900 million tons of 0.5 percent copper ore).

Limitations on Government: The agreement purported to restrict the government's general powers of regulation and control, both through clauses denying the government the right to alter the agreement and through restrictions on government power generally—for example, Clause 17:

> No Interference or Expropriation: The company shall be at all times entitled and permitted fully to enjoy all the rights, benefits and privileges granted or intended to be granted by or as a result of this agreement and at the same and also the rights of all past present and future members of the company and beneficial owners of shares in the company fully to enjoy the benefit of their shareholdings or their interests in shares or other rights arising therefrom shall at no time to the detriment of the company or such members or beneficial owners substantially altered or impaired, impeded or interfered with by executive or administrative action or in any other manner whatsoever (whether directly or indirectly) and without affecting the generality of the foregoing:
>
> (a) subject to the provisions of this agreement, the company's present freedom of choice of directors, executives, advisers, consultants, associates, employees, contractors, suppliers and customers, and its present freedom to declare credit and pay dividends and to grant other rights to its members shall continue without substantial interference; and
>
> (b) no discriminatory action, whether by way of industrial, fiscal or social legislation or otherwise shall be taken against the company. . . .
>
> So long as the company complies with this agreement . . . the administration shall not resume or expropriate or permit the resumption or expropriation of any asset . . . and of the products . . . the business of the company or any shares held or owned by any person in the company.[19]

At the time the agreement was introduced in the House of Assembly for ratification, Australian spokesmen asserted that the substantial

concessions to the company were necessary because of what was said
to be the marginal nature of the operation. (During the negotiations,
the company had presented financial estimates based on a throughput
of 30,000 tons of ore per day and involving a total investment of some
A$125 million. But even before the agreement had been presented to
the House of Assembly, the company had begun working on a much
larger operation—actual throughput is more than 90,000 tons per day—
with a much less marginal economic outlook.) Despite the substantially higher tax rates then in force in other copper-producing countries, there was no substantial opposition to the agreement in the
House.

The only controversy revolved around the successful effort by
the representative of the mine area in the House, Paul Lapun (since
1972 Minister of Mines) to ensure that a proportion of the royalty
payments went directly to the people living in the mine area.

Recently, however, criticism of the agreement grew steadily.
As it became apparent that the mine would generate substantial profits
even at low to moderate copper prices (the profit for nine months'
operation during the low-price year 1972 was A$28 million), both
academic critics and political leaders made public attacks on the
agreement.[20] The two leading Bougainville politicians, Lapun and
Fr. John Monis, both called for renegotiation in 1972.

With publication of the company's 1973 results showing a A$158
million profit, the moves toward renegotiation picked up speed. Chief
Minister Michael Somare announced on March 4 that he would seek renegotiation, with a major emphasis on securing a much greater share
of revenue for the government and cutting short the company's tax-
free period.

Economic Impact of the Mine

The Bougainville project had a significant impact upon the Papua
New Guinea economy during the 1969-72 construction period. It more
than doubled the annual rate of capital formation and doubled imports,
thus temporarily worsening the deficit on the current account of the
balance of payments. It increased wage incomes by 20 percent and
increased gross monetary sector product (GMSP) by 25 percent and
gross monetary expenditure by 38 percent (see Table 24.1).

It tended to raise aggregate demand faster than aggregate supply
(since construction necessarily preceded output from the project), and
it tended to push up wage rates in the monetary sector. Until 1974 it
also placed demands upon public sector expenditure in excess of its
contribution to public-sector income.

TABLE 24.1

Comparison of First Three Full Years of Bougainville Operations: Direct Impact with Monetary Sector
Activity in Papua New Guinea Immediately Prior to Construction
(millions of Australian dollars)

	Year Ending June 1969	Direct Impact of Project (year ending June)			Direct Impact Average for Three Years Ending June 1972 as Percent of 1968/69	Direct Impact of Project (year ending June)			Direct Impact Average for Three Years Ending June 1975 as Percent of 1968/69
		1970	1971	1972		1973	1974	1975	
a. Wages, salaries, and supplements	140	16	36	33	20	19	20	19	14
b. Other income	85	11	27	28	26	107	116	115	132
c. Depreciation allowances	16	5	11	13	60	23	24	27	154
d. GMSP at factor cost	241	32	74	74	25	149	160	161	65
e. Indirect taxes less subsidies	17	2	5	3	20	2	2	2	12
f. GMSP at market prices	258	34	79	77	25	151	162	163	62
g. Imports of goods and services	188	37	98	70	36	42	41	44	23
h. Market supplies	446	71	177	147	29	193	203	207	45
i. Exports of goods and services	93	—	—	35	13	149	185	191	188
j. Gross monetary expenditure	353	77	177	112	38	44	18	16	7
Personal consumption	162	—	—	—	—	—	—	—	—
Other net current expenditure	99	—	—	—	—	—	—	—	—
Gross capital formation	120	71	177	108	108	44	18	16	22
Statistical discrepancy	-28	—	—	—	—	—	—	—	—

Source: Department of External Territories, National Income Estimates for PNG 1960/61–1969/70 (Canberra, 1971).

In the three years to 1975, the impact of the project is changing. In this period, the project will have a reduced effect on the rate of gross capital formation in Papua New Guinea. Exports for which the project is responsible are vastly in excess of comparable imports, and hence it contributes to a positive balance of payments on current account. The project's rate of generation of wage income has fallen compared with the construction period. And the project's rate of contribution to GMSP has risen substantially (through output rather than investment), and contribution to supply is outstripping its boost to aggregate demand.

Without the recent renegotiation of the agreement, the future total contribution of the project to GMSP would have grown only very slowly (as local income grew) as a proportion of total income from the project; consequently, the project's total impact as a percentage of GMSP would have risen to about 40 percent in 1974 and 1975 and then progressively decline to about 20 percent by 1990. Even so, the project will remain the largest single component of GMSP for many years to come.

The direct effects of the project will continue to outweigh the indirect effects. This is so not only because of the high propensity of the project to import and because of the high import leakage from the project's local Papua New Guinea purchases and wage payments, but it is also because of the nature of the project's operations. Value added is a high proportion of its current receipts and capital contribution is responsible for a high proportion of total value added.

In addition to these gross effects, the Bougainville project also has demonstrable effects in the areas of environment and land use, labor and training, agriculture, and local business and industry.

ENVIRONMENT AND LAND USE

The total area occupied by the project and its infrastructure is about 60 square miles or approximately 1.5 percent of the area of Bougainville island. Seventy percent of this is the tailings disposal area. Because the island is in a seismically unstable zone, less extensive tailings disposal methods (for example, dike or dam building) were not considered sufficiently safe.

Atmospheric effects of the project's operations include discharge of oxides from the thermal power station and dust. Due to high rainfall and cyclone incidence, topography at the mine site cannot be yet considered finally stable. While the company has begun a major experimentation program to determine the suitability of the tailings disposal area for future agricultural use, there is as yet no clear indication of how and when the land affected may again be usable.

More generally, in relation to the land acquired from Papua New Guineans for the mine, there was a history of sharp conflict. The village communities affected gave the very highest importance to land as the source of their material standard of life. Land was also the basis of their feelings of security, and the focus of most of their religious attention.

Despite continuing compensation payments and rental fees, local resentment over the taking of the land remains high, and there is strong opposition to any expansion of mining on Bougainville, whether by the existing company, the government, or anyone else.

Labor

The mine is a relatively small employer of labor. As of mid-1974, total employment at the mine is some 4,200, or less than 2 percent of the country's full-time labor force. Of the 4,200, roughly 70 percent are Papua New Guineans, although only two of the 112 professional and managerial jobs in the company are currently held by nationals.

Labor demand for the project was considerably higher during the three-year construction period from 1969 to 1972, with a peak work force of more than 10,000. Large numbers of people who came from other parts of the country to participate in Bougainville in the construction boom remain unemployed. Local residents tend to blame these outsiders for crime and other problems.

The company began a large-scale training and extension program in 1968. Over A$1.6 million was spent (to June 1972) in buildings and equipment, and A$2.2 million for training staff, wages, salaries, subsidies, and stores and services.

Because of lack of data, it is difficult to assess the significance of these training figures in relation to the general qualifications of the country's indigenous work force. However, the fact that over 3,350 persons received technical or professional training by the end of 1972 is in itself significant. The training effort at the mine is somewhat in excess of its own immediate requirements, and labor turnover will distribute some added skills in the national labor force.

The company intends to increase the indigenous component of its work force as rapidly as possible. It was originally planned that 95 percent of the work force by 1980 would be indigenous. However, because of the difficulty in obtaining technical graduates and high turnover, a more realistic estimate would be that out of a forecast work force of approximately 4,700 in 1980, about 80 percent (or 3,800) will be Papua New Guineans.

A related problem is the upward push on national wage levels exerted by the mine. To date, company officials have held salaries of

TABLE 24.2

Bougainville Copper Project: Estimated Wages and Salaires Paid
(constant dollars)

Year Ending June	Monthly Wage Rate (A$)		Wages and Salaries Paid (U.S.$)		
	Indigenous	Nonindigenous	Indigenous	Nonindigenous	Total
1970	127.6	712	4	12	16
1971	127.2	747	8	28	36
1972	132.9	789	8	25	33
1973	138.3	810	5	14	19
1974	143.3	832	6	15	21
1975	148.2	358	6	14	20
1976	154.3	877	6	13	19
1977	159.6	895	7	12	19
1978	164.4	921	7	11	18
1979	169.7	1,002	8	14	22
1980	175.3	1,027	8	12	20
1981	180.6	1,053	8	10	18
1982	186.0	1,074	9	10	19
1983	191.6	1,095	10	9	19
1984	197.3	1,117	10	8	18
1985	203.2	1,139	10	8	18
1986	209.3	1,162	10	6	16
1987	215.6	1,185	11	6	17
1988	222.1	1,209	11	6	17
1989	228.8	1,233	11	6	17
1990	235.7	1,258	12	6	18

Source: W. D. Scott and Company, "The Impact of the Bougainville Copper Project" (Port Moresby, 1973).

local workers close to the level set by such other large employers as
the public service. With the emergence of a mineworkers union and
the obvious availability of large profits to meet wage claims, however,
it can be expected that the mineworkers will follow the pattern set in
countries like Zambia, Chile, and Peru and become a good deal more
highly paid than the rest of the country's work force.

Agriculture

The opening of the mine has had conflicting effects on the agricultural base of the Bougainville economy. Negative effects include
direct loss of agricultural land, diversion of production from traditional cash crops, and diversion of labor from agriculture. Positive
effects are centered around the increased market for locally produced
food and the opening up of a large portion of the island as a result of
road construction for the mine.

A major direct impact of the mine relates to the loss of crop
production from the 1,000-acre expatriate held Arawa plantation,
taken by the administration for Arawa township. The main direct
local impact was in the loss of wages to local workers. Other land
occupations—particularly in the tailings and port areas—have resulted
in loss of cash crop production and income for the indigenous residents.
However, the company has compensated the owners in each case,
either in cash terms, or by replanting of crops destroyed, paying
nearly A$500,000 in compensation for cocoa and coconuts destroyed.

During the construction phase of the project, up to 33 percent
of the Papua New Guinean labor employed, or a peak figure of 2,300,
was from Bougainville. Three thousand national workers are currently
employed, nearly half of them Bougainvilleans. This employment has
had a marked effect on village life, especially as young males leave
home. The immediate result of less manpower in the village is a deterioration of access routes (that is, grass not cut, washouts not re-
paired), a lowering in standard of care and attention of agricultural
plantings, and lack of labor to harvest and carry the crop. Some
plantings are not harvested at all. A significant amount of money
saved while working for wages, however, comes back to the villages
and is spent on clearing land and planting more cocoa. Total wages of
Bougainvillean workers amounted to A$2.6 million during construction
(1970-72) and to A$2-$2.5 million during the first full year of production (1972-73). This is equivalent to the total income from sales of
cash crops of cocoa and copra throughout the island.

Another indirect effect of mining activity on the agricultural
sphere is that the relatively high wages now available in the mine area
have had an effect on labor costs for agriculture in surrounding areas.

Until 1971 it was possible for a man clearing land to hire labor for 60-70 cents per day, but it is now difficult to obtain labor even at the current legal rate of A$1.60 per day.

The transisland road constructed by the mining company has provided access to market for cocoa and copra produced in the fertile, heavily populated southwest part of Bougainville. This has cut freight costs to the overseas terminal at Kieta by more than 50 percent, and greatly increased the availability and regularity of transport. Because of the long lead time required to increase production in these crops, however, it will be some years before the economic impact can be measured.

The mine has also provided strong demand for locally produced fruit and vegetables, as well as for beef cattle raised in other parts of Papua New Guinea. The estimated value of locally produced food used in the mine is more than A$500,000 per year, virtually all of it produced by Papua New Guinean small farmers.

Linkages

As is typical in most mining operations, the Bougainville project has produced few large-scale links to the rest of the economy. Forward linkages are nonexistent; all copper produced is to be shipped in the form of concentrates to overseas smelters at least until 1982. Backward linkages are of little importance. Of total purchasing for operational purposes by the mine of some A$15 million per year, only A$2 million is purchased locally, and the import content of those local purchases is more than 50 percent.

The company has sponsored a number of small businesses in the Bougainville area, providing finance for retail store and tavern owners, construction contractors, and other service enterprises. But, in broad economic terms, the mine continues to function as virtually a complete enclave.

GOVERNMENT POLICY CHOICES

The question now facing the Papua New Guinea Government is what to do about the fact of the industry's existence. The available policy choices appear to deal with four major issues:

1. The extent of desired linkages from the mine to the rest of the economy;

2. The use of manpower training within the mining industry for broader national purposes;

3. The desired amount of government revenue from the mine, and how that revenue should be managed; and

4. The extent of desired government equity and management control.

There is considerable debate within the Papua New Guinea government as to how desirable it is to establish links between major mining enterprises and the rest of the economy. A strong and vocal element within the government believes that these major projects inevitably disrupt the traditional way of life of the people, and that they should therefore be kept to a minimum, with their major function being to supply the government with the revenue it needs to undertake rural development and social service programs. Because of Papua New Guinea's relatively fortunate situation, with abundant natural resources, a small population, and a high standard of living in the subsistence sector, there is considerable support for this view, and in the short term it appears likely to be the position officially adopted by the government.

Under such a policy the rate of mining expansion would be strictly limited, and while there would be some requirements for local purchasing and the use of local labor, there would not be an effort to build a modern industrial economy around a growth pole provided by mining.

This approach is not wholly unsuitable to Papua New Guinea's own historical and economic position. A major transformation of the economy now, at a time before large cities have grown up, before half the people of the country have even become engaged in the money economy, might well be a possibility, but it is one that appears so far-reaching in its impact on the lives of people that one can understand the reluctance of the political leadership to embrace it. The alternative, of strictly limiting mining development to the amount needed to pay for rural projects and to the amount that can in some sense be managed and controlled by Papua New Guineans, appears the course likely to be chosen.

A narrower area of choice for the government concerns the emphasis to be placed on training within the mining industry. To the extent that the mines can provide a surplus of heavy equipment operators, electricians, mechanics, and other trained personnel, they will contribute to a general increase in the level of available skills, the ability of the country to carry out desired development projects on its own, and the ability to reduce dependence on foreigners in major enterprises. But this training program also has certain negative effects, including the upward pressure on wages likely to be exerted by those who have been employed at the mine and the more general problem of the absorption of Western values and lifestyles by mine-workers. On balance, the massive training program provided by the Bougainville mine is probably seen as a net benefit by most government leaders, but some worry about its disruptive effects and question whether it might not be better if the mine were to function as even more of an enclave.

On the question of revenue, the government's position thus far has simply been to say that the returns from the Bougainville project have been inadequate, and that, especially in periods of high metal prices (like the first half of 1974) the great majority of the surplus should go to the government. Profits from Bougainville are almost wholly dependent on the fluctuations in world copper prices; a 10 percent increase in metal prices means a 17 percent increase in profits. Since the company's management ability contributes little to the level of profits, once the mine has been established and is in operation, there has been strong support for a sharply graduated tax scale that at very high prices would take up to 80 percent of the marginal income for the government. The government has said that the mining tax formulas incorporated in the renegotiated concession agreement are designed to allow repayment of loans and the earning of a reasonable return (on the order of 15 to 20 percent on a discounted cash flow basis) before these higher marginal rates are imposed. In the case of Bougainville, the application of such a formula to its 1973 profits of $158 million would have produced tax revenue of about $90 million.

A related question is the timing of tax flows. Under the original agreement, no tax was payable until about 1978. Under the new terms, 1974 profits will be subject to tax. There has been pressure to institute a pay-as-you-go system, in which the company would make estimated tax payments during the course of the year. Under the existing system of tax collection, even though the 1974 income is subject to tax, the government will not receive the bulk of its money until late 1975.

A contrary factor, however, is the high level of Australian aid that has been promised for the period to 1976. In the short term, it is likely that a substantial portion of any additional revenue from Bougainville would merely substitute for Australian aid, as the aid commitment was reduced. While such a substitution has important values from the point of view of increased self-reliance, these facts have made Papua New Guinean political leaders a bit less firm on the question of immediate revenue from Bougainville than they otherwise might have been.

Finally, the less-than-satisfactory experience of some countries with 51 percent government ownership and management contracts has been closely watched from Papua New Guinea. Except that any government contribution to infrastructure is to be compensated for by a proportionate share of the equity in a mining project, there is relatively little current emphasis on majority ownership. Until sufficient numbers of Papua New Guineans are available to take over management positions in the mines, the emphasis is likely to be on revenue. Rather than trying to control profit leakages due to interaffiliate transactions and other causes by its ownership position, the government is enacting broad powers to review all contracts between foreign-owned enterprises and overseas firms. Under these powers, for example, the sale of concentrates at less than the London Metal Exchange price minus a fair margin for treatment costs will not be approved. Similarly, mine purchases from overseas suppliers will be subject to government approval, as will payment to parent firms for head office and administrative expenses, transfer of technology, and other items.

THE MINING COMPANIES

Unlike the CIPEC countries, Papua New Guinea does not have a long history of domination by one or two mining companies. While Rio Tinto Zinc (RTZ), the parent company of Bougainville Copper, is by far the largest single corporate influence on the country's economy at present, many other corporations hold prospecting authorities and are likely to be actively involved in mining in the future.

Bougainville is one of the few RTZ projects located in a developing country. If one excludes the Lornex mines and the Brinco hydroelectric power project in Canada, most of the company's widespread investments are in "safe" areas. Its other major copper mine is at Palabora in South Africa, and it has large iron ore, bauxite, uranium, and copper-lead-zinc interests in Australia. For a relatively low-profile, and not widely known company, the Bougainville venture represented a major gamble. The gamble, however, has paid off well; profits from Bougainville accounted for nearly 40 percent of total earnings in 1973, a figure much larger than the proportion of RTZ investment that the mine represents.[21]

The financing of the Bougainville mine is a typical example of approach of retaining control while exposing the corporation to the least possible financial risk. Initially, the project was developed by Conzinc Riotinto of Australia (CRA), an 80-percent-owned RTZ subsidiary. CRA in turn retains a 53 percent interest in Bougainville Copper Lts., the actual mining company. By selling off the minority interests in both CRA and Bougainville Copper to the public at premium prices (in the case of Bougainville, the public offering price was A$1.55, compared to a par value of 50 cents), RTZ retained effective control while minimizing its financial exposure.

In the relatively casual political atmosphere of Papua New Guinea (ministers are normally on a first-name basis with most people with whom they deal regularly, for example), it is difficult to judge the ex-

tent of RTZ political influence. The company's officials have reason-
ably easy access to political leaders, but, then, so does everyone
else. There is no evidence to date that the company has tried to manipulate domestic politics. Of course, until the recent renegotiation of
the 1967 agreement, there was no reason why the company should seek
anything more from the political arena. It remains to be seen what
weapons the company will bring to bear as government policy impinges
more harshly on profits.

On the island of Bougainville itself, however, the company has
played a much more active political role. Most Bougainvillean students at the country's two universities are sponsored by the company
and are likely to work for it after graduation. The company operates
its own business development and agricultural extension units, and
its public relations staff visit villages throughout the island, promoting the company's image and attracting crowds by showing movies.
The Panguna Development Foundation, a charitable offshoot of the
mine, has become the chief agency for financing Bougainvillean businessmen. It is not inconceivable that in a dispute with the central
government over renegotiation, the company could play on the already
present secessionist feeling on Bougainville.

The other companies active in mineral prospecting in Papua
New Guinea represent a fairly broad range of large mining corporations from around the world. The two projects currently closest to
development involve Kennecott Corporation and a partnership between
Mt. Isa Mines (50 percent owned by Asarco) and a Japanese group
headed by Sumitomo Mining. Other active prospecting operations are
being maintained by U.S. Steel, the Anglo-American Corporation group,
BHP (the largest Australian-owned corporation), and the Canadian
Placer mining group.

                          GOVERNMENT RESOURCES

Compared to the resources available to the mining corporations
the Papua New Guinea government's structure for dealing with major
mining interests is less than overwhelming. At the civil service level,
a small Office of Minerals and Energy has only been formed in the
past year. The office has a professional staff of less than a dozen—
all, except the director, expatriates—which is responsible for negotiating and enforcing mining agreements as well as for managing the
planned hydroelectric power development on the Purari River—which
will ultimately be as large as Zaire's Inga Project.

Because of the small size of the government as a whole, major
mining issues inevitably involve virtually the whole governmental apparatus. Thus, for example, the renegotiation of the Bougainville

agreement came under the direct supervision of a group of the six most senior ministers and was carried out at the day-to-day level by a team headed by the secretary for Finance and the director of the Planning Office.

The strain that coping with the mining industry places on the government is readily apparent. In the past year, half a dozen consultants, from the Commonwealth Secretariat, CIPEC, Harvard, and elsewhere have been brought in to assist with aspects of mining policy. The need for a vastly increased staff in the ministry is obvious, but because of the continuing shortage of university-trained Papua New Guineans, and the bureaucratic delays in recruiting of outsiders, it appears likely that the shortage will continue for some time to come.

Political pressures also create difficulties for the government in dealing strongly with the mining companies. The government itself is a coalition, one of whose major parties is clearly identified as a businessmen's party, despite its acceptance of the government's overall cautious policy toward foreign investment. There has been no strong independence movement, and as a result there is no clearly established vanguard party or leadership cadre that could be the vehicle for a strongly nationalistic economic policy to be carried out. Most members of the House of Assembly, accurately reflecting the view of their constituents, see foreign investment of virtually any sort as a source of desirable development for their areas. Only a few politicians have thus far raised the sort of questions about the impact of large foreign corporations that are now commonplace in other developing countries.

A final factor descreasing the likelihood of strong government action is the typical Papua New Guinean leadership style, embodied most clearly in Chief Minister Michael Somare himself. This style is one of compromise, accommodation, and delay. Every possible effort is made to avoid visible head-on conflicts.

CONCLUSIONS

This chapter brings out several key points relating to the existing state of the mining industry in Papua New Guinea.

First, Papua New Guinea is only now moving away from the conventional emphasis on growth toward a more rurally based vision of development.

Second, it is not yet clear how mining fits into this emerging development policy. In contrast to the CIPEC countries, the leadership of Papua New Guinea has not vigorously espoused policies of national ownership of the mines, nor of using mining as a major growth pole for industrialization.

Third, the capacity of the government for dealing with large mining companies is limited.

The overall situation, then, is an uncertain one, in which, despite movements toward positive action like the renegotiation of the Bougainville agreement, the outcome is by no means assured.

NOTES

1. Papua New Guinea Bureau of Statistics, Statistical Bulletin (Port Moresby, August 1974).

2. Commonwealth of Australia, Parliamentary Debates, June 1950, pp. 3638 and 3646.

3. As early as 1945, the Labour minister responsible for the territories said, "it would be wrong to encourage production in the territories of goods which would be competitive with commodities produced in Australia." Parliamentary Debates, July 1945, p. 4304.

4. International Bank for Reconstruction and Development, The Economic Development of the Territory of Papua and New Guinea (Baltimore, 1965), p. 20 (hereafter cited as IBRD Report).

5. Ibid., p. 438.

6. Territory of Papua and New Guinea, Programmes and Policies for the Economic Development of Papua and New Guinea (Port Moresby, 1968; hereafter cited as Five-Year Plan), p. 13.

7. IBRD Report, p. 431.

8. Ibid., p. 214.

9. Ibid., p. 29.

10. Ibid., p. 376.

11. Five-Year Plan, p. 119.

12. For the following data consequences of the first Five-Year Plan, see Papua New Guinea Central Planning Office, Papua New Guinea's Improvement Plan for 1973-74 (Port Moresby, 1973; hereafter cited as Improvement Plan), p. 16.

13. Ibid., p. 2.

14. A Report on Development Strategies for Papua New Guinea, prepared by a mission from the Overseas Development Group (University of East Anglia) (Port Moresby, 1973), p. 11.

15. Quoted in Australian Department of External Territories, Investing in Papua and New Guinea (Canberra, 1968), p. 47.

16. Bureau of Statistics, op. cit.

17. Papua New Guinea Office of Information, Press Release A0384, March 14, 1974.

18. House of Assembly Debates, June 1973, pp. 2080-81.

19. Mining (Bougainville Copper Agreement) Act 1967.

20. See S. Zorn, "Bougainville: Managing the Copper Industry," New Guinea 7, 3 (January 1973): 23-40.

21. The only generally available account of RTZ activities is R. West, <u>River of Tears</u> (London, 1972).

25

Research is a mansion with many rooms. One of its manifold purposes is to enable us to learn from experience, the better to deal with out own world and our own problems. Yet can Zambia, for example, learn from the Chilean experience of the 1950s and 1960s how to deal with copper-based underdevelopment in Africa in 1974?

It is logically impossible to argue from one set of facts directly to another set of facts. One must first find some higher-order generalization that subsumes both sets. Until we have articulated such higher-order generalizations, it is impossible consciously to learn from experience.

Such generalizations are not "laws" in the sense of the "laws" of the physical sciences. They do not and cannot express unvarying relationships. Propositions concerning human events can never be more than heurisms, for the relationships between human beings are never invariant. Such propositions can advise us as to the sorts of data to search for to explain particular events or to solve particular problems. They can suggest the likely consequences of action.

The research that this book contains would be incomplete without an attempt to state some general propositions of this sort. What follows is an attempt to do so. These propositions are, however, only in part based upon the cases reported here. Some of the propositions put forward seemed to me to be necessary to understand the troubled situation to be examined. So far as I know, however, these propositions accord with experience. Policy research finds its justification in the generation of policies to guide human activity. Precisely in that activity, too, do the products of such research find their ultimate warrant or diswarrant.

THE EXISTING STATE OF AFFAIRS

1. Chile, Peru, Zaire, and Zambia have rich copper resources, which have been extensively exploited in the past.

2. The economies of these countries are characterized by the following:

a. The export of semirefined or refined, but not processes, copper;

b. The export of large proportions of earned surpluses overseas;

c. The import of manufactured goods from overseas, with a very small domestic manufacturing sector;

d. High productivity with respect to copper, using advanced technologies becoming increasingly capital-intensive;

e. Low productivity based on relatively low technologies in most of the rural sector;

f. The rapid growth of the urban population, brought about mainly by migration from the rural to the urban sectors;

g. Very low wage rates;

h. High urban unemployment;

i. A set of economic institutions (banking, export-import trade, property laws, wholesale trade, and so on) that generate decisions supportive of these conditions and not of decisions looking to increased local industrialization, offering widespread job opportunities, out of the urban areas, in economic activities with linkages to the rest of the economy;

j. A markedly skewed income distribution curve in favor of the owners and managers of the copper industry, its satellite industries, the economic institutions mentioned above, and high civil servants and politicians, and against the vast mass of the urban and rural poor;

k. A dependency relationship with the industrialized, capitalist countries and multinational corporations based in those countries (that is, a range of choice open to these four countries to a great extent structured by decisions made by corporations, individuals, or other decision-makers in the industrialized, capitalist world).

3. Despite a series of measures taken by the governments of these countries aimed at alleviating these conditions, in the main they continue to exist.

GENERAL PARADIGM; DEFINITIONS

1. A society is comprised of actors who make choices they believe will maximize their interests within the constraints and resources

of their total environment as they perceive them. Of these constraints and resources, the institutions of the society are the factors most susceptible to conscious change through state action.

2. The phrase "political elite" shall mean those who occupy high political office, high office in the civil service, high army officers (in countries where they have political decision-making power), and high party officials (where they have political decision-making power).

3. The phrase "economic ruling class" shall mean the owners and managers of large economic enterprise (mines, industry, banks, estate agriculture, trading companies, and so forth), whether or not they are resident in the country.

4. The word "institutions" shall mean repetitive patterns of human action (that is, behavior consciously undertaken).

THE COLONIAL SITUATION

During the period of colonial overrule in Africa, the community of interest between the political elite, the colonial and imperial rulers, and the foreign entrepreneurs who constituted the economic ruling class was writ large. It was formally embodied in the political conditions of colonialism. The same situation existed for many years in Latin America, despite nominal independence. For convenience, I have labeled both of these conditions as "the colonial situation," even though the Latin American countries had formal political independence at the time:

1. Where:
  a. The following constraints and resources exist:
    i. Valuable deposits of copper, with existing processing factories in the industrialized, capitalist countries; and
    ii. The existing technology permits its profitable extraction in the less developed countries, using cheap, relatively untrained local labor; and
    iii. The existing technology requires a few, very highly trained managers and technicians; and
    iv. The existing economy is based upon a relatively low level of technology in agriculture, with correspondingly low levels of productivity, specialization, and exchange; and
  b. The interests of political elites and foreign investors are common or mutually interdependent; and
  c. The following institutions exist:
    i. A state structure possessing a sufficient reserve of political and coercive power to compel the labor force to work for low wages in mining; and

    ii. A legal order that through the laws of property and contract empowers those that own mining property to determine how the property and the surpluses it generates will be used;

2. The following activity by foreign entrepreneurs and the political elite will ensue:

a. Entrepreneurs from the developed, capitalist countries will take advantage of these conditions and institutions
    i. To invest in copper mining; and
    ii. To direct the surpluses earned as from time to time seems more desirable
        a) To investment in further copper mining;
        b) To high income and conspicuous consumption for themselves, the technocrats required to operate the mines, and the political elites;
        c) To profits, dividends, and interest exported to the industrialized capitalist countries; and
    iii. The copper industry having been established, to invest in the banks, import-export trade, estate agriculture, and other profitable activities aimed at either supporting the mining industry or supplying luxury goods to the high-income market; and
    iv. To pay very low wages to workers.

b. The political elite will use law and state power as follows:
    i. To maintain or create property and contract laws that support the power of the economic ruling class to act as above;
    ii. To maintain or create economic institutions
        a) That systematically make the rural areas less attractive as places to live and to work than the foreign-owned enterprises, despite very low wages in those enterprises (for example, land tenure rules, poll or hut taxes, coerced labor laws, and so on);
        b) That formally delegate to the economic ruling class the power to rule over selected areas of industry (industry boards, marketing boards, and so on);
        c) That weaken or destroy the power of the poor through collective action to seek to better their lot;
        d) That structure the range of choice for the political elite in their private capacities so that it is in their private interests to become and remain allied with, or part of, the economic ruling class (credit, bribery, and so on); and
        e) That provide infrastructural support for foreign economic enterprise (railroads and roads, credit, geological surveys, and so on); and

iii. To maintain or create political institutions as follows:

a) Governmental decision-making institutions that are highly compartmented and hierarchical, and therefore capable of generating decisions looking toward the maintenance of the existing structure, but incapable of generating decisions looking toward planned, radical change;

b) Institutions that restrict recruitment to the political elite to persons with ideologies supportive of the existing state of affairs;

c) Institutions that have in fact a high degree of participation in decision-making by the economic elite, and little or no participation in fact by the poor;

d) Governmental institutions that generate decisions improving the infrastructure as required by the interests of the economic ruling class personally and of their enterprises and of the political elite itself, but not of the mass of the poor;

e) Governmental institutions with a high degree of discretionary power in members of the bureaucracy.

3. As a consequence of opportunities created by the activity of foreign entrepreneurs and political elites described above, a few indigenous citizens may enter the economic ruling class or the class of highly paid technocrats in their service.

THE POSTCOLONIAL SITUATION

As a result of internal political processes in countries already possessing formal political independence, or of the struggle against the de jure colonial power, new political elites have come to power in all these four countries. They espouse a rhetoric of opposition to the economic ruling class, frequently couched in the vocabulary of socialism. Their personal interests are, at the time when they take power, to some degree inconsistent with the interests of the economic ruling class. This situation I denote as the postcolonial situation.

1. Immediately after a new political elite with interests to some extent contradictory to those of the economic ruling class takes power, the state of affairs described above continues, with the only change being the new contradiction between the interests of the new political elite and the economic ruling class.

2. New investment in the country whether locally generated or not will be based on considerations of short-term profitability so long as either

a. In fact investment decisions are made by private entrepreneurs in their own interest, whatever the formal legal structure of control over such decisions may be (for example, 51 percent ownership by government)—and in fact such decisions will be made by the private entrepreneurs so long as they have the right to do so by law, or so long as they have the power to do so by their control over marketing or technology; or

b. Investment decisions are made by employees of state enterprise or by public servants, whose careers and rewards as they perceive them are in law or in fact dependent upon their investment decisions being profitable, without regard to other social criteria for investment.

3. Given the existing social, political, economic, and legal institutions received from the colonial era, short-term profitability will be greatest in the following investments (regardless of the degree of infrastructural development in rural areas):

a. The further development of copper mining; or

b. In enterprises with some or all of the following characteristics:

   i. Designed to produce goods or services for the existing internal high-income markets;

   ii. Using large amounts of imported inputs;

   iii. Near existing urban areas;

   iv. Using capital-intensive, advanced technologies;

   v. Using foreign capital, or capital accumulated by existing high-income strata.

4. Multinational corporations, the chief source of foreign private capital, will make the decision, whether to invest in the country or elsewhere, depending upon their worldwide interests, and the worldwide opportunity costs of the proposed internal investment, and not upon the priorities of the host country.

5. If the independent government seeks to recapture the economic surplus (through taxation, joint ownership, or otherwise), private investors will increasingly find it in their worldwide interests to invest in

a. Downstream copper enterprises elsewhere; and

b. The production of substitute material elsewhere; and

c. Diversified enterprises elsewhere; and

d. Copper mines in countries where the political elites will not seek to recapture such a large share of the surplus.

6. The political elite will seek to change the institutions received from the colonial era, if, but only if, and to the extent that:

a. The political elite and the economic ruling class in fact have antagonistic interests, and continue to do so; and

b. The political elite are seized by an ideology that identifies the problems of the poor as the central concern of the political elite, and explains it in terms of the institutions that perpetuate underdevelopment; and

c. Governmental decision-making institutions are developed that are capable of generating decisions that are likely to induce radical institutional change; and

d. Governmental institutions are developed that have in fact a high degree of participation in decision-making by the poor, and little participation in fact by the economic ruling class; and

e. The political elites have both the knowledge and resources to change institutions against the interests of the economic elites.

7. Change in the composition of the economic ruling class from foreign nationals to citizens, or changes in the formal ownership of economic enterprise, will not alone change repetitive patterns of behavior by the persons involved, and hence will not change the institutions they comprise.

8. To the extent that the political elite continues to receive very high salaries, and to the extent that they join the economic ruling class by making investments, and to the extent that they create new institutions that give to particular members the power over the disposition of surpluses earned in enterprise (for example, through parastatal corporations), the political elite will develop alliances or common interests with the economic ruling class.

CHARACTERISTICS OF A SUCCESSFUL DEVELOPMENT STRATEGY IN THE POSTCOLONIAL PERIOD

In fact, the government of the four countries have, each in their own way, behaved in the main as described in the preceding section. The results are the troubled situation whose main characteristics have been described above in the continued underdevelopment of the four countries. A successful development strategy flows from these explanations. A political elite committed to development in the interests of the mass of the population must undertake the following programs:

1. It must ensure that the political elite does not have or develop interests that are in fact common or mutually interdependent with the economic ruling class; and therefore

    a. It must change existing institutions so that it will not be in the interests of members of the political elite to become or to remain allied with economic elite (for example, through a leadership code), and which deprive them of opportunities to do so; and

    b. Since in the long run any economic elite will likely wield sufficient power to coopt or coerce any political elite, the new institutions created or development programs undertaken must be designed to have low potential for creating new members of the economic ruling class or enhancing its power, and high potential for reducing its size and power.

    2. It must create institutions of recruitment to the political elite, which ensure that the elite will be comprised of persons seized by an ideology that identifies the problems of the poor as the central concern of the political elite, and explains poverty in terms of the institutions that perpetuate underdevelopment; and

    3. It must create institutions capable of providing knowledge sufficient to enable the political elite to change the institutional structure effectively; and

    4. It must create institutions to provide the political elite with sufficient resources (money, trained manpower, armaments, and so on) to change the institutional structure and to overcome any resistance by the economic ruling class; and

    5. It must create new governmental institutions capable of generating decisions looking toward fundamental institutional changes (that is, planning and implementing institutions); and

    6. It must create institutions by which the mass rather than the economic ruling class can participate in governmental and economic decision-making (through developing channels of access to the political elite and through actual participation in decision-making on the local and enterprise level); and

    7. It must create institutions that permit no more discretion to bureaucrats than is required by the matter at issue, provides means for requiring that discretion be structured by rules, and for confiding bureaucrats to the limits defined by those rules; and

    8. It must create institutions to recapture the surpluses exported overseas; and

    9. It must create institutions so that the power of the economic ruling class over investment decisions, over the operation of productive enterprise, and over the disposition of the surplus will be replaced by participatory governmental decision-making institutions; and

    10. It must create institutions to ensure that the recaptured surpluses are invested in productive enterprise with the following characteristics, pursuant to a long-term plan for economic development:

    a. Acceptable rates of profitability over time;

    b. Located in previously underdeveloped areas of the country;

    c. Increasing the productivity of the rural areas both in agriculture and in industry;

    d. Using an appropriate technology, in the sense that

        i. It increases as much as possible the number of employment opportunities; and

        ii. It requires so far as possible for operation and servicing skills and material inputs that the country itself can provide;

    e. Uses local raw materials so far as possible;

    f. So far as possible has a favorable effect upon the foreign exchange position;

    g. Produces outputs that contribute to increased productivity and to raising the levels of living of the lowest income group; and

11. It must create institutions that reduce the power of the multinational corporations to control the marketing of copper, and their monopoly over the expertise relating to copper technology; and

12. It must create institutions to develop and implement an incomes policy progressively reducing the gap between the lowest and highest incomes (including profits as well as salaries and wages); and

13. It must create institutions capable of improving the infrastructure (roads, schools, governmental services generally in depressed rural and urban areas); and

14. It must enact and implement laws likely to create and maintain the changed behavior required by these institutional changes.

ABOUT THE EDITOR AND CONTRIBUTORS

The views expressed by the authors are not necessarily those of the institutions where they are employed; their positions at the time of the conference are indicated here for identification only.

CLAES BRUNDENIUS has been conducting research on the role of multinational corporations in Latin America at the Institute of Economic History, University of Lund, Sweden. This study was financed by the Nordic Cooperation Committee for Internal Politics including Conflict and Peace Research, Stockholm. It is reprinted with permission from the Journal of Peace Research no. 3 (1972).

MANUEL CABIESES BARRERA is an economist and a member of the Ministry of Planning in Peru.

AL GEDICKS is the research coordinator, Community Action on Latin America (CALA), and a Social Organization Trainee in the Department of Sociology, University of Wisconsin, Madison.

RICARDO FFRENCH-DAVIS is Research Professor at the Centrode Estudios de Planificacion Nacional (Ceplan), and Economics Professor at the University of Chile and the Catholic University of Chile.

DANILO JELENC is a member of the UN Advisory Team to the Planning Division of the Ministry of Planning and Finance of the Government of Zambia.

ILUNGU ILUNKAMBA is an Assistant, Economics Faculty, University of Zaire-Kinshasa.

LUABEYA KABEYA is a Professor of Economics in Kinshasa, Zaire.

E. ALEXANDER, an economist employed by Anglo-American Corporation (Central Africa), Ltd., did not prepare a paper for the Lusaka Conference, although he participated in the discussions and added the comment to Mr. Ushewokunze's paper at a later date.

LUMBUNGU KAMANDA is a Professor in the Faculties of Business, Economics, and Agronomy, at the University of Zaire-Kinshasa.

MULUMBA LUKOJI is a Professor of Law, University of Zaire-Kinshasa.

USELE MAWASI is an Assistant, Faculty of Economics, University of Zaire-Kinshasa and a member of the Planning Service Bureau of the President of the Republic.

DIAMBOMBA MU KANDA MIALA is Professor in the Economics Faculty, University of Zaire-Kinshasa.

DOROTHEA MEZGER is conducting research at the Max Planck Institute of Starnberg, Federal Republic of Germany. The conference organizers wish to express their appreciation to the Institute for having made it possible for Ms. Mezger to make this contribution.

LUIS PASARA is a Professor of Law at the Catholic University, Lima, Peru and a researcher at Desco.

MANZILA LUTUMBA SAL 'ASAL is Chairman of the Law School, University of Zaire-Kinshasa.

DIEGO GARCIA-SAYAN is a Professor in the Law School at the Catholic University, Lima, Peru, and a researcher at Desco.

ANN SEIDMAN, an Economist, is currently Project Director, Center for Research on Women, Wellesley College.

RAJ SHARMA, an Economist, is Economic Adviser to the Ministry of Planning and Finance, Zambia.

R. M. K. SILITSHENA is a Lecturer in Economic Geography, University of Botswana, Lesotho and Swaziland in Botswana.

GEORGE SIMWINGA is a Lecturer in Political and Administrative Studies, University of Zambia.

KENJI TAKEUCHI is Senior Economist, Commodities and Export Projections Division, Economic Analysis and Projections Department, International Bank for Reconstruction and Development, Washington, D.C.

ERNESTO TIRONI is Professor of Economics at the Centro de Estudios de Planificacion Nacional (Ceplan) and the School of Economics at the Catholic University of Chile.

C. M. USHEWOKUNZE is an Assistant Lecturer in Law, University of Zambia.

KATANGA MUKUMADI YAMATUMBA is Chairman, Economics Department, University of Zaire-Kinshasa.

STEPHEN ZORN is an adviser to the Chief Minister of Papua New Guinea on mining and investment matters.

ARAB OIL: Impact on Arab Nations and Global Implications
> edited by Naiem A. Sherbiny and
> Mark A. Tessler

ECONOMIC ANALYSIS AND THE MULTINATIONAL
ENTERPRISE
> edited by John H. Dunning

EXPROPRIATION OF U.S. PROPERTY IN SOUTH
AMERICA: Nationalization of Oil and Copper Companies in Peru, Bolivia, and Chile
> George M. Ingram

THE MULTINATIONAL CORPORATION AS A FORCE
IN LATIN AMERICAN POLITICS: A Case Study of
the International Petroleum Company in Peru
> Adalberto J. Pinelo

MULTINATIONAL CORPORATIONS AND GOVERN-
MENTS: Business-Government Relations in an International Context
> edited by Patrick M. Boarman and
> Hans Schollhammer

THE NATION-STATE AND TRANSNATIONAL COR-
PORATIONS IN CONFLICT: With Special Reference
to Latin America
> edited by Jon P. Gunnemann

THE PRICING OF CRUDE OIL: Economic and Strategic Guidelines for an International Energy Policy
(expanded and updated edition)
> Taki Rifai